T0330972

PROCEEDINGS OF THE SIXTH INTERNATIONAL SCHOOL

ADVANCED COURSES OF MATHEMATICAL ANALYSIS VI

EDITORS

FRANCISCO JAVIER MARTÍN-REYES
PEDRO ORTEGA SALVADOR
MARÍA LORENTE
CRISTÓBAL GONZÁLEZ

Universidad de Málaga, Spain

PROCEEDINGS OF THE SIXTH INTERNATIONAL SCHOOL

ADVANCED COURSES OF MATHEMATICAL ANALYSIS VI

Universidad de Málaga, Málaga, Spain 8 – 12 September 2014

World Scientific

NEW JERSEY • LONDON • SINGAPORE • BEIJING • SHANGHAI • HONG KONG • TAIPEI • CHENNAI • TOKYO

Published by

World Scientific Publishing Co. Pte. Ltd.
5 Toh Tuck Link, Singapore 596224
USA office: 27 Warren Street, Suite 401-402, Hackensack, NJ 07601
UK office: 57 Shelton Street, Covent Garden, London WC2H 9HE

British Library Cataloguing-in-Publication Data
A catalogue record for this book is available from the British Library.

ADVANCED COURSES OF MATHEMATICAL ANALYSIS VI
Proceedings of the Sixth International School

ISBN 978-981-3147-63-8

Printed in Singapore

Preface

These Proceedings contain the four courses and six plenary lectures presented at the *VI International Course of Mathematical Analysis in Andalucía* held in Antequera (Málaga, Spain) during the period 8-12 September, 2014. The event was organized by the groups of Mathematical Analysis from the University of Málaga, and, as its name suggests, is a continuation of the previous five editions. Each of these editions took place in different cities of Andalucía. It started in Cádiz in 2002, then passed to Granada in 2004, La Rábida (Huelva) in 2007, Cádiz again in 2009, and Almería in 2011. These meetings are possible thanks to the joint effort of the different research groups in the areas of Mathematical Analysis from all the Universities of Andalucía. They stand as a great opportunity for all generations of mathematicians, especially the younger ones, to enrich their knowledge with the latest results in the field. Therefore, it is crucial that we continue to work together on the celebration of this conference.

In 2011, at the meeting of Almería, the Analysis group from the University of Málaga agreed to organize this sixth edition of the course. Let us specially thank Juan Carlos Navarro and El Amin Kaidi Lhachmi from the University of Almería, and Juan Francisco Mena and María Victoria Velasco from the University of Granada who helped and encouraged us from the very beginning to run the conference. Finally, the Course was celebrated and now we are able to offer you these Proceedings with all the excellent courses and talks that were delivered in it.

We would like to thank everyone who supported and helped us in the organization of the event. Special thanks go to the various Analysis research groups from Andalucía and to the Mathematical Analysis Department from the University of Málaga, for their scientific and financial contributions. We would also like to acknowledge the financial support provided by the Spanish Government through its Ministry of Economy and Competitiveness (Grant MTM2011-28149-C02-02), by the University of Málaga (Campus of International Excellence - Andalucía Tech) through its Office for Research and Research Transference, and also by the *Unicaja* Bank. Thanks also go to the Royal Spanish Mathematical Society that scientifically sponsored this meeting, and to the City Hall of Antequera for its collaboration in making it possible.

We also thank all attendees, especially those who presented short talks and posters, for selecting this event to show their latest work. Last but not least, we are very grateful to the speakers for accepting our invitations and taking their time to write the papers for these Proceedings.

The editors,
May 2016.

Organizing committees

Scientific organizing members

Univ. Almería:
– El Amin KAIDI LHACHMI
– Juan Carlos NAVARRO PASCUAL

Univ. Cádiz:
– Fernando LEÓN SAAVEDRA
– Francisco Javier PÉREZ FERNÁNDEZ

Univ. Granada:
– Juan Francisco MENA JURADO
– Rafael PAYÁ ALBERT
– Ángel RODRÍGUEZ PALACIOS
– María Victoria VELASCO COLLADO

Univ. Huelva:
– Cándido PIÑEIRO GÓMEZ
– Ramón J. RODRÍGUEZ ÁLVAREZ

Univ. Jaen:
– Miguel MARANO CALZOLARI
– José María QUESADA TERUEL
– Francisco ROCA RODRÍGUEZ

Univ. Málaga:
– Daniel GIRELA ÁLVAREZ
– Francisco Javier MARTÍN REYES

Univ. Pablo Olavide:
– Antonio VILAR NOTARIO

Univ. Sevilla:
– Tomás DOMÍNGUEZ BENAVIDES
– Antonio FERNÁNDEZ CARRIÓN
– Carlos PÉREZ MORENO
– Luis RODRÍGUEZ PIAZZA

Local organizing members

Francisco Javier MARTÍN REYES (Coord.)	– Univ. Málaga
Venancio ÁLVAREZ GONZÁLEZ	– Univ. Málaga
Daniel GIRELA ÁLVAREZ	– Univ. Málaga
Cristóbal GONZÁLEZ ENRÍQUEZ	– Univ. Málaga
Antonio JIMÉNEZ MELADO	– Univ. Málaga
María LORENTE DOMÍNGUEZ	– Univ. Málaga
María Auxiliadora MÁRQUEZ FERNÁNDEZ	– Univ. Málaga
Pedro ORTEGA SALVADOR	– Univ. Málaga
Consuelo RAMÍREZ TORREBLANCA	– Univ. Córdoba
Alberto de la TORRE RODRÍGUEZ	– Univ. Málaga

Contents

PART A

Courses

Convex inequalities, isoperimetry and spectral gap*

David Alonso-Gutiérrez

Departamento de Matemáticas and IUMA,
Universidad de Zaragoza,
Pedro Cerbuna 12, 50009 Zaragoza, Spain
E-mail: alonsod@unizar.es

Jesús Bastero

Departamento de Matemáticas and IUMA,
Universidad de Zaragoza,
Pedro Cerbuna 12, 50009 Zaragoza, Spain
E-mail: bastero@unizar.es

The main idea of these notes is to present the Kannan-Lovász-Simonovits spectral gap conjecture on the correct estimate for the spectral gap of the Laplace-Beltrami operator associated to any log-concave probability on $\mathbb{R}^n$.

Keywords: Log-concave probability, Brunn-Minkowski inequality, concentration phenomena, convex bodies, spectral gap, isoperimetric inequality, Poincaré inequality.

1. Introduction and Notation

These notes are the content of the three lectures explained by the second author in the "VI International Course of Mathematical Analysis in Andalucía", held in Antequera in September, 2014. The authors want to express their gratitude to the organizers of this meeting for giving to one of them the possibility of presenting this quite new theory to young researchers. The content of this course is included in the reference [2], where a more detailed and complete information appears.

The main idea of these notes is to present the Kannan-Lovász-Simonovits spectral gap conjecture (KLS). This question was originally posed in relation with some problems in theoretical computer science, but it has a well-understood analytic-geometrical meaning: give the correct estimate for the

*This work has been done with the financial support of MTM2013-42105-P, DGA E-64 and P1-1B2014-35 projects and of Institut Universitari de Matemàtiques i Aplicacions de Castelló.

spectral gap (first non trivial eigenvalue) of the Laplace-Beltrami operator associated to any log-concave probability in $\mathbb{R}^n$. The KLS conjecture can also be expressed in terms of a type of Cheeger's isoperimetric inequality and, in this way, is related to Poincare's inequalities and to the concentration of measure phenomenon. In the meanwhile this conjecture is now one of the central points in asymptotic geometric analysis which is the new branch of functional analysis coming from the geometry of Banach spaces when it interplays with classical convex geometry and probability.

The notes are divided in three parts. In Section 2 we will present Prékopa-Leindler inequality as a certain reverse of classical Hölder's inequality. We will also deduce from it Brunn-Minkowski, isoperimetric inequality and Borell's inequality on concentration of mass for log-concave probabilities on $\mathbb{R}^n$. Section 3 is devoted to Cheeger-type isoperimetric inequalities, its relation to Poincaré-type inequalities and E. Milman's recent result on the role of convexity in this framework. In Section 4 we will present the KLS conjecture and the main results in this subject, up to now. We will also relate KLS to other conjectures and present our own results on these topics. The references list papers used for the preparation of this work.

Let us introduce some notation. A convex body K in $\mathbb{R}^n$ is a convex, compact subset of $\mathbb{R}^n$ having the origin in its interior. Given an n-dimensional convex body K, we will denote by $|K|_n$ (or simply $|K|$) its volume. ($\widetilde{K} = |K|^{-\frac{1}{n}} K$, so that $|\widetilde{K}|_n = 1$). The volume of the n-dimensional Euclidean ball will be denoted by ω_n. We also use $|\cdot|$ for the modulus of a real number or the Euclidean norm for a vector in $\mathbb{R}^n$. When we write $a \sim b$, for $a, b > 0$, it means that the quotient of a and b is bounded from above and from below by absolute constants. $O(n)$ and $SO(n)$ will always denote the orthogonal and the symmetric orthogonal group on $\mathbb{R}^n$. The Haar probability on them will be denoted by σ. We will also denote by σ_{n-1} or just σ the Haar probability measure on S^{n-1}.

2. Convex inequalities

In this section we focus on Prékopa-Leindler inequality, presenting it as a kind of reverse of classical Hölder's inequality. Next we present Brunn-Minkowski inequality, which really is the version of Prékopa-Leindler's for charateristic functions. As a consequence we prove the classical isoperimetric inequality in $\mathbb{R}^n$ and Borell's inequality which is the main tool for working with log-concave probabilities.

2.1. *Hölder's and reverse Hölder's inequalities*

It is well known that given two measurable functions $f, g : \mathbb{R}^n \to \mathbb{R}_+$ and $0 \le \lambda \le 1$, Hölder's inequality says that

$$\int_{\mathbb{R}^n} f(x)^{1-\lambda} g(x)^\lambda dx \le \left(\int_{\mathbb{R}^n} f(x)dx\right)^{1-\lambda} \left(\int_{\mathbb{R}^n} g(x)dx\right)^\lambda.$$

In the case that we take characteristic functions $f = \chi_A$, $g = \chi_B$, $(A, B \subseteq \mathbb{R}^n)$, we have

$$|A \cap B|_n \le |A|_n^{1-\lambda} |B|_n^\lambda, \qquad \forall\, 0 \le \lambda \le 1,$$

or, equivalently,

$$|A \cap B|_n \le \min\{|A|_n, |B|_n\}.$$

It is also well known that we cannot reverse in general these two inequalities even by adding some constant, i.e., in general we cannot affirm for any constant greater than one that

$$\min\{|A|_n, |B|_n\} \not\le C|A \cap B|_n,$$

or

$$\left(\int_{\mathbb{R}^n} f\right)^{1-\lambda} \left(\int_{\mathbb{R}^n} g\right)^\lambda \not\le C \int_{\mathbb{R}^n} f^{1-\lambda} g^\lambda.$$

However, we can reverse the inequality if we consider some other expression instead of $\int_{\mathbb{R}^n} f^{1-\lambda} g^\lambda$. For that, we use the sup-convolution of these two functions, which is defined in the following way. Given $f, g : \mathbb{R}^n \to \mathbb{R}_+$ measurable, and $0 \le \lambda \le 1$, we define

$$f^{1-\lambda} *_{\sup} g^\lambda(z) := \sup_{z=(1-\lambda)x+\lambda y} f(x)^{1-\lambda} g^\lambda(y).$$

(In this definition we consider all possible couples (x, y) such that z is convex combination of x, y for λ, in particular, we have $f^{1-\lambda}(z)g^\lambda(z) \le f^{1-\lambda} *_{\sup} g^\lambda(z)$).

This function is not necessarily measurable, but we can consider its exterior Lebesgue integral defined by

$$\int_{\mathbb{R}^n}^* f^{1-\lambda} *_{\sup} g^\lambda(z)dz = \inf\left\{\int_{\mathbb{R}^n} h(z)dz : f^{1-\lambda} *_{\sup} g^\lambda(z) \le h(z)\right\},$$

and we have the following result:

Theorem 2.1 (Prékopa-Leindler's inequality). *Let $f, g, h : \mathbb{R}^n \to \mathbb{R}_+$ be three measurable functions such that, for some $0 \le \lambda \le 1$,*

$$f(x)^{1-\lambda} g(y)^\lambda \le h((1-\lambda)x + \lambda y), \qquad \forall\, x, y \in \mathbb{R}^n.$$

Then

$$\left(\int_{\mathbb{R}^n} f(x)dx\right)^{1-\lambda}\left(\int_{\mathbb{R}^n} g(y)dy\right)^{\lambda} \le \int_{\mathbb{R}^n} h(z)dz.$$

Moreover, we have the reverse Hölder's inequality,

$$\left(\int_{\mathbb{R}^n} f(x)dx\right)^{1-\lambda}\left(\int_{\mathbb{R}^n} g(y)dy\right)^{\lambda} \le \int_{\mathbb{R}^n}^{*} f^{1-\lambda} *_{sup} g^{\lambda}(z)dz.$$

Proof. *Dimension* $n = 1$.

Let $A, B \subseteq \mathbb{R}$ be non-empty compact sets. Then we have

$$|A+B|_1 \ge |A|_1 + |B|_1.$$

(This inequality is the one-dimensional case of Brunn-Minkowski inequality, which we will present later.) Indeed,

$$A+B \supseteq (\min A + B) \cup (A + \max B),$$

and

$$(\min A + B) \cap (A + \max B) = \min A + \max B,$$

which implies

$$|A+B|_1 \ge |\min A + B|_1 + |A + \max B|_1 = |A|_1 + |B|_1.$$

For the rest of Borel sets we can use an approximation argument. Given two bounded Borel measurable functions f, g we can assume without loss of generality that $\|f\|_\infty = \|g\|_\infty = 1$.

For any $0 \le t < 1$, since

$$\{x \in \mathbb{R} : h(x) \ge t\} \supseteq (1-\lambda)\{x \in \mathbb{R} : f(x) \ge t\} + \lambda\{x \in \mathbb{R} : g(x) \ge t\},$$

we have that

$$\begin{aligned}\int_{\mathbb{R}} h(x)dx &\ge \int_0^1 |\{h \ge t\}|dt \ge (1-\lambda)\int_0^1 |\{f \ge t\}|dt + \lambda\int_0^1 |\{g \ge t\}|dt \\ &\ge \text{ (by the arithmetic-geometric mean inequality)} \\ &\ge \left(\int_{\mathbb{R}} f(x)dx\right)^{1-\lambda}\left(\int_{\mathbb{R}} g(x)dx\right)^{\lambda}.\end{aligned}$$

For general measurable functions we apply approximation arguments and the monotone convergence theorem.

Induction for $n > 1$.

Fix $x_1 \in \mathbb{R}$, let $f_{x_1} : \mathbb{R}^{n-1} \to [0,\infty)$ be $f_{x_1}(x_2,\dots,x_n) = f(x_1,\dots,x_n)$. Whenever $z_1 = (1-\lambda)x_1 + \lambda y_1$, and $(x_2,\dots,x_n), (y_2,\dots,y_n) \in \mathbb{R}^{n-1}$, we have,

$$h_{z_1}\big((1-\lambda)(x_2,\dots,x_n) + \lambda(y_2,\dots,y_n)\big) \ge f_{x_1}(x_2,\dots,x_n)^{1-\lambda} g_{y_1}(y_2,\dots,y_n)^{\lambda}.$$

By the induction hypothesis,

$$\int_{\mathbb{R}^{n-1}} h_{z_1}(\overline{z})d\overline{z} \geq \left(\int_{\mathbb{R}^{n-1}} f_{x_1}(\overline{x})d\overline{x}\right)^{1-\lambda}\left(\int_{\mathbb{R}^{n-1}} g_{y_1}(\overline{y})d\overline{y}\right)^{\lambda}.$$

Applying again the inequality for $n = 1$ and Fubini's theorem we obtain the result. □

If we consider characteristic functions, we have $f = \chi_A$, $g = \chi_B$, $A, B \subseteq \mathbb{R}^n$ and we obtain the following consequence.

Corollary 2.1 (Brunn-Minkowski inequality). *Let A, B two Borel sets in $\mathbb{R}^n$. For any $0 \leq \lambda \leq 1$*

$$|A|^{1-\lambda}|B|^{\lambda} \leq |(1-\lambda)A + \lambda B|, \tag{1}$$

or, equivalently,

$$|A|^{\frac{1}{n}} + |B|^{\frac{1}{n}} \leq |A+B|^{\frac{1}{n}}, \tag{2}$$

whenever $A \neq \emptyset \neq B$.

It is clear that (1) and (2) are equivalent:

- (1) $\Longrightarrow$ (2). We take

$$A' = \frac{A}{|A|^{\frac{1}{n}}}, \qquad B' = \frac{B}{|B|^{\frac{1}{n}}}, \qquad \lambda = \frac{|B|^{\frac{1}{n}}}{|A|^{\frac{1}{n}} + |B|^{\frac{1}{n}}}.$$

- (2) $\Longrightarrow$ (1). By the arithmetic-geometric mean inequality,

$$|(1-\lambda)A + \lambda B|^{\frac{1}{n}} \geq (1-\lambda)|A|^{\frac{1}{n}} + \lambda|B|^{\frac{1}{n}} \geq |A|^{\frac{1-\lambda}{n}}|B|^{\frac{\lambda}{n}}.$$

2.2. *Log-concave measures*

A measure (or probability measure) $d\mu(x)$ in $\mathbb{R}^n$ is log-concave if

$$d\mu(x) = e^{-V(x)}dx,$$

where $V : \mathbb{R}^n \to (-\infty, \infty]$ is a convex function (and the support of V is an n-dimensional convex subset in $\mathbb{R}^n$).

Examples.

- The Lebesgue measure in $\mathbb{R}^n$.
- The uniform measure on K, convex body in $\mathbb{R}^n$ (compact, convex with non empty interior).
- The exponential measure, $d\mu(x) = e^{-|x|}dx$ in $\mathbb{R}^n$.
- The classical Gaussian measure in $\mathbb{R}^n$, $d\mu(x) = \frac{1}{(\sqrt{2\pi})^n}\exp\left(-\frac{|x|^2}{2}\right)dx$.

2.2.1. *Brunn-Minkowski inequality for log-concave probabilities*

Theorem 2.2. *Any log-concave probability μ on $\mathbb{R}^n$ satisfies the Brunn-Minkowski inequality, i.e.,*

$$\mu\big((1-\lambda)A+\lambda B\big) \geq \mu(A)^{1-\lambda}\mu(B)^{\lambda},$$

for any $A, B \subseteq \mathbb{R}^n$ Borel sets and any $0 \leq \lambda \leq 1$.

Proof. We take $f(x) = \chi_A(x)e^{-V(x)}$, $g(y) = \chi_B(y)e^{-V(y)}$, and

$$h(z) = \chi_{(1-\lambda)A+\lambda B}(z)e^{-V(z)}.$$

Then we apply Prékopa-Leindler inequality. □

2.3. *Isoperimetric inequality in* $\mathbb{R}^n$

The classical isoperimetric inequality says that among all the Borel sets having the same volume the corresponding Euclidean ball is the one with the smallest perimeter or, reciprocally, among all the Borel sets having the same perimeter the corresponding Euclidean ball is the one with the greatest volume. This fact can be expressed in the following way.

Theorem 2.3. *Let A be any bounded Borel set in $\mathbb{R}^n$, then*

$$\frac{|\partial A|^{\frac{1}{n-1}}}{|A|^{\frac{1}{n}}} \geq \frac{|S^{n-1}|^{\frac{1}{n-1}}}{|B_2^n|^{\frac{1}{n}}},$$

where

$$|\partial A| = \liminf_{t\to 0}\frac{|A^t|-|A|}{t},$$

and A^t, the t-dilation of A, is $A^t = \{x \in \mathbb{R}^n; d(x, A) \leq t\} = A + tB_2^n$.

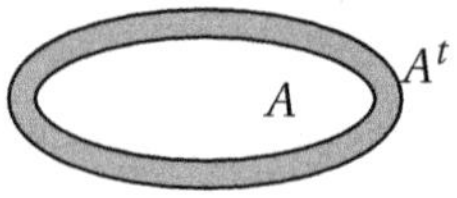

Proof. Observe that

$$|A^t|-|A| = |A+tB_2^n|-|A| \geq \left(|A|^{\frac{1}{n}} + t\omega_n^{\frac{1}{n}}\right)^n - |A| = nt|A|^{\frac{n-1}{n}}\omega_n^{\frac{1}{n}} + o(t).$$

Hence

$$|\partial A| = \liminf_{t\to 0}\frac{|A^t|-|A|}{t} \geq n|A|^{\frac{n-1}{n}}\omega_n^{\frac{1}{n}},$$

and then

$$\frac{|\partial A|^{\frac{1}{n-1}}}{|A|^{\frac{1}{n}}} \geq \frac{|S^{n-1}|^{\frac{1}{n-1}}}{|B_2^n|^{\frac{1}{n}}}. \qquad \square$$

2.4. *C. Borell's inequality*

Theorem 2.4. *Let μ be a log-concave probability in $\mathbb{R}^n$. Then for any symmetric convex set $A \subseteq \mathbb{R}^n$ with $\mu(A) \geq \theta \geq \frac{1}{2}$ we have*

$$\mu(tA)^c \leq \theta\Big(\frac{1-\theta}{\theta}\Big)^{1+\frac{t}{2}} \qquad \forall\, t > 1.$$

For instance, if $\mu(A) \geq 2/3$,

$$\mu(tA)^c \leq \frac{1}{2}\exp\Big(-\frac{t\log 2}{2}\Big) \qquad \forall\, t > 1.$$

This inequality means that there is an exponential decay of the mass for $(t > 1)$-dilations of A symmetric, with absolute constants.

Proof. It is a consequence of the fact that

$$A^c \supseteq \frac{2}{t+1}(tA)^c + \frac{t-1}{t+1}A,$$

and Brunn-Minkowski inequality,

$$1-\theta \geq \mu(A^c) \geq \mu((tA)^c)^{\frac{2}{t+1}}\mu(A)^{\frac{t-1}{t+1}}. \qquad \square$$

As a consequence we have a type of reverse Hölder's inequality and exponential decay of semi-norms.

Corollary 2.2. *There exist absolute constants, $C_1, C_2 > 0$, such that for any log-concave probability on $\mathbb{R}^n$ and for any semi-norm $f : \mathbb{R}^n \to [0,\infty)$, we have,*

(i) $\Big(\int_{\mathbb{R}^n} f^p\, d\mu\Big)^{\frac{1}{p}} \leq C_1 p \int_{\mathbb{R}^n} f\, d\mu, \qquad \forall\, p > 1,$

(ii) $\mu\Big\{x \in \mathbb{R}^n : f(x) \geq C_2 t \int_{\mathbb{R}^n} f\, d\mu\Big\} \leq 2\exp\Big(-t\log 2\Big), \qquad \forall\, t > 0.$

Proof. (i) Since any semi-norm is integrable we can assume that $\int_{\mathbb{R}^n} f\, d\mu = 1$. Let $A = \{f < 3\}$. By Markov's inequality $\mu(A) \geq 2/3$. Then

$$\mu\Big\{f \geq 3t\Big\} = \mu(tA)^c \leq \frac{1}{2}\exp\Big(-t\frac{\log 2}{2}\Big), \qquad t > 1.$$

Let $p > 1$,

$$\int_{\mathbb{R}^n} f^p d\mu = \int_0^3 p t^{p-1} \mu\{f > t\} dt + \int_3^\infty p t^{p-1} \mu\{f > t\} dt$$
$$\leq 3^p + 3^p \int_1^\infty p s^{p-1} e^{-2s} ds \leq (C_1 p)^p,$$

for some absolute constant $C_1 > 0$, and (i) follows.

(ii) Assume that $\int_{\mathbb{R}^n} f d\mu = 1$. By Markov's inequality, taking $p = t \geq 1$,

$$\mu\{f > 2C_1 t\} = \mu\left\{\frac{f}{2C_1 t} > 1\right\} \leq \int \frac{f^t}{(2C_1 t)^t} d\mu \leq \frac{(C_1 t)^t}{(2C_1 t)^t} = e^{-t \log 2}.$$

For $0 < t \leq 1$, the trivial bound 1 does the job. □

Remark 2.1. Inequalities for the moments:
Consider the case in which $f(x) = |x|$, and use that $\mathbb{E}_\mu |x|^p = \int_{\mathbb{R}^n} |x|^p d\mu$.

- By Borell's inequality,

$$\left(\mathbb{E}_\mu |x|^p\right)^{\frac{1}{p}} \leq C_1 p \mathbb{E}_\mu |x|, \qquad \forall p > 1,$$

  and

$$\mu\left\{|x| \geq C_2 t \mathbb{E}_\mu |x|\right\} \leq 2 \exp(-t), \qquad \forall t > 0.$$

- Paouris (2006) improved the inequality to

$$\left(\mathbb{E}_\mu |x|^p\right)^{\frac{1}{p}} \leq C \max\left\{\mathbb{E}_\mu |x|, p\lambda_\mu\right\},$$

  where $\lambda_\mu = \sup_{\theta \in S^{n-1}} \left(\mathbb{E}_\mu |\langle x, \theta\rangle|^2\right)^{\frac{1}{2}}$. We also have, for $t \geq 1$,

$$\mu\left\{|x| \geq C t \mathbb{E}_\mu |x|\right\} \leq \exp\left(-3 \frac{t \mathbb{E}_\mu |x|}{\lambda_\mu}\right),$$

  which is stronger than Borell's inequality.

3. Isoperimetric inequalities

In this section we will study isoperimetric inequalities with respect to log-concave probabilities. We will consider their corresponding functional Poincaré's inequalities and will show E. Milman's result on the role on convexity.

3.1. *Isoperimetric versus functional inequalities*

As we said before, the classical isoperimetric inequality on $\mathbb{R}^n$ says that

$$|\partial A| \geq C|A|^{1-\frac{1}{n}}, \qquad \forall \text{ bounded Borel set } A \subseteq \mathbb{R}^n,$$

where $C = \frac{|S^{n-1}|}{|B_2^n|^{1-\frac{1}{n}}}$, and $|\partial A|$ is the outer Minkowski content of A:

$$|\partial A| = \liminf_{\varepsilon\to 0} \frac{|A^\varepsilon| - |A|}{\varepsilon},$$

with $A^\varepsilon = \{a + x; a \in A, |x| < \varepsilon\} = A + \varepsilon B$, being the ε-dilation of A. The outer Minkowski content coincides with the $(n-1)$-dimensional Hausdorff measure of the boundary for bounded Borel sets with smooth enough boundary.

We know that another approach for proving the classical isoperimetric inequality is to establish the corresponding Sobolev inequality in the extreme.

Theorem 3.1 (Federer-Fleming). *The following statements are equivalent with the same constant C:*

- *For any bounded Borel set $A \subset \mathbb{R}^n$,*

$$|\partial A| \geq C|A|^{1-\frac{1}{n}}.$$

- *For any locally Lipschitz compactly supported function $f: \mathbb{R}^n \to \mathbb{R}$,*

$$\| |\nabla f| \|_1 \geq C \|f\|_{\frac{n}{n-1}}.$$

Here

$$\|f\|_{\frac{n}{n-1}} = \left(\int_{\mathbb{R}^n} |f(x)|^{\frac{n}{n-1}} dx\right)^{\frac{n-1}{n}},$$

and

$$|\nabla f(x)| = \limsup_{y\to x} \frac{|f(y) - f(x)|}{|y - x|},$$

which is defined for every $x \in \mathbb{R}^n$ and coincides almost everywhere with the classical modulus of the gradient due to Rademacher's theorem.

3.2. *Isoperimetric inequalities for log-concave probabilities*

Kannan, Lovász and Simonovits (see [17]) posed the following question which originally arose in relation with some problems in theoretical computer sciences, i.e., an algorithmic question about the complexity of volume computation for convex bodies: Given a convex body $K \subset \mathbb{R}^n$ find a surface S which divide K into two parts, K_1, K_2, whose measure is minimal relative to the volume of the two parts. Namely, which is the greatest constant $C = C(K)$ that makes the following formula true for all surfaces S dividing K into two parts K_1 and K_2?

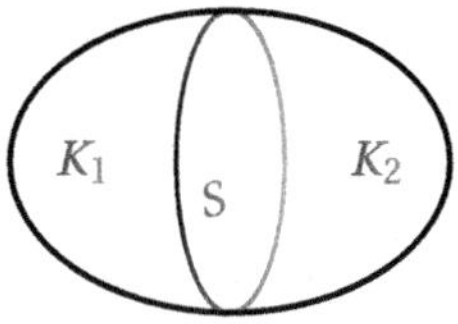

$$\mathrm{vol}_{n-1}(S) \geq C \frac{\mathrm{vol}(K_1)\cdot \mathrm{vol}(K_2)}{\mathrm{vol}(K)}.$$

If we normalize, $\mu(A) = \dfrac{|A|}{|K|}$, we have

$$\mathrm{vol}_{n-1}(\partial_K K_1) \geq C\, \frac{\mathrm{vol}(K_1)\cdot \mathrm{vol}(K_2)}{\mathrm{vol}(K)}, \qquad \Longleftrightarrow \qquad \mu^+(A) \geq C\, \mu(A)\mu(A^c),$$

which is known as Cheeger-type isoperimetric inequality. Hence the problem is, given μ (the uniform probability on a convex body K or more generally any log-concave probability), estimate the best constant $C > 0$ for which

$$\mu^+(A) \geq C\mu(A)\mu(A^c) \qquad \forall \text{ Borel set } A \subseteq \mathbb{R}^n,$$
$$\Updownarrow\ (C \geq C' \geq C/2)$$
$$\mu^+(A) \geq C' \min\{\mu(A), \mu(A^c)\} \qquad \forall \text{ Borel set } A \subseteq \mathbb{R}^n,$$

where

$$\mu^+(A) := \liminf_{\varepsilon\to 0} \frac{\mu(A^\varepsilon) - \mu(A)}{\varepsilon}, \qquad \text{and} \qquad A^\varepsilon = \{a + x; a \in A, |x| < \varepsilon\}.$$

In a similar way to the classical one we have the following result

Theorem 3.2 (Maz'ja, Cheeger). *Let μ be a Borel probability measure in $\mathbb{R}^n$. The following statements are equivalent:*

(i) For any Borel set $A \subseteq \mathbb{R}^n$, $\mu^+(A) \geq C_1 \min\{\mu(A), \mu(A^c)\}$.
(ii) For any f, integrable and locally Lipschitz, $C_2\|f - \mathbb{E}_\mu f\|_1 \leq \|\,|\nabla f|\,\|_1$.

Moreover $C_2 \leq C_1 \leq 2C_2$.

Here we have used the following notation:

$$\mathbb{E}_\mu f = \int_{\mathbb{R}^n} f\, d\mu, \qquad \text{and} \qquad \|g\|_1 = \mathbb{E}_\mu |g| = \int_{\mathbb{R}^n} |g| d\mu.$$

Proof of (i) $\Longrightarrow$ (ii). We shall use the coarea formula, and a result for positive locally Lipschitz functions.

Lemma 3.1 (Federerer). *Assume that $f : \mathbb{R}^n \to \mathbb{R}$ is smooth, and that $g : \mathbb{R}^n \to [0,\infty)$ is measurable. Then*

$$\int_{\mathbb{R}^n} g(x)\cdot|\nabla f(x)|dx = \int_{-\infty}^{\infty}\int_{\{f(x)=t\}} g(f^{-1}(t))d\mathcal{H}_{n-1}\,dt.$$

In particular, for $g=1$, we have

$$\int_{\mathbb{R}^n} |\nabla f(x)|\,dx = \int_{-\infty}^{\infty}\int_{\{f(x)=t\}} d\mathcal{H}_{n-1}\,dt = \int_{-\infty}^{\infty} \mathcal{H}_{n-1}\{x\in\mathbb{R}^n; f(x)=t\}\,dt.$$

Lemma 3.2 (see [8]). *Assume that $f>0$ is locally Lipschitz and μ is a log-concave probability. Then*

$$\int_{\mathbb{R}^n} |\nabla f(x)|d\mu(x) \geq \int_0^{\infty} \mu^+(A_t)\,dt,$$

where $A_t = \{x\in\mathbb{R}^n : f(x)>t\}$.

Now, assume that $f>0$. Then

$$\begin{aligned}
\int_{\mathbb{R}^n} |\nabla f(x)|d\mu(x) &\geq \int_0^{\infty} \mu^+\{f>t\}dt \\
&\overset{(i)}{\geq} C_1\int_0^{\infty} \min\{\mu(A_t),\mu(A_t^c)\}\,dt \\
&\geq C_1\int_0^{\infty} \mu(A_t)\mu(A_t^c)\,dt \\
&= \frac{C_1}{2}\int_0^{\infty} \|\chi_{A_t}-\mathbb{E}_\mu\chi_{A_t}\|_1\,dt \\
&= \frac{C_1}{2}\int_0^{\infty} \sup_{\|g\|_\infty=1}\int_{\mathbb{R}^n}(\chi_{A_t}(x)-\mathbb{E}_\mu\chi_{A_t})g(x)\,d\mu(x)\,dt \\
&\geq \frac{C_1}{2}\sup_{\|g\|_\infty=1}\int_0^{\infty}\int_{\mathbb{R}^n}(\chi_{A_t}(x)-\mathbb{E}_\mu\chi_{A_t})g(x)\,d\mu(x)\,dt \\
&= \frac{C_1}{2}\sup_{\|g\|_\infty=1}\int_0^{\infty}\int_{\mathbb{R}^n}\chi_{A_t}(x)\big(g(x)-\mathbb{E}_\mu g\big)\,d\mu(x)\,dt \\
&= \frac{C_1}{2}\sup_{\|g\|_\infty=1}\int_{\mathbb{R}^n}\big(g(x)-\mathbb{E}_\mu g\big)f(x)\,d\mu \\
&= \frac{C_1}{2}\|f-\mathbb{E}_\mu f\|_1.
\end{aligned}$$

In the general case we proceed for functions bounded from below and then by an approximation argument. □

Proof of (ii) $\Longrightarrow$ (i). Let A be a Borel set in $\mathbb{R}^n$. For $0<\varepsilon<1$, set

$$f_\varepsilon(x)=\max\Big\{0,1-\frac{d(x,A^{\varepsilon^2})}{\varepsilon-\varepsilon^2}\Big\}.$$

- $0\le f_\varepsilon(x)\le 1$,
- $f_\varepsilon(x)=1$ if $x\in A^{\varepsilon^2}\left(\supseteq A\right)$,
- $f(x)=0$, whenever $d(x,A)>\varepsilon$,
- $\lim_{\varepsilon\to 0} f_\varepsilon=\chi_{\overline{A}}$,
- $|f_\varepsilon(x)-f_\varepsilon(y)|\le \frac{1}{\varepsilon(1-\varepsilon)}\left|d(x,A^{\varepsilon^2})-d(y,A^{\varepsilon^2})\right|\le \frac{|x-y|}{\varepsilon(1-\varepsilon)}$,
- f_ε is locally Lipschitz and, for $x\in\mathbb{R}^n$, $|\nabla f_\varepsilon(x)|\le \frac{1}{\varepsilon-\varepsilon^2}$.

It is true that $|\nabla f_\varepsilon(x)|=0$ whenever $x\in\{x\in\mathbb{R}^n; d(x,A)>\varepsilon\}\cup A^{\varepsilon^2}\supseteq\{x\in\mathbb{R}^n; d(x,A)>\varepsilon\}\cup A$. Thus,

$$\int_{\mathbb{R}^n}|\nabla f_\varepsilon(x)|d\mu(x)\le\frac{\mu(A^{\varepsilon+\varepsilon^2})-\mu(A)}{\varepsilon-\varepsilon^2}.$$

By (ii),

$$C_2\|f_\varepsilon-\mathbb{E}_\mu f_\varepsilon\|_1\le\frac{\mu(A^{\varepsilon+\varepsilon^2})-\mu(A)}{\varepsilon-\varepsilon^2},$$

and so, letting $\varepsilon\to 0^+$, we obtain

$$2C_2\mu(A)\mu(A^c)\le\mu^+(A),$$

which gives (i). □

3.3. *Poincaré inequalities associated to a log-concave μ*

Given $1\le p\le q\le\infty$, we introduce

Definition 3.1. $D_{p,q}(\mu)$ is the greatest constant that makes the following inequality true

$$D_{p,q}\|f-\mathbb{E}_\mu f\|_p\le\|\,|\nabla f|\,\|_q,$$

for any locally Lipschitz integrable function $f\in L^p(d\mu)$.

- Case $p=q=1$: $D_{1,1}$ is equivalent to the isoperimetric Cheeger constant for μ
- Case $p=q=2$: is the Poincaré inequality for $d\mu=e^{V(x)}\,dx$:

$$D_{2,2}^2\underbrace{\int_{\mathbb{R}^n}\left|f-\int_{\mathbb{R}^n}f\,d\mu\right|^2d\mu}_{Var_\mu(f)}\le\int_{\mathbb{R}^n}|\nabla f|^2\,d\mu.$$

- $D_{2,2}^2 = \lambda_1$ is known as the spectral gap of μ, which is the first eigenvalue of the Laplace-Beltrami operator, associated to μ,

$$L = \Delta - \langle \nabla V, \nabla \rangle.$$

Proposition 3.1. *The following relations hold:*

- *By Hölder's inequality, if* $1 \le p \le q \le \infty$,

$$D_{p,q} \le D_{p,\infty} \le D_{1,\infty}, \qquad \text{and} \qquad D_{p,q} \le D_{1,q} \le D_{1,\infty}.$$

- *Maz'ja and Cheeger (1960):*

$$D_{1,1} \le C D_{2,2},$$

where C is an absolute constant.

- *Easy modification of Hölder's inequality:*

$$D_{p,p} \le C p' D_{p',p'}, \qquad \forall 1 \le p \le p' \le \infty,$$

where C is an absolute constant

Much more important and difficult are the following.

Theorem 3.3. *If* μ *is a log-concave probability we have*

(i) *Ledoux (1994):* $D_{2,2} \le C D_{1,1}$.
(ii) *E. Milman (2010):* $D_{1,\infty} \le C D_{1,1}$.

Consequently,

$$D_{p,q} \le D_{1,\infty} \le C D_{1,1} \le C p D_{p,q}.$$

Here, C is an absolute constant (independent of μ *and even of the dimension).*

Part (ii) of Theorem 3.3 is a consequence of the following result.

Theorem 3.4 (E. Milman, see [22]). *Let* μ *be a log-concave probability in* $\mathbb{R}^n$. *Assume that*

$$D_{1,\infty} \mathbb{E}_\mu |f - \mathbb{E}_\mu f| \le \| |\nabla f| \|_\infty, \qquad \forall f \text{ locally Lipschitz.}$$

Then

$$\mu^+(A) \ge C D_{1,\infty} \mu(A)^2, \qquad \forall \text{ borelian } A \text{ with } \mu(A) \le \frac{1}{2},$$

where $C > 0$ *is an absolute constant. Moreover,*

$$C D_{1,\infty} \mathbb{E}_\mu |f - \mathbb{E}_\mu f| \le \mathbb{E}_\mu |\nabla f|, \qquad \text{and} \qquad C D_{1,\infty} \le D_{1,1}.$$

Proof. The "moreover" part: $\mu^+(A) \geq CD_{1,\infty}\mu(A)^2$ is enough. Indeed,

$$\mu(A) = \frac{1}{2} \Longrightarrow \mu^+(A) \geq C\frac{D_{1,\infty}}{4}.$$

The isoperimetric profile, defined by

$$I_\mu(t) := \inf\{\mu^+(A) : \mu(A) = t\}, \qquad 0 \leq t \leq \frac{1}{2},$$

is a concave function (this result is true both in Riemannian geometry and for log-concave probabilities due to the work of a lot of people: Bavard-Pansu, Bérard-Besson-Gallot, Gallot, Morgan-Johnson, Sternberg-Zumbrun, Kuwert, Bayle-Rosales, Bayle, Morgan, Bobkov).

Given $0 \leq t \leq \frac{1}{2}$,

$$I_\mu(t) \geq 2tI_\mu\left(\frac{1}{2}\right) \geq 2tCD_{1,\infty}\frac{1}{4} \geq CD_{1,\infty}\frac{t}{2}.$$

Then

$$\mu^+(A) \geq CD_{1,\infty}\mu(A), \qquad \text{whenever } \mu(A) \leq \frac{1}{2}.$$

Hence, Cheeger's theorem implies

$$\mathbb{E}_\mu|\nabla f| \geq CD_{1,\infty}\mathbb{E}_\mu|f - \mathbb{E}_f|.$$

Next we will use the semigroups technique introduced by Ledoux: Given $d\mu = e^{-V(x)}dx$, V convex and smooth, let L be the associated Laplace-Beltrami operator, given by $L = \Delta - \langle\nabla V, \nabla\rangle$. Let $(P_t)_{t\geq 0}$ be the semigroup generated by L. It is characterized by the heat diffusion given by the following system of differential equations of second order:

$$\begin{cases} \frac{d}{dt}P_t(f) &= L(P_t(f)), \\ P_0(f) &= f, \end{cases}$$

for every bounded smooth function f.

The main properties are:

(1) $P_t(1) = 1$,
(2) $f \geq 0 \Longrightarrow P_t(f) \geq 0$,
(3) $\mathbb{E}_\mu P_t(f) = \mathbb{E}_\mu f$,
(4) $\mathbb{E}_\mu|P_t(f)|^p \leq \mathbb{E}_\mu|f|^p$, $\forall p \geq 1$,
(5) (Bakry-Ledoux). If $2 \leq q \leq \infty$ and f is bounded and smooth,

$$\| |\nabla P_t(f)| \|_{L^q(\mu)} \leq \frac{1}{\sqrt{2t}}\|f\|_{L^q(\mu)}.$$

(6) (Ledoux). If f is bounded and smooth,

$$\|f - P_t(f)\|_{L^1(\mu)} \le \sqrt{2t}\| |\nabla f| \|_{L^1(\mu)}.$$

Assume now that A is closed, $\mu(A) \le \frac{1}{2}$. Given $\varepsilon > 0$, let $A^\varepsilon = \{x \in \mathbb{R}^n : d(x, A) < \varepsilon\}$. The function $\chi_{A,\varepsilon}(x) = \max\{1 - \frac{1}{\varepsilon} d(x, A), 0\}$ is Lipschitz and

- $\lim_{\varepsilon \to 0} \chi_{A,\varepsilon}(x) = 1 \iff x \in A$,
- $|\nabla \chi_{A,\varepsilon}(x)| = \begin{cases} = 0, & \text{if } \ x \in \operatorname{int} A, \\ = 0, & \text{if } \ d(x, A) > \varepsilon, \\ \le \frac{1}{\varepsilon}, & \text{if } \ x \notin \operatorname{int} A, 0 \le d(x, A) \le \varepsilon. \end{cases}$

Then

$$\frac{\mu(A^\varepsilon) - \mu(A)}{\varepsilon} \ge \int_{\mathbb{R}^n} |\nabla \chi_{A,\varepsilon}(x)| d\mu(x) \ge (\text{by } 6) \ge \frac{1}{\sqrt{2t}} \mathbb{E}_\mu |\chi_{A,\varepsilon} - P_t(\chi_{A,\varepsilon})|.$$

When $\varepsilon \to 0$, we have

$$\begin{aligned} \sqrt{2t}\,\mu^+(A) &\ge \mathbb{E}_\mu |\chi_A - P_t(\chi_A)| = 2\Big(\mu(A) - \int_A P_t(\chi_A)(x) d\mu(x)\Big) \\ &= 2\Big(\mu(A)\mu(A^c) - \mathbb{E}_\mu\big(\chi_A - \mu(A)\big)\big(P_t(\chi_A) - \mu(A)\big)\Big) \\ (\text{by Hölder's ineq.}) &\ge 2\Big(\mu(A)\mu(A^c) - \|\chi_A - \mu(A)\|_\infty \mathbb{E}_\mu |P_t(\chi_A) - \mu(A)|\Big). \end{aligned}$$

We use the hypothesis and we have

$$\mathbb{E}_\mu |P_t(\chi_A) - \mu(A)| \le \frac{1}{D_{1,\infty}} \mathbb{E}_\mu |\nabla P_t(\chi_A)|,$$

since $\mathbb{E}_\mu P_t(\chi_A) = \mu(A)$. Also $\nabla P_t\big(\chi_A - \mu(A)\big) = \nabla P_t(\chi_A)$, so

$$\begin{aligned} \mathbb{E}_\mu |\nabla P_t(\chi_A)| &\le \| |\nabla P_t(\chi_A)| \|_2 \\ (\text{by } 5) &\le \frac{1}{\sqrt{2t}} \|P_t \chi_A - \mu(A)\|_2 \\ (\text{by } 4) &\le \frac{1}{\sqrt{2t}} \|\chi_A - \mu(A)\|_2 \\ &\le \frac{1}{\sqrt{2t}} \|\chi_A - \mu(A)\|_\infty \le \frac{1}{\sqrt{2t}}. \end{aligned}$$

Since $\mu(A) \le \frac{1}{2}$,

$$\sqrt{2t}\mu^+(A) \ge \mu(A) - \frac{2}{\sqrt{2t} D_{1,\infty}}.$$

Choose $\sqrt{2t} D_{1,\infty} = \frac{4}{\mu(A)}$ and we get

$$\mu^+(A) \ge 8 D_{1,\infty} \mu(A)^2. \qquad \square$$

4. K-L-S spectral gap conjecture

Given μ a log-concave probability on $\mathbb{R}^n$, the Kannan-Lovász-Simonovits problem is to estimate the greatest constant in the inequality

$$\mu^+(A) \geq C\mu(A), \qquad \forall \text{ Borel set } A \text{ with } \mu(A) \leq \frac{1}{2}.$$

We know that this problem is equivalent, up to absolute constants, to estimate the best constant in Poincaré's inequality,

$$\lambda_2 \mathbb{E}_\mu |f - \mathbb{E}_\mu f|^2 \leq \mathbb{E}_\mu |\nabla f|^2, \qquad \forall \text{ locally Lipschitz } f,$$

and also equivalent to estimate the best constant in

$$C' \mathbb{E}_\mu |f - \mathbb{E}_\mu f|^2 \leq 1, \qquad \forall \text{ 1-Lipschitz } f.$$

KLS Conjecture. *The greatest constant is attained for affine functions, up to an absolute constant.*

Let f be an affine function on $\mathbb{R}^n$, $f(x) = t + \langle a, x\rangle$, $t \in \mathbb{R}$, $a \in \mathbb{R}^n$. Its Lipschitz constant is $|a|$. If we assume that f is affine and 1-Lipschitz, then $f(x) = t + \langle \theta, x\rangle$, $t \in \mathbb{R}$, $\theta \in S^{n-1}$. Thus, $\mathbb{E}_\mu f = t + \langle \theta, \mathbb{E}_\mu x\rangle$ and

$$\mathbb{E}_\mu |f - \mathbb{E}_\mu f|^2 = \mathbb{E}_\mu \langle \theta, x - \mathbb{E}_\mu x\rangle^2.$$

Let λ_μ be greatest eigenvalue of the covariance matrix of μ, i.e.,

$$\lambda_\mu^2 := \sup_{\theta \in S^{n-1}} \mathbb{E}_\mu \langle \theta, x - \mathbb{E}_\mu x\rangle^2.$$

Hence, an equivalent formulation for the KLS conjecture is as follows.

KLS Conjecture. *Let μ be a log-concave probability on $\mathbb{R}^n$. Then*

$$\mathbb{E}_\mu |f - \mathbb{E}_\mu f|^2 \leq C\lambda_\mu^2 E_\mu |\nabla f|^2, \qquad \forall \text{ 1-Lipschitz } f,$$

for some absolute constant $C > 0$.

4.1. *Known results*

Theorem 4.1 (KLS estimate, see [17]). *Given μ (log-concave) and f Lipschitz integrable, we have*

$$\mathbb{E}_\mu |f - \mathbb{E}_\mu f|^2 \leq C\mathbb{E}_\mu |x - \mathbb{E}_\mu x|^2 \cdot \mathbb{E}_\mu |\nabla f|^2,$$

where $C > 0$ is an absolute constant.

Proof.

$$\begin{aligned}
\mathbb{E}_\mu|f-\mathbb{E}_\mu f|^2 &\le \mathbb{E}_\mu\big(|f(x)-f(\mathbb{E}_\mu x)|+|f(\mathbb{E}_\mu x)-\mathbb{E}_\mu f|\big)^2\\
&= \mathbb{E}_\mu\big(|f(x)-f(\mathbb{E}_\mu x)|+|\mathbb{E}_\mu(f(\mathbb{E}_\mu x)-f)|\big)^2\\
\text{(by Minkowski's ineq.)} &\le 4\mathbb{E}_\mu|f(x)-f(\mathbb{E}_\mu x)|^2\\
&\le 4\mathbb{E}_\mu|x-\mathbb{E}_\mu x|^2\cdot \||\nabla f|\|_\infty^2\\
&\le C\mathbb{E}_\mu|x-\mathbb{E}_\mu x|^2\cdot \mathbb{E}_\mu|\nabla f|^2.
\end{aligned}$$

This inequality is worse than the one conjectured by the authors since $\mathbb{E}_\mu|x-\mathbb{E}_\mu x|^2=\sum_{i=1}^n \mathbb{E}_\mu(x_i-\mathbb{E}_\mu x_i)^2\le n\lambda_\mu^2$. □

Theorem 4.2 (Payne-Weinberger (1960)). *If μ is the normalized uniform measure on a convex body K, then*

$$\mathbb{E}_\mu|f-\mathbb{E}_\mu f|^2\le \frac{4}{\pi^2}\,diam(K)^2\cdot\mathbb{E}_\mu|\nabla f|^2.$$

If B_2^n is the Euclidean ball and μ the normalized uniform measure on it, the sharp estimate is

$$\mathbb{E}_\mu|f-\mathbb{E}_\mu f|^2\le \frac{C}{n}\cdot\mathbb{E}_\mu|\nabla f|^2.$$

The first estimate in Payne-Weinberger inequality is a trivial consequence of KLS estimate since

$$E_\mu|x-\mathbb{E}_\mu x|^2\le \text{ (by Jensen)} \le \mathbb{E}_{\mu\otimes\mu}|x-y|^2\le \text{ (diam } K)^2.$$

The second one is much more acurate.

Theorem 4.3 (Talagrand, 1991). *Let $d\mu(x)=\frac{1}{2^n}e^{-\sum_{i=1}^n|x_i|}\,dx$. Then*

$$\mathbb{E}_\mu|f-\mathbb{E}_\mu f|^2\le C\cdot\mathbb{E}_\mu|\nabla f|^2.$$

(This is the classical Talagrand's inequality for the exponential probability.)

Theorem 4.4 (Gaussian case). *Let $d\mu(x)=(2\pi)^{-n/2}e^{-|x|^2/2}\,dx$. Then*

$$Var_\mu f=\mathbb{E}_\mu|f-\mathbb{E}_\mu f|^2\le\mathbb{E}_\mu|\nabla f|^2.$$

Proof. It is easy to see that $\lambda_\mu=1$. Let $u\in\mathscr{D}(\mathbb{R}^n)$ be a test functions. Consider the associated Laplace-Beltrami operator L

$$Lu(x)=\Delta u(x)-\langle x,\nabla u(x)\rangle.$$

We know that $\{Lu;u\in\mathscr{D}\}$ is dense in $\{f\in L^2(\mathbb{R}^n,d\mu):\mathbb{E}_\mu f=0\}$. Then, $\inf_{u\in\mathscr{D}}\left\{\mathbb{E}_\mu(Lu-f)^2\right\}=0$, and integrating by parts,

- $\mathbb{E}_\mu f\, Lu = -\mathbb{E}_\mu\langle \nabla f, \nabla u\rangle$, (Green's formula),
- $\mathbb{E}_\mu (Lu)^2 = \mathbb{E}_\mu\langle \nabla u, \nabla u\rangle + \mathbb{E}_\mu \sum_{i,j}(\partial_{ij}u(x))^2 \geq \mathbb{E}_\mu |u|^2$.

Assume that $\mathbb{E}_\mu f = 0$. Since

$$\mathrm{Var}_\mu f = \mathbb{E}_\mu |f - \mathbb{E}_\mu f|^2 = \mathbb{E}_\mu f^2,$$

we have,

$$\begin{aligned}\mathbb{E}_\mu f^2 - \mathbb{E}_\mu (Lu - f)^2 &= 2\mathbb{E}_\mu f\, Lu - \mathbb{E}_\mu (Lu)^2\\ &\leq -2\mathbb{E}_\mu\langle \nabla f, \nabla u\rangle - \mathbb{E}_\mu |u|^2\\ &\leq \mathbb{E}_\mu |\nabla f|^2.\end{aligned}$$

Taking the infimum in u we obtain the result,

$$\mathrm{Var}_\mu f \leq \lambda_\mu^2 \cdot \mathbb{E}_\mu |\nabla f|^2. \qquad \square$$

Theorem 4.5. *The normalized measure on the previous classes verify the KLS conjecture.*

- *p-balls, $1 \leq p \leq \infty$ (Sodin (2008) , Łatala and Wojtaszczyk (2008)).*
- *The simplex (Barthe and Wolff, 2009).*
- *Some revolution bodies (Bobkov 2003, Hue 2011).*
- *Unconditional bodies (Klartag, 2009) with a $\log n$ constant, i.e.,*

$$Var_\mu f = \mathbb{E}_\mu \left| f - \mathbb{E}_\mu f \right|^2 \leq C \log n\, \lambda_\mu^2 \mathbb{E}_\mu |\nabla f|^2.$$

(A convex body K is unconditional if $(x_1, \dots, x_n) \in K$ if and only if $(|x_1|, \dots, |x_n|) \in K$.)

A general upper bound which is the best known estimate, up to now, is

Theorem 4.6 (Guédon-Milman (2011), Eldan (2013)). *For any log-concave probability μ in $\mathbb{R}^n$, Poincaré's inequality is true in the following way*

$$Var_\mu f = \mathbb{E}_\mu |f - \mathbb{E}_\mu f|^2 \leq C n^{2/3} (\log n)^2 \lambda_\mu^2 \mathbb{E}_\mu |\nabla f|^2,$$

for any locally Lipschitz integrable function f.

4.2. *Relations with other conjectures*

There are other well-known geometric and probabilistic conjectures related with the KLS conjecture. They are the variance or thin shell conjecture and the slicing problem.

4.2.1. *The slicing problem, Bourgain (1986):*

There exists an absolute constant $C > 0$ such that every convex body K in $\mathbb{R}^n$ with volume 1 has, at least, one $(n-1)$-dimensional section such that

$$|K \cap H|_{n-1} \geq C.$$

The slicing problem or hyperplane conjecture was introduced by Bourgain when he was proving the boundedness in L^p of the Hardy-Littlewood maximal function on convex bodies. It is known to be true in the following families:

- Unconditional convex bodies,
- Zonoids,
- Random polytopes,
- Polytopes in which the number of vertices is proportional to the dimension, i.e., for instance, $N/n \leq 2$,
- The unit balls of finite dimensional Schatten classes,for $1 \leq p \leq \infty$,
- $(n-1)$-orthogonal projection of the classes above,
- ...and more.

And a general estimate is as follows.

Theorem 4.7. *There exists an absolute constant $C > 0$ such that for every convex body K in $\mathbb{R}^n$ with volume 1 at least one $(n-1)$-dimensional section satisfies:*

- *(Bourgain, 1986).* $|K \cap H|_{n-1} \geq \frac{C}{n^{1/4} \log n}$.
- *(Klartag, 2006).* $|K \cap H|_{n-1} \geq \frac{C}{n^{1/4}}$.

4.2.2. *Thin shell width or variance conjecture*
(Bobkov-Koldobsky, Anttila-Ball-Perissinaki, 2003)

There exists an absolute constant $C > 0$ such that for every log-concave probability μ in $\mathbb{R}^n$,

$$\sigma_\mu = \sqrt{\mathbb{E}_\mu \big| |x| - \mathbb{E}_\mu |x| \big|^2} \leq C \lambda_\mu.$$

It is not difficult to prove that the thin shell conjecture is just the KLS conjecture to be true for the function $|x|$ or for $|x|^2$. The name is due to the following fact.

Theorem 4.8. *If the thin shell width conjecture were true, we would have a stronger concentration of the mass around the mean for log-concave probabil-*

ities:

$$\mu\Big\{\big||x|-\mathbb{E}_\mu|x|\big|>t\mathbb{E}_\mu|x|\Big\}\le 2\exp\Big(-C' t^{\frac{1}{2}}\frac{(\mathbb{E}_\mu|x|)^{\frac{1}{2}}}{\lambda_\mu^{\frac{1}{2}}}\Big),\qquad \forall\, t>0.$$

The thin shell width conjecture is true for the uniform probability on:

- finite dimensional p-balls, $1\le p\le\infty$,
- finite dimensional Orlicz-balls,
- revolution bodies,
- $(n-1)$-dimensional orthogonal projections of the cross-polytope (1-ball),
- $(n-1)$-dimensional orthogonal projections of the cube and even all their linear deformations,
- although this conjecture is not linear invariant, in a random sense, more than half of linear deformations of the classes above also satisfy this conjecture.

The best known estimate is the one given by O. Guédon and E. Milman.

Theorem 4.9 (Guédon-Milman, 2010). *There exists an absolute constant $C>0$ such that for every log-concave probability μ in $\mathbb{R}^n$,*

$$\sigma_\mu\le Cn^{1/3}\lambda_\mu.$$

The relation among the three conjectures is the following:

- Eldan-Klartag (2010) proved that if the thin shell conjecture is true for all log-concave probabilities then the slicing problem is also true for any convex body.
- Eldan (2013) proved that if the thin shell width conjecture were true for any log-concave probability then the Kannan-Lovász-Simonovits spectral gap conjecture would be true, up to a $\log n$ factor.
- In a parallel way Ball-Nguyen (2013) proved that if the KLS conjecture were true for a family of convex bodies the slicing problem would be true for this family.

Finally, let us mention that the contribution of the authors with respect to the thin shell width conjecture are contained in the references [1–3], and they are as follows.

- The $(n-1)$-dimensional orthogonal projections of the cross-polytope (1-ball) verify this conjecture.
- The $(n-1)$-dimensional orthogonal projections of the cube and even all their linear deformations verify this conjecture.

- If μ verifies the thin shell width conjecture then $\nu = \mu \circ T$ also verifies the thin shell width conjecture at least for half of T's (T linear map) in a probabilistic meaning and 'at random' if the Schatten norm of $\| T \|_{C_4}$ satisfies

$$\frac{\| T \|_{HS}}{\| T \|_{C_4}} = o(n^{\frac{1}{4}}).$$

- The $(n-1)$-dimensional orthogonal projections of the p-balls verify this conjecture if $p > n$ and up to a $\log(1+p)$ factor whenever $1 \leq p \leq n$.

References

[1] D. Alonso-Gutiérrez and J. Bastero, The variance conjecture on some polytopes, in *Asymptotic geometric analysis*, Fields Inst. Commun., Vol. 68, pp. 1–20 (Springer, New York, 2013).

[2] D. Alonso-Gutiérrez and J. Bastero, *Approaching the Kannan-Lovász-Simonovits and variance conjectures*, Lecture Notes in Mathematics, Vol. 2131, pp. x+148 (Springer, Cham, 2015).

[3] D. Alonso-Gutiérrez and J. Bastero. The variance conjecture on hyperplane projections of the ℓ_p^n balls. Forthcoming paper.

[4] K. Ball and V. H. Nguyen, Entropy jumps for isotropic log-concave random vectors and spectral gap, *Studia Math.* **213**, pp. 81–96 (2012).

[5] F. Barthe and D. Cordero-Erausquin, Invariances in variance estimates, *Proc. Lond. Math. Soc. (3)* **106**, pp. 33–64 (2013).

[6] F. Barthe and P. Wolff, Remarks on non-interacting conservative spin systems: the case of gamma distributions, *Stochastic Process. Appl.* **119**, pp. 2711–2723 (2009).

[7] S. G. Bobkov and A. Koldobsky, On the central limit property of convex bodies, in *Geometric aspects of functional analysis*, Lecture Notes in Math., Vol. 1807, pp. 44–52 (Springer, Berlin, 2003).

[8] S. G. Bobkov and C. Houdré, Some connections between isoperimetric and Sobolev-type inequalities, *Mem. Amer. Math. Soc.* **129**, pp. viii+111 (1997).

[9] C. Borell, Convex measures on locally convex spaces, *Ark. Mat.* **12**, pp. 239–252 (1974).

[10] S. Brazitikos, A. Giannopoulos, P. Valettas and B.-H. Vritsiou, *Geometry of isotropic convex bodies*, Mathematical Surveys and Monographs, Vol. 196, pp. xx+594 (American Mathematical Society, Providence, RI, 2014).

[11] J. Cheeger, A lower bound for the smallest eigenvalue of the Laplacian, in *Problems in analysis (Papers dedicated to Salomon Bochner, 1969)*, pp. 195–199 (Princeton Univ. Press, Princeton, N. J., 1970).

[12] R. Eldan, Thin shell implies spectral gap up to polylog via a stochastic localization scheme, *Geom. Funct. Anal.* **23**, pp. 532–569 (2013).

[13] R. Eldan and B. Klartag, Approximately Gaussian marginals and the hyperplane conjecture, in *Concentration, functional inequalities and isoperimetry*, Contemp. Math., Vol. 545, pp. 55–68 (Amer. Math. Soc., Providence, RI, 2011).

[14] M. Gromov and V. D. Milman, Generalization of the spherical isoperimetric inequality to uniformly convex Banach spaces, *Compositio Math.* **62**, pp. 263–282 (1987).

[15] O. Guédon and E. Milman, Interpolating thin-shell and sharp large-deviation estimates for isotropic log-concave measures, *Geom. Funct. Anal.* **21**, pp. 1043–1068 (2011).

[16] N. Huet, Spectral gap for some invariant log-concave probability measures, *Mathematika* **57**, pp. 51–62 (2011).

[17] R. Kannan, L. Lovász and M. Simonovits, Isoperimetric problems for convex bodies and a localization lemma, *Discrete Comput. Geom.* **13**, pp. 541–559 (1995).

[18] B. Klartag, Power-law estimates for the central limit theorem for convex sets, *J. Funct. Anal.* **245**, pp. 284–310 (2007).

[19] B. Klartag, A Berry-Esseen type inequality for convex bodies with an unconditional basis, *Probab. Theory Related Fields* **145**, pp. 1–33 (2009).

[20] R. Latała and J. O. Wojtaszczyk, On the infimum convolution inequality, *Studia Math.* **189**, pp. 147–187 (2008).

[21] M. Ledoux, *The concentration of measure phenomenon*, Mathematical Surveys and Monographs, Vol. 89, pp. x+181 (American Mathematical Society, Providence, RI, 2001).

[22] E. Milman, On the role of convexity in isoperimetry, spectral gap and concentration, *Invent. Math.* **177**, pp. 1–43 (2009).

[23] V. D. Milman and A. Pajor, Isotropic position and inertia ellipsoids and zonoids of the unit ball of a normed n-dimensional space, in *Geometric aspects of functional analysis (1987–88)*, Lecture Notes in Math., Vol. 1376, pp. 64–104 (Springer, Berlin, 1989).

[24] V. D. Milman and G. Schechtman, *Asymptotic theory of finite-dimensional normed spaces*, Lecture Notes in Mathematics, Vol. 1200, pp. viii+156 (Springer-Verlag, Berlin, 1986), With an appendix by M. Gromov.

[25] G. Paouris, Concentration of mass on convex bodies, *Geom. Funct. Anal.* **16**, pp. 1021–1049 (2006).

[26] S. Sodin, An isoperimetric inequality on the l_p balls, *Ann. Inst. Henri Poincaré Probab. Stat.* **44**, pp. 362–373 (2008).

[27] M. Talagrand, A new isoperimetric inequality and the concentration of measure phenomenon, in *Geometric aspects of functional analysis (1989–90)*, Lecture Notes in Math., Vol. 1469, pp. 94–124 (Springer, Berlin, 1991).

[28] S. S. Vempala, Recent progress and open problems in algorithmic convex geometry, in *30th International Conference on Foundations of Software Technology and Theoretical Computer Science*, LIPIcs. Leibniz Int. Proc. Inform., Vol. 8, pp. 42–64 (Schloss Dagstuhl. Leibniz-Zent. Inform., Wadern, 2010).

Two weight inequalities for fractional integral operators and commutators*

David Cruz-Uribe, OFS

Department of Mathematics, University of Alabama, Tuscaloosa, AL, 35401 USA
E-mail: dcruzuribe@ua.edu

Keywords: Fractional integral operators, commutators, dyadic operators, weights, testing conditions, A_p bump conditions.

1. Introduction

In these lecture notes we describe some recent work on two weight norm inequalities for fractional integral operators, also known as Riesz potentials, and for commutators of fractional integrals. Our point of view is strongly influenced by the groundbreaking work on dyadic operators that led to the proof of the A_2 conjecture by Hytönen [42] and the simplification of that proof by Lerner [58, 59]. (See also [43] for a more detailed history and bibliography of this problem.) Fractional integrals are of interest in their own right and have important applications in the study of Sobolev spaces and PDEs. They are positive operators and in many instances proofs are much easier for fractional integrals than they are for Calderón-Zygmund singular integrals. But as we will see, in some cases they are more difficult to work with, and we will give several examples of results which are known to hold for singular integrals but remain conjectures for fractional integrals.

After giving some preliminary results in Section 2, in Section 3 we lay out the abstract theory of dyadic grids and show how inequalities for fractional integrals and commutators can be reduced to the study of dyadic operators. All of these ideas were implicit in the classical Calderón-Zygmund decomposition but in recent years the essentials have been extracted, yielding a substantially new perspective.

In Section 4 we show how the dyadic approach can be used to simplify the proof of one weight norm inequalities for fractional maximal and integral oper-

*The author is supported by NSF grant DMS 1362425. While these lecture notes were being written, he was also supported by the Stewart-Dorwart faculty development fund and the Dean of Faculty at Trinity College, Hartford, CT.

ators. The purpose of this digression is two-fold. First, it provides a nice illustration of the power of these dyadic methods, as the proofs are markedly simpler than the classical proofs. Second, we will use these proofs to illustrate the technical obstacles we will encounter in trying to prove two weight inequalities.

There are two approaches to two weight inequalities for fractional integrals: the testing conditions, first introduced by Sawyer [88, 91], and the "A_p bump" conditions introduced by Neugebauer [76] and Pérez [79]. Both approaches have their advantages. In Section 5 we consider testing conditions. The fundamental result we discuss is due to Lacey, Sawyer and Uriarte-Tuero [55], but we will present a beautiful simplification of their proof due to Hytönen [43]. We conclude this section with a conjecture concerning testing conditions for commutators of fractional integrals.

In Sections 6 and 7 we will discuss bump conditions. Besides the work of Pérez cited above, the contents of these sections are based on recent work by the author and Moen [24–26]. We conclude the last section with several open problems.

Throughout these lecture notes we assume that the reader is familiar with real analysis (e.g., as presented by Royden [86]) and with classical harmonic analysis including the basics of the theory of Muckenhoupt A_p weights and one weight norm inequalities (e.g., the first seven chapters of Duoandikoetxea [35]). Additional references include the classic books by Stein [94] and García-Cuerva and Rubio de Francia [38] and the more recent books by Grafakos [40, 41]. Many of the results we give for weighted norm inequalities for fractional integrals are scattered through the literature—there is unfortunately no single source for this material. We will provide copious references throughout, including historical ones. Some of the material in these notes is new and has not appeared in the literature before.

These notes are based on three lectures delivered at the *6th International Course of Mathematical Analysis in Andalucía*, held in Antequera, Spain, September 8–12, 2014. They are, however, greatly expanded to include both new results and many details that I did not present in my lectures due to time constraints. In addition, I have taken this opportunity to correct some (relatively minor) mistakes in the proofs I sketched in the lectures. I am grateful to the organizers for the invitation to present this work. I would also like to thank Kabe Moen, my principal collaborator on fractional integrals (or Riesz potentials, as he prefers), and Carlos Pérez, who introduced me to bump conditions and has shared his insights with me for many years. It has been a privilege to work with both of them.

2. Preliminaries

In this section we gather some essential definitions and a few background results. Hereafter, we will be working in $\mathbb{R}^n$ and n will always denote the dimension. We will denote constants by C, c, etc. and the value may change at each appearance. If necessary, we will denote the dependence of the constants parenthetically: e.g., $C = C(n, p)$. The letters P and Q will be used to denote cubes in $\mathbb{R}^n$. By a weight we will always mean a non-negative, measurable function that is positive on a set of positive measure.

Averages of functions will play a very important role in these notes, so we introduce some useful notation. Given any set E, $0 < |E| < \infty$, we define

$$\fint_E f(x)\,dx = \frac{1}{|E|}\int_E f(x)\,dx.$$

More generally given a non-negative measure μ, we define

$$\fint_E f(x)\,d\mu = \frac{1}{\mu(E)}\int_E f(x)\,d\mu.$$

In other words, an average is always with respect to the measure. If we have a measure of the form $\sigma\,dx$, where σ is a weight, we will write $d\sigma$, as in $\fint_E f\,d\sigma$, to emphasize this fact. We will also use the following more compact notation, particularly when the set is a cube Q:

$$\fint_Q f(x)\,dx = \langle f\rangle_Q, \qquad \fint_Q f(x)\,d\sigma = \langle f\rangle_{Q,\sigma}.$$

We now define the two operators we will be focusing on. Given $0 < \alpha < n$ and a measurable function f, we define the fractional integral operator I_α by

$$I_\alpha f(x) = \int_{\mathbb{R}^n} \frac{f(y)}{|x-y|^{n-\alpha}}\,dy.$$

Given a function $b \in BMO$, the space of functions of bounded mean oscillation, we define the commutator

$$[b, I_\alpha]f(x) = b(x)I_\alpha f(x) - I_\alpha(bf)(x) = \int_{\mathbb{R}^n} \big(b(x) - b(y)\big)\frac{f(y)}{|x-y|^{n-\alpha}}\,dy.$$

The fractional integral operator is classical: it was introduced by M. Riesz [85]. The commutators are more recent and were first considered by Chanillo [7]. The following are some of the basic properties of these operators; unless otherwise noted, see Stein [94], Chapter V, for details.

(1) I_α is a positive operator: if $f(x) \geq 0$ a.e., then $I_\alpha f(x) \geq 0$. Note, however, that $[b, I_\alpha]$ is not positive.

(2) For $1 < p < \frac{n}{\alpha}$, if we define q by $\frac{1}{p} - \frac{1}{q} = \frac{\alpha}{n}$, then

$$I_\alpha : L^p \to L^q,$$

and (see Chanillo [7]) for all $b \in BMO$,

$$[b, I_\alpha] : L^p \to L^q.$$

(3) When $p = 1$, $q = \frac{n}{n-\alpha}$, then I_α satisfies the weak type inequality

$$I_\alpha : L^1 \to L^{q,\infty},$$

but commutators are more singular and do not satisfy a weak $(1, \frac{n}{n-\alpha})$ inequality. For a counter-example and a substitute inequality, see [16].

(4) We can define fractional powers of the Laplacian via the Fourier transform using the fractional integral operator: for all Schwartz functions f and $0 < \alpha < n$,

$$(-\Delta)^{-\frac{\alpha}{2}} f(x) = cI_\alpha f(x).$$

We also have that for all $f \in C_c^\infty$,

$$|f(x)| \le cI_1(|\nabla f|)(x).$$

Fractional integrals have found wide application in the study of PDEs. Here we mention a few results. Recall the Sobolev embedding theorem (see Stein [94], Chapter V): if f is contained in the Sobolev space $W^{1,p}$, then for $1 \le p < n$ and $p^* = \frac{np}{n-p}$,

$$\|f\|_{L^{p^*}} \le C\|\nabla f\|_{L^p}.$$

When $p > 1$ this is an immediate consequence of the inequality relating I_1 and the gradient, and the strong type norm inequality for I_1. When $p = 1$ it can be proved using the weak type inequality for I_1 and a decomposition argument due to Maz'ya [65], p. 110 (see also Long and Nie [64] and [22], Lemma 4.31).

Two weight norm inequalities for I_α also yield weighted Sobolev embeddings. In particular, they can be used to prove inequalities of the form

$$\|f\|_{L^p(u)} \le C\|\nabla f\|_{L^p}.$$

These were introduced by Fefferman and Phong (see [36]) in the study of the Schrödinger operator. Such inequalities can also be used to prove that weak solutions of elliptic equations with non-smooth coefficients are strong solutions: see, for example, Chiarenza and Franciosi [8] and [28]. For additional applications we refer to the paper by Sawyer and Wheeden [93] and the many references it contains. (We remark in passing that this paper has been extremely influential in the study of two weight norm inequalities for fractional integrals.)

Closely related to the fractional integral operator is the fractional maximal operator: given $0 < \alpha < n$ and $f \in L^1_{loc}$, define

$$M_\alpha f(x) = \sup_Q |Q|^{\frac{\alpha}{n}} \fint_Q |f(y)|\, dy \cdot \chi_Q(x),$$

where the supremum is taken over all cubes with sides parallel to the coordinate axes. The fractional maximal operator was introduced by Muckenhoupt and Wheeden [67] in order to prove one weight norm inequalities for I_α via a good-λ inequality. This result is the analog of the one linking the Hardy-Littlewood maximal operator and Calderón-Zygmund singular integrals proved by Coifman and Fefferman [11].

For $1 < p < \frac{n}{\alpha}$, M_α satisfies the same strong (p, q) inequality as I_α. In addition, it satisfies the upper endpoint estimate $M_\alpha : L^\infty \to L^{\frac{n}{\alpha}}$. In contrast, if $f \in L^\infty$, then $I_\alpha f$ need not be bounded, but does satisfy an exponential integrability condition. See, for instance, Ziemer [105], Theorem 2.9.1.

Our approach to norm inequalities for the fractional integral operator will avoid M_α; however, we will use it as a model operator since it has many features in common with I_α but is usually easier to work with. We note in passing that there is an Orlicz fractional maximal operator that plays a similar role for commutators of fractional integrals: see [16]. (This operator also plays a role in the study of two weight, weak $(1,1)$ inequalities for I_α: see Section 7.)

3. Dyadic operators

In this section we explain the machinery of dyadic grids and dyadic operators. These ideas date back to the 1950's and the seminal work of Calderón and Zygmund [3], and have played a prominent role in harmonic analysis since then. In the past fifteen years they have been reformulated and taken on a new prominence because of their connection with the A_2 conjecture. An important early presentation of this new point of view was the lecture notes on dyadic harmonic analysis by Pereyra [78]. As she described them:

> These notes contain what I consider are the main actors and universal tools used in this area of mathematics. They also contain an overview of the classical problems that lead mathematicians to study these objects and to develop the tools that are now considered the *abc* of harmonic analysis. The modern twist is the connection to a parallel dyadic world where objects, statements and sometimes proofs are simpler, but yet illuminated enough to guarantee that one can translate them into the non-dyadic world.

The major advance since this was written was the realization that not only could dyadic operators illuminate what was going on with their non-dyadic counterparts, but in fact the solution of non-dyadic problems could be reduced to proving the corresponding results for dyadic operators. Our understanding of this approach continues to evolve: see for instance, the very recent work on dyadic approximation by Lerner and Nazarov [60] and Conde-Alonso and Rey [13].

This philosophy of dyadic operators can be summarized by paraphrasing the title of the hit song from Irving Berlin's 1946 musical, *Annie Get Your Gun* (see Figure 1):

Anything you can do, I can do better (dyadically)!

Fig. 1. Ethel Merman as Annie Oakley, 1946

Dyadic grids

We begin by recalling the classical dyadic grid. This is the countable collection of cubes that are dyadic translates and dilations of the unit cube, $[0,1)^n$:

$$\Delta = \{Q = 2^k([0,1)^n + m) : k \in \mathbb{Z}, m \in \mathbb{Z}^n\}.$$

These cubes have a number of important properties: any cube in Δ has side-length a power of two; any two cubes in Δ are disjoint or one is contained in the other; given any $k \in \mathbb{Z}$, the subcollection Δ_k of cubes with side-length 2^k forms a partition of $\mathbb{R}^n$.

The importance of dyadic cubes lies in the Calderón-Zygmund cubes, which give a very powerful decomposition of a function. For proof of this result, see García-Cuerva and Rubio de Francia [38], Chapter II, and [22], Appendix A.

Proposition 3.1. *Let $f \in L^1_{loc}$ be such that $\langle f\rangle_Q \to 0$ as $|Q| \to \infty$ (e.g., $f \in L^p$, $1 \le p < \infty$). Then for each $\lambda > 0$ there exists a collection of disjoint dyadic cubes $\{Q_j\} \subset \Delta$ such that*

$$\lambda < \fint_{Q_j} |f(x)|\,dx \le 2^n \lambda.$$

Moreover, given $a \ge 2^{n+1}$, for each $k \in \mathbb{Z}$ let $\{Q_j^k\}$ be the collection of cubes obtained by taking $\lambda = a^k$ above. Define

$$\Omega_k = \bigcup_j Q_j^k, \qquad E_j^k = Q_j^k \setminus \Omega_{k+1}.$$

Then for all j and k, the sets E_j^k are pairwise disjoint and $|E_j^k| \ge \frac{1}{2}|Q_j^k|$.

These cubes are closely related to the dyadic maximal operator: given $f \in L^1_{loc}$, define the operator M^d ([a]) by

$$M^d f(x) = \sup_{Q\in\Delta} \fint_Q |f(y)|\,dy \cdot \chi_Q(x).$$

Then for each $\lambda > 0$, if we form the cubes Q_j from the first part of Proposition 3.1,

$$\{x \in \mathbb{R}^n : M^d f(x) > \lambda\} = \bigcup_j Q_j.$$

The Calderón-Zygmund cubes were introduced by Calderón and Zygmund [3]. The essential idea underlying the sets E_j^k from the second half of Proposition 3.1 is due to Calderón [2] (working with balls in a space of homogeneous type). This idea was then applied to Calderón-Zygmund cubes by García-Cuerva and Rubio de Francia [38, Chapter IV] in their proof of the reverse Hölder inequality. It appears to have first been explicitly stated and proved as a property of Calderón-Zygmund cubes by Pérez [81].

Given the specific example of the Calderón-Zygmund cubes, we make the following two definitions that extract their fundamental properties.

Definition 3.1. A collection of cubes $\mathscr{D}$ in $\mathbb{R}^n$ is a dyadic grid if:

---

[a] In the notation we will introduce below, we would call this operator M^Δ. Here we prefer to use the classical notation. As Emerson said, "*Foolish consistency is the hobgoblin of little minds.*"

(1) If $Q \in \mathscr{D}$, then $\ell(Q) = 2^k$ for some $k \in \mathbb{Z}$.
(2) If $P, Q \in \mathscr{D}$, then $P \cap Q \in \{P, Q, \emptyset\}$.
(3) For every $k \in \mathbb{Z}$, the cubes $\mathscr{D}_k = \{Q \in \mathscr{D} : \ell(Q) = 2^k\}$ form a partition of $\mathbb{R}^n$.

Definition 3.2. Given a dyadic grid $\mathscr{D}$, a set $\mathscr{S} \subset \mathscr{D}$ is sparse if for every $Q \in \mathscr{S}$,

$$\Big| \bigcup_{\substack{P \in \mathscr{S} \\ P \subsetneq Q}} P \Big| \leq \frac{1}{2}|Q|.$$

Equivalently, if we define

$$E(Q) = Q \setminus \bigcup_{\substack{P \in \mathscr{S} \\ P \subsetneq Q}} P,$$

then the sets $E(Q)$ are pairwise disjoint and $|E(Q)| \geq \frac{1}{2}|Q|$.

It is immediate that the classical dyadic cubes Δ are a dyadic grid. By Proposition 3.1, given a function $f \in L^1_{loc}$, if we form the cubes $\{Q_j^k\}$, then they are a sparse subset of Δ with $E(Q_j^k) = E_j^k$. Because of this fact, given a fixed dyadic grid $\mathscr{D}$, we will often refer to cubes in it as dyadic cubes.

Clearly, we can get dyadic grids by taking translations of the cubes in Δ. The importance of this is that every cube in $\mathbb{R}^n$ is contained in a cube from a fixed, finite collection of such dyadic grids.

Theorem 3.1. *There exist dyadic grids $\mathscr{D}^k$, $1 \leq k \leq 3^n$, such that given any cube Q, there exists k and $P \in \mathscr{D}^k$ such that $Q \subset P$ and $\ell(P) \leq 3\ell(Q)$.*

The origin of Theorem 3.1 is obscure but we believe that credit should be given to Okikiolu [77] and, for a somewhat weaker version, to Chang, Wilson and Wolff [6].([b]) The total number of dyadic grids needed can be reduced, though at the price of increasing the constant C relating the size of the cubes.

---

[b]Theorem 3.1 and variations of it have recently been attributed to Christ in [70] and also to Garnett and Jones in [54], Section 2.2. In particular, some people suggested that it was in the paper by Garnett and Jones [39] on dyadic BMO. It is not. Moreover, these authors have told me and others that this result did not originate with them, though they knew and shared it. The earliest appearance of a version of Theorem 3.1 in print seems to be in Okikiolu [77], Lemma 1b. Earlier, Chang, Wilson and Wolff [6], Lemma 3.2, had a weaker but substantially similar version. They showed that given the set $\Delta_A = \{Q \in \Delta, \ell(Q) \leq 2^A\}$, then there exists a finite collection of translates of Δ such that given any $Q \in \Delta_A$, $3Q$ is contained in a cube of comparable size from one of these translated grids. A refined version of this lemma later appeared in Wilson [103], Lemma 2.1.
The basic idea underlying the proof of Theorem 3.1 is sometimes referred to as the "one-third trick" (e.g. in Refs. [56, 62]). This idea has been variously attributed (cf. Refs. [56, 66]) to Garnett or Garnett and Jones, Davis, and Wolff. The earliest unambiguous appearance seems to have been Wolff [104], Lemma 1.4; Wolff attributes this lemma to S. Janson.

Hytönen and Pérez [44], Theorem 1.10, showed that 2^n dyadic grids suffice, with $C = 6$. (For details of the proof, see [59], Proposition 2.1.) Conde [12] proved that only $n+1$ grids are necessary, and this bound is sharp, but with a constant $C \approx n$.

Proof. We will use the following 3^n translates of the standard dyadic grid Δ:

$$\mathscr{D}^t = \{2^j([0,1)^n + m + t) : j \in \mathbb{Z}, m \in \mathbb{Z}^n\}, \quad t \in \{0, \pm 1/3\}^n. \tag{1}$$

Now fix a cube Q; then there exists a unique $j \in \mathbb{Z}$ such that

$$\frac{2^j}{3} \leq \ell(Q) < \frac{2^{j+1}}{3}.$$

At most 2^n cubes in Δ of sidelength 2^j intersect Q; let P be one such that $|P \cap Q|$ is maximal. Note that maximality implies P contains the center of Q.

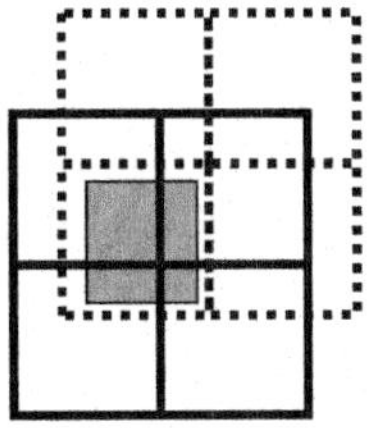

Fig. 2. The construction of P' containing Q

To get the desired cube we translate P, acting on each coordinate in succession (see Figure 2). If a face of P (i.e. a $n-1$ dimensional hyper-plane on the boundary) perpendicular to the j-th coordinate axis intersects the interior of Q, translate P parallel to the j-th coordinate axis in the direction of the closest face of Q a distance $\frac{2^j}{3}$. Because of the maximality of P, this direction is away from the interior of P. Hence, this moves the face out of Q, and the opposite face remains outside as well, so more of Q is contained in the interior of P. Thus, after at most n steps we will have a cube P' that is contained in one of the grids $\mathscr{D}^t$, $\ell(P') = \ell(P) \leq 3\ell(Q)$, and such that $Q \subset P'$. □

Though we do not use it here, we want to note that there is another important approach to dyadic grids. Nazarov, Treil and Volberg [71, 72, 74] have shown that random dyadic grids (i.e., translates of Δ where the translation is taken according to some probability distribution) are very well behaved "on average." This approach was central to Hytönen's original proof of the A_2 conjecture in [42].

Dyadic operators

We can now introduce the dyadic operators that we will use in place of the fractional maximal and integral operators and commutators. We begin with the fractional maximal operator. Given $0 < \alpha < n$, a dyadic grid $\mathscr{D}$ and $f \in L^1_{loc}$, define

$$M_\alpha^{\mathscr{D}} f(x) = \sup_{Q\in\mathscr{D}} |Q|^{\frac{\alpha}{n}} \fint_Q |f(y)|\,dy \cdot \chi_Q(x).$$

Proposition 3.2. *There exists a constant $C(n,\alpha)$ such that for every function $f \in L^1_{loc}$ and $1 \leq t \leq 3^n$,*

$$M_\alpha^{\mathscr{D}^t} f(x) \leq M_\alpha f(x) \leq C(n,\alpha) \sup_t M_\alpha^{\mathscr{D}^t} f(x),$$

where the grids $\mathscr{D}^t$ are defined by (1).

Proposition 3.2 is stated in [25] without proof; when $\alpha = 0$ this was proved in [44, Theorem 1.10], and the proof we give for $\alpha > 0$ is essentially the same.

Proof. The first inequality is immediate. To prove the second, fix x and a cube Q containing x. Then by Theorem 3.1 there exists t and $P \in \mathscr{D}^t$ such that $Q \subset P$ and $|P| \leq 3^n|Q|$. Therefore,

$$\begin{aligned} |Q|^{\frac{\alpha}{n}} \fint_Q |f(y)|\,dy &\leq 3^{n-\alpha}|P|^{\frac{\alpha}{n}} \fint_P |f(y)|\,dy \\ &\leq C(n,\alpha) M_\alpha^{\mathscr{D}^t} f(x) \leq C(n,\alpha) \sup_t M_\alpha^{\mathscr{D}^t} f(x). \end{aligned}$$

If we take the supremum over all cubes Q containing x, we get the desired inequality. □

Because we are working with a finite number of dyadic grids, we have that

$$\sup_t M_\alpha^{\mathscr{D}^t} f(x) \approx \sum_{t=1}^{3^n} M_\alpha^{\mathscr{D}^t} f(x),$$

and the constants depend only on n. In other words, we can dominate any sublinear expression for M_α by a sum of expressions involving $M_\alpha^{\mathscr{D}^t}$. The same will be true for I_α. Hereafter, we will use this equivalence without comment.

The dyadic analog of the fractional integral operator is defined as an infinite sum: given $0 < \alpha < n$ and a dyadic grid $\mathscr{D}$, for all $f \in L^1_{loc}$ let

$$I_\alpha^{\mathscr{D}} f(x) = \sum_{Q\in\mathscr{D}} |Q|^{\frac{\alpha}{n}} \langle f\rangle_Q \cdot \chi_Q(x).$$

The dyadic fractional integral operator (with $\mathscr{D} = \Delta$) was introduced by Sawyer and Wheeden [93] who showed that averages over an infinite family of dyadic grids dominated I_α. Here we show that only a finite number of grids is necessary; this was proved in [26], Proposition 2.2.

Proposition 3.3. *There exist constants* $c(n,\alpha)$, $C(n,\alpha)$ *such that for every non-negative function* $f \in L^1_{loc}$ *and* $1 \le t \le 3^n$,

$$c(n,\alpha) I_\alpha^{\mathscr{D}^t} f(x) \le I_\alpha f(x) \le C(n,\alpha) \sup_t I_\alpha^{\mathscr{D}^t} f(x),$$

where the grids $\mathscr{D}^t$ *are defined by* (1).

Proof. To prove the first inequality, fix a dyadic grid $\mathscr{D} = \mathscr{D}^t$, a non-negative function f, and $x \in \mathbb{R}^n$. Without loss of generality we may assume that f is bounded: since I_α and $I_\alpha^{\mathscr{D}}$ are positive operators, the inequality for unbounded f follows by the monotone convergence theorem.

Let $\{Q_k\}_{k\in\mathbb{Z}} \subset \mathscr{D}$ be the unique sequence of dyadic cubes such that $\ell(Q_k) = 2^k$ and $x \in Q_k$. Then for every integer $N > 0$,

$$\begin{aligned}
&\sum_{\substack{Q\in\mathscr{D}\\ \ell(Q)\le 2^N}} |Q|^{\frac{\alpha}{n}} \langle f\rangle_Q \cdot \chi_Q(x)\\
&= \sum_{k=-\infty}^{N} |Q_k|^{\frac{\alpha}{n}-1} \int_{Q_k\setminus Q_{k-1}} f(y)\,dy + \sum_{k=-\infty}^{N} |Q_k|^{\frac{\alpha}{n}-1} \int_{Q_{k-1}} f(y)\,dy\\
&\le c(n,\alpha) \sum_{k=-\infty}^{N} \int_{Q_k\setminus Q_{k-1}} \frac{f(y)}{|x-y|^{n-\alpha}}\,dy + 2^{\alpha-n} \sum_{\substack{Q\in\mathscr{D}\\ \ell(Q)\le 2^N}} |Q|^{\frac{\alpha}{n}} \langle f\rangle_Q \cdot \chi_Q(x)\\
&= c(n,\alpha) \int_{Q_N} \frac{f(y)}{|x-y|^{n-\alpha}}\,dy + 2^{\alpha-n} \sum_{\substack{Q\in\mathscr{D}\\ \ell(Q)\le 2^N}} |Q|^{\frac{\alpha}{n}} \langle f\rangle_Q \cdot \chi_Q(x).
\end{aligned}$$

Because f is bounded, the last sum is finite. Therefore, since $2^{\alpha-n} < 1$, we can rearrange terms and take the limit as $N \to \infty$ to get

$$c(n,\alpha) I_\alpha^{\mathscr{D}} f(x) \le I_\alpha f(x).$$

To prove the second inequality, let $Q(x,r)$ be the cube of side-length $2r$ centered at x. Then

$$\begin{aligned}
I_\alpha f(x) &= \sum_{k\in\mathbb{Z}} \int_{Q(x,2^k)\setminus Q(x,2^{k-1})} \frac{f(y)}{|x-y|^{n-\alpha}}\,dy\\
&\le 2^{n-\alpha} \sum_{k\in\mathbb{Z}} 2^{-k(n-\alpha)} \int_{Q(x,2^k)} f(y)\,dy.
\end{aligned}$$

By Theorem 3.1, for each $k \in \mathbb{Z}$ there exists a grid $\mathscr{D}^t$, $1 \le t \le 3^n$, and $Q_t \in \mathscr{D}^t$ such that $Q(x,2^k) \subset Q_t$ and

$$2^{k+1} = \ell(Q(x,2^k)) \le \ell(Q_t) \le 6\ell(Q(x,2^k)) = 12\cdot 2^k.$$

Since $\ell(Q_t) = 2^j$ for some j, we must have that $2^{k+1} \le \ell(Q_t) \le 2^{k+3}$. Hence,

$$\begin{aligned}
2^{n-\alpha} \sum_{k\in\mathbb{Z}} (2^{-k})^{n-\alpha} \int_{Q(x,2^k)} f(y)\,dy
&\le C(n,\alpha) \sum_{k\in\mathbb{Z}} \sum_{t=1}^{3^n} \sum_{\substack{Q\in\mathscr{D}^t \\ 2^{k+1}\le\ell(Q)\le 2^{k+3}}} |Q|^{\frac{\alpha}{n}} \langle f\rangle_Q \cdot \chi_Q(x) \\
&\le C(n,\alpha) \sum_{t=1}^{3^n} \sum_{Q\in\mathscr{D}^t} |Q|^{\frac{\alpha}{n}} \langle f\rangle_Q \cdot \chi_Q(x) \\
&\le C(n,\alpha) \sum_{t=1}^{3^n} I_\alpha^{\mathscr{D}^t} f(x) \\
&\le C(n,\alpha) \sup_t I_\alpha^{\mathscr{D}^t} f(x).
\end{aligned}$$

If we combine these two estimates we get the second inequality. □

Intuitively, the dyadic version of the commutator $[b, I_\alpha]$ is the operator $[b, I_\alpha^{\mathscr{D}}]$. However, recall that this operator is not positive: we cannot prove the pointwise bound

$$\big|[b, I_\alpha] f(x)\big| \le C \sup_t \big|[b, I_\alpha^{\mathscr{D}^t}] f(x)\big|,$$

even for f non-negative. (We are not certain whether this inequality is in fact true.) But if we pull the absolute values inside the integral we do get a useful dyadic approximation of the commutator. The following result was implicit in [24]; the proof is essentially the same as the proof of the second inequality in Proposition 3.3.

Proposition 3.4. *There exists a constant $C(n,\alpha)$ such that for every non-negative function $f \in L^1_{loc}$ and $b \in BMO$,*

$$\big|[b, I_\alpha] f(x)\big| \le C(n,\alpha) \sup_t C_b^{\mathscr{D}^t} f(x),$$

where the grids $\mathscr{D}^t$ are defined by (1) *and*

$$C_b^{\mathscr{D}^t} f(x) = \sum_{Q\in\mathscr{D}^t} |Q|^{\frac{\alpha}{n}} \fint_Q |b(x) - b(y)| f(y)\,dy \cdot \chi_Q(x).$$

Sparse operators

We now come to another important reduction: we can replace the dyadic operators $M_\alpha^{\mathscr{D}}$ and $I_\alpha^{\mathscr{D}}$ with operators defined on sparse families. For the fractional maximal operator we replace it with a linear operator that resembles the fractional integral operator. Given a dyadic grid $\mathscr{D}$, a sparse set $\mathscr{S} \subset \mathscr{D}$ and $f \in L^1_{loc}$, define the operator $L^{\mathscr{S}}_\alpha$ by

$$L^{\mathscr{S}}_\alpha f(x) = \sum_{Q\in\mathscr{S}} |Q|^{\frac{\alpha}{n}} \langle f\rangle_Q \cdot \chi_{E(Q)}(x).$$

The idea for this linearization was implicit in Sawyer [87]; for the maximal operator see also de la Torre [98]. The following result was given without proof in [25].

Proposition 3.5. *Given a dyadic grid $\mathscr{D}$ and a non-negative function f such that $|Q|^{\frac{\alpha}{n}}\langle f\rangle_Q \to 0$ as $|Q| \to \infty$, there exists a sparse set $\mathscr{S} = \mathscr{S}(f) \subset \mathscr{D}$ and a constant $C(n,\alpha)$ independent of f such that for every $x \in \mathbb{R}^n$,*

$$L^{\mathscr{S}}_\alpha f(x) \le M^{\mathscr{D}}_\alpha f(x) \le C(n,\alpha) L^{\mathscr{S}}_\alpha f(x).$$

Proof. The sets $E(Q)$ are pairwise disjoint and for every $x \in E(Q)$, $|Q|^{\frac{\alpha}{n}}\langle f\rangle_Q \le M^{\mathscr{D}}_\alpha f(x)$, so the first inequality follows at once. To prove the second inequality, fix $a = 2^{n+1-\alpha}$ and for each $k \in \mathbb{Z}$, let

$$\Omega_k = \{x \in \mathbb{R}^n : M^{\mathscr{D}}_\alpha f(x) > a^k\}.$$

For every $x \in \Omega_k$ there exists $Q \in \mathscr{D}$ such that $|Q|^{\frac{\alpha}{n}}\langle f\rangle_Q > a^k$. Let $\mathscr{S}_k$ be the collection of maximal, disjoint cubes with this property. Such maximal cubes exist by our assumption on f. Further, by maximality we must also have that for each $P \in \mathscr{S}_k$, $a^k < |P|^{\frac{\alpha}{n}}\langle f\rangle_P \le 2^{n-\alpha}a^k$, and

$$\Omega_k = \bigcup_{P\in\mathscr{S}_k} P.$$

Let $\mathscr{S} = \bigcup_k \mathscr{S}_k$; we claim that $\mathscr{S}$ is sparse. Again by maximality these cubes are nested: if $P' \in \mathscr{S}_{k+1}$, then there exists $P \in \mathscr{S}_k$ such that $P' \subsetneq P$. Therefore, if we fix $k \in \mathbb{Z}$ and $P \in \mathscr{S}_k$, and consider the union of cubes $P' \in \mathscr{S}$ with $P' \subsetneq P$, we may restrict the union to $P' \in \mathscr{S}_{k+1}$. Clearly these cubes satisfy $|P'| \le 2^{-n}|P|$. Hence,

$$\begin{aligned}\Bigg|\bigcup_{\substack{P'\in\mathscr{S}\\ P'\subsetneq P}} P'\Bigg| &= \sum_{\substack{P'\in\mathscr{S}_{k+1}\\ P'\subsetneq P}} |P'| < \frac{1}{a^{k+1}} \sum_{\substack{P'\in\mathscr{S}_{k+1}\\ P'\subsetneq P}} |P'|^{\frac{\alpha}{n}} \int_{P'} f(y)\,dy \\ &\le \frac{2^{-\alpha}}{a^{k+1}}|P|^{\frac{\alpha}{n}} \int_P f(y)\,dy \le \frac{2^{n-2\alpha}}{a}|P| \le 2^{-1}|P|.\end{aligned} \tag{2}$$

To get the desired estimate, note first that by the definition of the cubes in $\mathscr{S}$, for each $k \in \mathbb{Z}$,

$$\Omega_k \setminus \Omega_{k+1} = \bigcup_{P\in\mathscr{S}_k} E(P).$$

Therefore, we have that for each $x \in \mathbb{R}^n$, there exists k such that $x \in \Omega_k \setminus \Omega_{k+1}$, and so there exists $P \in \mathscr{S}_k$ such that

$$M_\alpha^{\mathscr{D}} f(x) \le a^{k+1} \le a|P|^{\frac{\alpha}{n}} \langle f \rangle_P \cdot \chi_{E(P)}(x) = C(n,\alpha) \sum_{P \in \mathscr{S}} |P|^{\frac{\alpha}{n}} \langle f \rangle_P \cdot \chi_{E(P)}(x). \quad \square$$

The sparse operator associated with $I_\alpha^{\mathscr{D}}$ is nearly the same as $L_\alpha^{\mathscr{S}}$ except that the characteristic function is for the entire cube Q. Given a dyadic grid $\mathscr{D}$ and a sparse set $\mathscr{S} \subset \mathscr{D}$, we define

$$I_\alpha^{\mathscr{S}} f(x) = \sum_{Q \in \mathscr{S}} |Q|^{\frac{\alpha}{n}} \langle f \rangle_Q \cdot \chi_Q(x).$$

When $\alpha = 0$, this operator becomes the sparse Calderón-Zygmund operator that plays a central role in Lerner's proof of the A_2 conjecture: see Refs. [58, 59]. The operators $I_\alpha^{\mathscr{S}}$ were implicit in Sawyer and Wheeden [93], Pérez [79] and Lacey, *et al.* [51], and first appeared explicitly in [26], where the following result was proved.

Proposition 3.6. *Given a dyadic grid $\mathscr{D}$ and a non-negative function f such that $\langle f \rangle_Q \to 0$ as $|Q| \to \infty$, there exists a sparse set $\mathscr{S} = \mathscr{S}(f) \subset \mathscr{D}$ and a constant $C(n,\alpha)$ independent of f such that for every $x \in \mathbb{R}^n$,*

$$I_\alpha^{\mathscr{S}} f(x) \le I_\alpha^{\mathscr{D}} f(x) \le C(n,\alpha) I_\alpha^{\mathscr{S}} f(x).$$

Proof. The first inequality is immediate for any subset $\mathscr{S}$ of $\mathscr{D}$. To prove the second inequality, we first construct the sparse set $\mathscr{S}$. The argument is very similar to the construction in Proposition 3.2. Let $a = 2^{n+1}$. For each $k \in \mathbb{Z}$ define

$$\mathscr{Q}_k = \{Q \in \mathscr{D} : a^k < \langle f \rangle_Q \le a^{k+1}\}.$$

Then for every $Q \in \mathscr{D}$ such that $\langle f \rangle_Q \neq 0$, there exists a unique k such that $Q \in \mathscr{Q}_k$.

Now define $\mathscr{S}_k$ to be the maximal disjoint cubes contained in

$$\{P \in \mathscr{D} : \langle f \rangle_P > a^k\}.$$

Such maximal cubes exist by our hypothesis on f. It follows that given any $Q \in \mathscr{Q}_k$, there exists $P \in \mathscr{S}_k$ such that $Q \subset P$. Furthermore, these cubes are nested: if $P' \in \mathscr{S}_{k+1}$, then it is contained in some $P \in \mathscr{S}_k$. If we let $\mathscr{S} = \bigcup_k \mathscr{S}_k$, then arguing as in inequality (2) we have that $\mathscr{S}$ is sparse.

We now prove the desired inequality. Fix $x \in \mathbb{R}^n$; then

$$I_\alpha^{\mathscr{D}} f(x) = \sum_k \sum_{Q \in \mathscr{Q}_k} |Q|^{\frac{\alpha}{n}} \langle f \rangle_Q \cdot \chi_Q(x) \le \sum_k a^{k+1} \sum_{P \in \mathscr{S}_k} \sum_{\substack{Q \in \mathscr{Q}_k \\ Q \subseteq P}} |Q|^{\frac{\alpha}{n}} \cdot \chi_Q(x).$$

The inner sum can be evaluated:

$$\sum_{\substack{Q\in\mathscr{Q}_k\\ Q\subseteq P}} |Q|^{\frac{\alpha}{n}}\cdot\chi_Q(x) = \sum_{r=0}^{\infty} \sum_{\substack{Q\in\mathscr{Q}_k:Q\subseteq P\\ \ell(Q)=2^{-r}\ell(P)}} |Q|^{\frac{\alpha}{n}}\cdot\chi_Q(x) = \frac{1}{1-2^{-\alpha}}|P|^{\frac{\alpha}{n}}\cdot\chi_P(x).$$

Thus we have that

$$\begin{aligned}\sum_k a^{k+1}\sum_{P\in\mathscr{S}_k}\sum_{\substack{Q\in\mathscr{Q}_k\\ Q\subseteq P}} |Q|^{\frac{\alpha}{n}}\cdot\chi_Q(x) &\le C(\alpha)\sum_k a^{k+1}\sum_{P\in\mathscr{S}_k}|P|^{\frac{\alpha}{n}}\cdot\chi_P(x)\\ &\le C(n,\alpha)\sum_k\sum_{P\in\mathscr{S}_k}|P|^{\frac{\alpha}{n}}\langle f\rangle_P\cdot\chi_P(x) = C(n,\alpha)I_\alpha^{\mathscr{S}}f(x).\end{aligned}$$

If we combine these estimates we get the desired inequality. □

We conclude this section with a key observation:

In light of Propositions 3.2, 3.3, 3.5 and 3.6, when proving necessary and/or sufficient conditions for weighted norm inequalities for fractional maximal or integral operators, it suffices to prove the analogous inequalities for either the associated dyadic or sparse operators.

In the subsequent sections we will use this fact repeatedly. The ability to pass to a dyadic operator will considerably simplify the proofs. The choice to use the dyadic or sparse operator will be determined by the details of the argument.

Matters are more complicated for commutators. It is possible to reduce estimates for the dyadic commutator, or more precisely, the dyadic operator $C_b^{\mathscr{D}}$ defined in Proposition 3.4, to estimates for a sum defined over a sparse set. However, this reduction does not yield a pointwise inequality and is dependent on the particular result to be proved. For an example of this argument, we refer the reader to Theorem 6.4 below and [24], Theorem 1.6. This difficultly plays a role in some of the open problems which we will discuss below.

4. Digression: one weight inequalities

In this section we briefly turn away from the main topic of these notes, two weight norm inequalities, to present some basic results on one weight norm inequalities. We do so for two reasons. First, in this setting it is easier to see the advantages of the reduction to dyadic operators; second, a closer examination of the proofs in the one weight case will highlight where the major obstacles will be in the two weight case.

The fractional maximal operator

We first consider the fractional maximal operator. The governing weight class is a generalization of the Muckenhoupt A_p weights, and was introduced by Muckenhoupt and Wheeden [67].

Definition 4.1. Given $0 < \alpha < n$, $1 < p < \frac{n}{\alpha}$, and q such that $\frac{1}{p} - \frac{1}{q} = \frac{\alpha}{n}$, we say that a weight w such that $0 < w(x) < \infty$ a.e. is in $A_{p,q}$ if

$$[w]_{A_{p,q}} = \sup_Q \left(\fint_Q w^q \, dx \right)^{\frac{1}{q}} \left(\fint_Q w^{-p'} \, dx \right)^{\frac{1}{p'}} < \infty,$$

where the supremum is taken over all cubes Q. When $p = 1$ we say $w \in A_{1,q}$ if

$$[w]_{A_{1,q}} = \sup_Q \operatorname*{ess\,sup}_{x \in Q} \left(\fint_Q w^q \, dx \right)^{\frac{1}{q}} w(x)^{-1} < \infty.$$

The $A_{1,q}$ condition is equivalent to assuming that $M_q w(x) = M(w^q)(x)^{1/q} \leq [w]_{A_{1,q}} w(x)$, that is, $w^q \in A_1$. (For a proof of this when $q = 1$, see [38], Section 5.1.) More generally, if $p > 1$, we have that $w \in A_{p,q}$ if and only if $w^q \in A_{1+\frac{q}{p'}}$; this follows at once from the definition. By symmetry we have that $w \in A_{p,q}$ if and only if $w^{-1} \in A_{q',p'}$, and this is equivalent to $w^{-p'} \in A_{1+\frac{p'}{q}}$.

In our proofs we will generally not keep track of the dependence on the constant $[w]_{A_{p,q}}$ since our proofs do not yield sharp results. For the exact dependence, see [24, 51].

Theorem 4.1. *Given $0 < \alpha < n$, $1 \leq p < \frac{n}{\alpha}$, q such that $\frac{1}{p} - \frac{1}{q} = \frac{\alpha}{n}$, and a weight w, the following are equivalent:*

(1) $w \in A_{p,q}$;
(2) for any $f \in L^p(w^p)$,

$$\sup_{t>0} t\, w^q(\{x \in \mathbb{R}^n : M_\alpha f(x) > t\})^{\frac{1}{q}} \leq C(n,\alpha)[w]_{A_{p,q}} \left(\int_{\mathbb{R}^n} |f(x)|^p w(x)^p \, dx \right)^{\frac{1}{p}}.$$

The sufficiency of the $A_{p,q}$ condition was first proved in [67]. Our proof is basically the same as theirs, but using the sparse operator $L_\alpha^{\mathcal{S}}$ obviates the need for a covering lemma argument—this is "hidden" in the construction of the sparse operator. The necessity of the $A_{p,q}$ condition was not directly considered but was implicit in their results for the fractional integral. Our argument below is adapted from the case $\alpha = 0$ in [38], Section 5.1.

Proof. To show the sufficiency of the $A_{p,q}$ condition, without loss of generality we may assume f is non-negative. It is straightforward to show that if the sequence $\{f_k\}$ increases pointwise a.e. to f, then $M_\alpha f_k$ increases to $M_\alpha f$, so we may assume that f is bounded and has compact support. (For the details of this argument when $\alpha = 0$, see [17], Lemma 3.30.) Further, it will suffice to fix a dyadic grid $\mathscr{D}$ and prove the weak type inequality for $M_\alpha^{\mathscr{D}}$.

We first consider the case when $p > 1$. Fix $t > 0$. If $x \in \mathbb{R}^n$ is such that $M_\alpha^{\mathscr{D}} f(x) > t$, then there exists a cube $Q \in \mathscr{D}$ such that $|Q|^{\frac{\alpha}{n}} \langle f \rangle_Q > t$. Let $\mathscr{Q}$ be the set of maximal disjoint cubes in $\mathscr{D}$ with this property. (Such cubes exist by our assumptions on f.) Then by Hölder's inequality,

$$\begin{aligned}
t^q w^q(\{x \in \mathbb{R}^n : M_\alpha^{\mathscr{D}} f(x) > t\}) \\
&= t^q \sum_{Q \in \mathscr{Q}} w^q(Q) \\
&\leq \sum_{Q \in \mathscr{Q}} w^q(Q) \big(|Q|^{\frac{\alpha}{n}} \langle f \rangle_Q\big)^q \\
&= \sum_{Q \in \mathscr{Q}} |Q|^{q\frac{\alpha}{n} - q} w^q(Q) \Big(\int_Q f(y) w(y) w(y)^{-1}\, dy \Big)^q \\
&\leq \sum_{Q \in \mathscr{Q}} |Q|^{q\frac{\alpha}{n} - q} w^q(Q) \Big(\int_Q w(y)^{-p'}\, dy \Big)^{\frac{q}{p'}} \Big(\int_Q f(y)^p w(y)^p\, dy \Big)^{\frac{q}{p}};
\end{aligned}$$

by our choice of q, $q - q\frac{\alpha}{n} = 1 + \frac{q}{p'}$, so by the $A_{p,q}$ condition,

$$\begin{aligned}
&\leq [w]_{A_{p,q}}^q \sum_{Q \in \mathscr{Q}} \Big(\int_Q f(y)^p w(y)^p\, dy \Big)^{\frac{q}{p}} \\
&\leq [w]_{A_{p,q}}^q \Big(\sum_{Q \in \mathscr{Q}} \int_Q f(y)^p w(y)^p\, dy \Big)^{\frac{q}{p}} \\
&\leq [w]_{A_{p,q}}^q \Big(\int_{\mathbb{R}^n} f(y)^p w(y)^p\, dy \Big)^{\frac{q}{p}}.
\end{aligned}$$

The second to last inequality holds because $\frac{q}{p} \geq 1$ and the final inequality since the cubes in $\mathscr{Q}$ are pairwise disjoint by maximality. This completes the proof of the weak type inequality when $p > 1$.

When $p = 1$ the same proof works, omitting Hölder's inequality and using the pointwise inequality in the $A_{1,q}$ condition.

To prove the necessity of the $A_{p,q}$ condition, we again first consider the case $p > 1$. Fix a cube Q and let $f = w^{-p'} \chi_Q$. Then for $x \in Q$, $M_\alpha f(x) \geq |Q|^{\frac{\alpha}{n}} \langle w^{-p'} \rangle_Q$, and so for all $t < |Q|^{\frac{\alpha}{n}} \langle w^{-p'} \rangle_Q$, the weak type inequality implies that

$$t^q w^q(Q) \leq C \Big(\int_Q f(x)^p w(x)^p\, dx \Big)^{\frac{q}{p}} = C|Q|^{\frac{q}{p}} \Big(\fint_Q w(x)^{-p'}\, dx \Big)^{\frac{q}{p}}.$$

Taking the supremum over all such t yields

$$|Q|^{q\frac{\alpha}{n}} \int_Q w(x)^q\,dx \Big(\fint_Q w(x)^{-p'}\,dx \Big)^q \leq C|Q|^{\frac{q}{p}} \Big(\fint_Q w(x)^{-p'}\,dx \Big)^{\frac{q}{p}},$$

and rearranging terms we get the $A_{p,q}$ condition on Q with a uniform constant.

When $p = 1$ we repeat the above argument but now with $f = \chi_P$, where $P \subset Q$ is any cube. Then we get

$$\fint_Q w(x)^q\,dx \leq C \Big(\fint_P w(x)^p\,dx \Big)^{\frac{q}{p}}.$$

Let x_0 be a Lebesgue point of w^p in Q, and take the limit as $P \to \{x_0\}$; by the Lebesgue differentiation theorem we get

$$\fint_Q w(x)^q\,dx \leq C w(x_0)^q.$$

The $A_{1,q}$ condition follows at once. □

The weak type inequality and its proof have two consequences. First, the proof when $p = 1$, holds for all $p > 1$ and we can replace the cube P by any measurable set $E \subset Q$. Doing this yields an A_∞ type inequality:

$$\frac{|E|}{|Q|} \leq [w]_{A_{p,q}} \Big(\frac{w^q(E)}{w^q(Q)} \Big)^{\frac{1}{q}}. \tag{3}$$

Second, though we assumed *a priori* in the definition of the $A_{p,q}$ condition that $0 < w(x) < \infty$ a.e., we can use this inequality to show that in fact this is a consequence of the weak type inequality. For the details of the proof when $\alpha = 0$, see [38], Section 5.1. We note in passing that the usual A_∞ condition, which exchanges the roles of w^q and Lebesgue measure in (3), is more difficult to prove since it also requires the reverse Hölder inequality.

To prove the strong type inequality we could use the fact that $w^q \in A_{1+\frac{p'}{q}}$ implies $w^q \in A_{1+\frac{p'}{q}-\epsilon}$ for some $\epsilon > 0$ to apply Marcinkiewicz interpolation. This is the approach used in [67] and it requires the reverse Hölder inequality.

Instead, here we are going to give a direct proof that avoids the reverse Hölder inequality. It is based on an argument for the Hardy-Littlewood maximal operator due to Christ and Fefferman [9] that only uses (3). For the proof we also introduce an auxiliary operator, a weighted dyadic fractional maximal operator. Such weighted operators when $\alpha = 0$ have played an important role in the proof of sharp constant inequalities: see Refs. [21, 57, 59]. Given a non-negative Borel measure σ and a dyadic grid $\mathscr{D}$, define

$$M^{\mathscr{D}}_{\sigma,\alpha} f(x) = \sup_{Q\in\mathscr{D}} \sigma(Q)^{\frac{\alpha}{n}} \fint_Q |f(y)|\,d\sigma \cdot \chi_Q(x).$$

If $\alpha = 0$ we simply write $M_\sigma^{\mathscr{D}}$.

Lemma 4.1. *Given $0 \le \alpha < n$, $1 \le p < \frac{n}{\alpha}$, q such that $\frac{1}{p} - \frac{1}{q} = \frac{\alpha}{n}$, a dyadic grid $\mathscr{D}$, and a non-negative Borel measure σ,*

$$\sup_{t>0} t\sigma(\{x \in \mathbb{R}^n : M_{\sigma,\alpha}^{\mathscr{D}} f(x) > t\})^{\frac{1}{q}} \le \left(\int_{\mathbb{R}^n} |f(x)|^p \, d\sigma\right)^{\frac{1}{p}}.$$

Furthermore, if $p > 1$,

$$\left(\int_{\mathbb{R}^n} M_{\sigma,\alpha}^{\mathscr{D}} f(x)^q \, d\sigma\right)^{\frac{1}{q}} \le C(p,q)\left(\int_{\mathbb{R}^n} |f(x)|^p \, d\sigma\right)^{\frac{1}{p}}.$$

Proof. The proof of the weak $(1,q)$ inequality for $M_{\sigma,\alpha}^{\mathscr{D}}$ is essentially the same as the proof of Theorem 4.1. By Hölder's inequality we have for any cube $Q \in \mathscr{D}$,

$$\sigma(Q)^{\frac{\alpha}{n}} \fint_Q |f(x)| \, d\sigma \le \sigma(Q)^{\frac{\alpha}{n}} \left(\fint_Q |f(x)|^{\frac{n}{\alpha}} \, d\sigma\right)^{\frac{\alpha}{n}} \le \|f\|_{L^{\frac{n}{\alpha}}(\sigma)},$$

which immediately implies that $M_{\sigma,\alpha}^{\mathscr{D}} : L^{\frac{n}{\alpha}}(\sigma) \to L^\infty$. The strong (p,q) inequality then follows from off-diagonal Marcinkiewicz interpolation: see [96], Chapter V, Theorem 2.4. □

Theorem 4.2. *Given $0 < \alpha < n$, $1 < p < \frac{n}{\alpha}$, q such that $\frac{1}{p} - \frac{1}{q} = \frac{\alpha}{n}$, and a weight w, the following are equivalent:*

(1) $w \in A_{p,q}$;
(2) for any $f \in L^p(w^p)$,

$$\left(\int_{\mathbb{R}^n} M_\alpha f(x)^q w(x)^q \, dx\right)^{\frac{1}{q}} \le C(n,\alpha,p,[w]_{A_{p,q}})\left(\int_{\mathbb{R}^n} |f(x)|^p w(x)^p \, dx\right)^{\frac{1}{p}}.$$

Proof. Since the strong type inequality implies the weak type inequality, necessity follows from Theorem 4.1. To prove sufficiency we can again assume f is non-negative, bounded and has compact support, and so it is enough to prove the strong type inequality for $L_\alpha^{\mathscr{S}} f$, where $\mathscr{S}$ is any sparse subset of a dyadic grid $\mathscr{D}$.

Let $\sigma = w^{-p'}$. Since $E(Q)$, $Q \in \mathscr{S}$, are disjoint, then, by inequality (3), the properties of sparse cubes, the definition of $A_{p,q}$ and Lemma 4.1,

$$\begin{aligned}
\|(L_\alpha^{\mathcal{S}} f)w\|_q^q &= \sum_{Q\in\mathcal{S}} |Q|^{q\frac{\alpha}{n}} \langle f\rangle_Q^q w^q(E(Q)) \\
&\leq \sum_{Q\in\mathcal{S}} \big(\sigma(Q)^{\frac{\alpha}{n}} \langle f\sigma^{-1}\rangle_{Q,\sigma}\big)^q |Q|^{q\frac{\alpha}{n}-q} w^q(Q)\sigma(Q)^{q-q\frac{\alpha}{n}} \\
&= \sum_{Q\in\mathcal{S}} \big(\sigma(Q)^{\frac{\alpha}{n}} \langle f\sigma^{-1}\rangle_{Q,\sigma}\big)^q |Q|^{-\frac{q}{p'}-1} w^q(Q)\sigma(Q)^{\frac{q}{p'}}\sigma(Q) \\
&\leq C([w]_{A_{p,q}}) \sum_{Q\in\mathcal{S}} \big(\sigma(Q)^{\frac{\alpha}{n}} \langle f\sigma^{-1}\rangle_{Q,\sigma}\big)^q \sigma(E(Q)) \\
&\leq C([w]_{A_{p,q}}) \sum_{Q\in\mathcal{S}} \int_{E(Q)} M_{\sigma,\alpha}^{\mathcal{D}}(f\sigma^{-1})(x)^q\,d\sigma \\
&\leq C([w]_{A_{p,q}}) \int_{\mathbb{R}^n} M_{\sigma,\alpha}^{\mathcal{D}}(f\sigma^{-1})(x)^q\,d\sigma \\
&\leq C(p,q,[w]_{A_{p,q}}) \Big(\int_{\mathbb{R}^n} f(x)^p\sigma(x)^{-p}\sigma(x)\,dx\Big)^{\frac{q}{p}} \\
&= C(p,q,[w]_{A_{p,q}}) \Big(\int_{\mathbb{R}^n} f(x)^p w(x)^p\,dx\Big)^{\frac{q}{p}}. \qquad \square
\end{aligned}$$

The above proof has several features that we want to highlight. First, since the sets $E(Q)$, $Q\in\mathcal{S}$, are pairwise disjoint, we are able to pull the power q inside the summation. For dyadic fractional integrals (even sparse ones) this is no longer the case. As we will see below, one technique for avoiding this problem is to use duality. Second, a central obstacle is that we have a sum over cubes Q that are not themselves disjoint, so we need some way of reducing the sum to the sum of integrals over disjoint sets. Here we use that the cubes in $\mathcal{S}$ are sparse, and then use the A_∞ property given by inequality (3). In the two weight setting we will no longer have this property. To overcome this we will pass to a carefully chosen subfamily of cubes that are sparse with respect to some measure induced by the weights (e.g., $d\sigma$ in the proof above).

The fractional integral operator

We now turn to one weight norm inequalities for the fractional integral operator. We will give a direct proof of the strong type inequality that appears to be new, though it draws upon ideas already in the literature: in particular, the two weight bump conditions for the fractional integral due to Pérez [79] (see Theorem 6.3 below). The original proof of this result by Muckenhoupt and Wheeden [67] uses a good-λ inequality; another proof using sharp maximal function estimates and extrapolation was given in [18] (see also [22], Chapter 9). One important feature of these approaches is that they also yield the endpoint weak type inequality for the fractional integral. It would be very interesting to give a proof of this inequality using the techniques of this section as it might shed light on several open problems: see Section 7. ([c])

Theorem 4.3. *Given $0 < \alpha < n$, $1 < p < \frac{n}{\alpha}$, q such that $\frac{1}{p} - \frac{1}{q} = \frac{\alpha}{n}$, and a weight w, the following are equivalent:*

(1) $w \in A_{p,q}$;
(2) for any $f \in L^p(w^p)$,

$$\Big(\int_{\mathbb{R}^n} I_\alpha f(x)^q w(x)^q\,dx\Big)^{\frac{1}{q}} \le C(n,\alpha,p,[w]_{A_{p,q}})\Big(\int_{\mathbb{R}^n} |f(x)|^p w(x)^p\,dx\Big)^{\frac{1}{p}}.$$

Proof. By the pointwise inequality $M_\alpha^{\mathscr{D}} f(x) \le I_\alpha^{\mathscr{D}} f(x)$, the necessity of the $A_{p,q}$ condition follows from Theorem 4.2.

To prove sufficiency, we may assume f is non-negative. Furthermore, by the monotone convergence theorem, if $\{f_k\}$ is any sequence of functions that increases pointwise a.e. to f, then for each $x \in \mathbb{R}^n$, $I_\alpha f_k(x)$ increases to $I_\alpha f(x)$. Therefore, we may also assume that f is bounded and has compact support. Thus, it will suffice to prove this result for the sparse operator $I_\alpha^{\mathscr{S}}$, where $\mathscr{S}$ is any sparse subset of a dyadic grid $\mathscr{D}$.

Let $v = w^q$ and $\sigma = w^{-p'}$ and estimate as follows: there exists $g \in L^{q'}(w^{-q'})$, $\|gw^{-1}\|_{q'} = 1$, such that

$$\begin{aligned}\|(I_\alpha^{\mathscr{S}} f)w\|_q &= \int_{\mathbb{R}^n} I_\alpha^{\mathscr{S}} f(x)g(x)\,dx = \sum_{Q\in\mathscr{S}} |Q|^{\frac{\alpha}{n}}\langle f\rangle_Q \int_Q g(x)\,dx\\ &= \sum_{Q\in\mathscr{S}} |Q|^{\frac{\alpha}{n}-1}\sigma(Q)v(Q)^{1-\frac{\alpha}{n}}\langle f\sigma^{-1}\rangle_{Q,\sigma}\, v(Q)^{\frac{\alpha}{n}}\langle gv^{-1}\rangle_{Q,v}.\end{aligned}$$

Since $1 - \frac{\alpha}{n} = \frac{1}{p'} + \frac{1}{q}$, by the definition of the $A_{p,q}$ condition and inequality (3) (applied to both v and σ), we have that

$$\begin{aligned}|Q|^{\frac{\alpha}{n}-1}\sigma(Q)v(Q)^{1-\frac{\alpha}{n}} &\le [w]_{A_{p,q}}\sigma(Q)^{\frac{1}{p}}v(Q)^{\frac{1}{p'}}\\ &\le C([w]_{A_{p,q}})\sigma(E(Q))^{\frac{1}{p}}v(E(Q))^{\frac{1}{p'}}.\end{aligned}$$

[c]Note added in proof: after these lecture notes were written such a proof of the weak type inequality when $p = 1$ was found. See [14].

If we combine these two estimates, by Hölder's inequality and Lemma 4.1 we get that

$$\begin{aligned}
\|(I_\alpha^{\mathscr{S}} f)w\|_q &\le C([w]_{A_{p,q}}) \sum_{Q\in\mathscr{S}} \langle f\sigma^{-1}\rangle_{Q,\sigma}\sigma(E(Q))^{\frac{1}{p}} v(Q)^{\frac{\alpha}{n}} \langle gv^{-1}\rangle_{Q,v} v(E(Q))^{\frac{1}{p'}} \\
&\le C([w]_{A_{p,q}})\Big(\sum_{Q\in\mathscr{S}} \langle f\sigma^{-1}\rangle_{Q,\sigma}^p \sigma(E(Q))\Big)^{\frac{1}{p}}\Big(\sum_{Q\in\mathscr{S}} [v(Q)^{\frac{\alpha}{n}} \langle gv^{-1}\rangle_{Q,v}]^{p'} v(E(Q))\Big)^{\frac{1}{p'}} \\
&\le C([w]_{A_{p,q}})\Big(\sum_{Q\in\mathscr{S}} \int_{E(Q)} M_\sigma^{\mathscr{D}}(f\sigma^{-1})(x)^p\,d\sigma\Big)^{\frac{1}{p}}\Big(\sum_{Q\in\mathscr{S}} \int_{E(Q)} M_{v,\alpha}^{\mathscr{D}}(gv^{-1})(x)^{p'}\,dv\Big)^{\frac{1}{p'}} \\
&\le C([w]_{A_{p,q}})\Big(\int_{\mathbb{R}^n} M_\sigma^{\mathscr{D}}(f\sigma^{-1})(x)^p\,d\sigma\Big)^{\frac{1}{p}}\Big(\int_{\mathbb{R}^n} M_{v,\alpha}^{\mathscr{D}}(gv^{-1})(x)^{p'}\,dv\Big)^{\frac{1}{p'}} \\
&\le C(p,q,[w]_{A_{p,q}})\Big(\int_{\mathbb{R}^n} (f(x)\sigma(x)^{-1})^p\,d\sigma\Big)^{\frac{1}{p}}\Big(\int_{\mathbb{R}^n} (g(x)v(x)^{-1})^{q'}\,dv\Big)^{\frac{1}{q'}} \\
&= C(p,q,[w]_{A_{p,q}})\|fw\|_p\|gw^{-1}\|_{q'} \\
&= C(p,q,[w]_{A_{p,q}})\|fw\|_p. \qquad \square
\end{aligned}$$

Commutators

We conclude this section with the statement of the one weight norm inequality for the commutator $[b, I_\alpha]$. This was proved in [24] using a Cauchy integral formula technique due to Chung, Pereyra and Pérez [10]. We refer the reader there for the details of the proof.([d])

Theorem 4.4. *Given $0<\alpha<n$, $1<p<\frac{n}{\alpha}$, q such that $\frac{1}{p}-\frac{1}{q}=\frac{\alpha}{n}$, $b\in BMO$ and a weight w, then for any $f\in L^p(w^p)$,*

$$\Big(\int_{\mathbb{R}^n} [b,I_\alpha]f(x)^q w(x)^q\,dx\Big)^{\frac{1}{q}} \le C(n,\alpha,p,[w]_{A_{p,q}},\|b\|_{BMO})\Big(\int_{\mathbb{R}^n} |f(x)|^p w(x)^p\,dx\Big)^{\frac{1}{p}}.$$

5. Testing conditions

In this section we turn to our main topic: two weight norm inequalities for fractional maximal and integral operators and for commutators. We will consider one of the two dominant approaches to this problem: the Sawyer testing conditions.

[d]Note added in proof: a new proof using the dyadic approach of these lecture notes can be found in [14].

Two weight inequalities

Before discussing characterizations of two weight inequalities, we first reformulate the inequalities in a way that works well with arbitrary weights. We are interested in weak and strong type inequalities of the form

$$\sup_{t>0} t\, u(\{x \in \mathbb{R}^n : |Tf(x)| > t\})^{\frac{1}{q}} \le C\Big(\int_{\mathbb{R}^n} |f(x)|^p u(x)\, dx\Big)^{\frac{1}{p}},$$

$$\Big(\int_{\mathbb{R}^n} |Tf(x)|^q v(x)\, dx\Big)^{\frac{1}{q}} \le C\Big(\int_{\mathbb{R}^n} |f(x)|^p v(x)\, dx\Big)^{\frac{1}{p}},$$

where $1 < p \le q < \infty$ and T is one of M_α, I_α, or $[b, I_\alpha]$. For the weak type inequality we can also consider the (more difficult) endpoint inequality when $p = 1$. For two weight inequalities we no longer assume that there is a relationship among p, q and α. This allows us to consider "diagonal" inequalities: e.g., $I_\alpha : L^p(v) \to L^p(u)$. For this reason it is more convenient to write the weights as measures (e.g., "$u\,dx$") rather than as "multipliers" as we did in the previous section for one weight norm inequalities.

However, there are some problems with this formulation. For instance, since I_α is self-adjoint, a strong type inequality is equivalent to the dual inequality. For instance, the dual inequality to

$$I_\alpha : L^p(v) \to L^q(u) \qquad \text{is} \qquad I_\alpha : L^{q'}(u^{1-q'}) \to L^{p'}(v^{1-p'}).$$

To make sense of this we need to assume either that $0 < v(x) < \infty$ a.e. (which precludes weights that have compact support) or deal with weights that are measurable functions but equal infinity on sets of positive measure. This is possible, but it requires some care to consistently evaluate expressions of the form $0 \cdot \infty$. For a careful discussion of the details in one particular setting, see [22], Section 7.2.

To avoid these problems we adopt a point of view first introduced by Sawyer [88, 89]. We introduce a new weight $\sigma = v^{1-p'}$ and replace f by $f\sigma$; then we can restate the weak and strong type inequalities as

$$\sup_{t>0} t\, u(\{x \in \mathbb{R}^n : |T(f\sigma)(x)| > t\})^{\frac{1}{q}} \le C\Big(\int_{\mathbb{R}^n} |f(x)|^p \sigma(x)\, dx\Big)^{\frac{1}{p}},$$

$$\Big(\int_{\mathbb{R}^n} |T(f\sigma)(x)|^q u(x)\, dx\Big)^{\frac{1}{q}} \le C\Big(\int_{\mathbb{R}^n} |f(x)|^p \sigma(x)\, dx\Big)^{\frac{1}{p}}.$$

With this formulation, the dual inequality becomes much more natural: for example, for I_α, the dual inequality to

$$I_\alpha(\cdot\sigma) : L^p(\sigma) \to L^q(u) \tag{4}$$

is

$$I_\alpha(\cdot u) : L^{q'}(u) \to L^{p'}(\sigma). \tag{5}$$

Hereafter, in a slight abuse of terminology, we will refer to an inequality like (5) as the dual of (4) even if the operator involved (e.g., M_α) is not self-adjoint or even linear.

Another advantage of this formulation (though not one we will consider here) is that in this form one can take u and σ to be non-negative measures. See for instance, Sawyer [88], or more recently, Lacey [49].

Finally, we note in passing that two weight inequalities when $q < p$ are much more difficult and we will not discuss them. For more information on such inequalities for I_α, we refer the reader to Verbitsky [101] and the recent paper by Tanaka [97]. We are not aware of any analogous results for M_α or $[b, I_\alpha]$.

Testing conditions for fractional maximal operators

Our first approach to characterizing the pairs of weights (u,σ) for which a two weight inequality holds is via testing conditions. The basic idea of a testing condition is to show that an operator T satisfies the strong (p,q) inequality $T(\cdot\sigma) : L^p(\sigma) \to L^q(u)$ if and only if T satisfies it when restricted to a family of test functions: for instance, the characteristic functions of cubes, χ_Q. This approach to the problem is due to Sawyer [87–89, 91], who first proved testing conditions for maximal operators, the Hardy operator, and fractional integrals. For this reason, these are often referred to as Sawyer testing conditions.

Testing conditions received renewed interest in the work of Nazarov, Treil and Volberg [73, 75, 102]; they first made explicit the conjecture that testing conditions were necessary and sufficient for singular integral operators, beginning with the Hilbert transform. (Even this case is an extremely difficult problem which was only recently solved by Lacey, Sawyer, Shen and Uriarte-Tuero [50, 53].) They also pointed out (see [102]) the close connection between testing conditions and the David-Journé $T1$ theorem that characterizes the boundedness of singular integrals on L^2. This was not immediately obvious in the original formulation of the $T1$ theorem, but became clear in the version given by Stein [95].

We first consider the testing condition that characterizes the strong (p,q) inequality for the fractional maximal operator. As we noted, this was first proved by Sawyer [87]. Here we give a new proof based on ideas of Hytönen [43] and Lacey, *et al.* [53]. For a related proof that avoids duality and is closer in spirit to the proof of Theorem 4.2, see Kairema [45].

Theorem 5.1. *Given $0 \le \alpha < n$, $1 < p \le q < \infty$, and a pair of weights (u,σ), the following are equivalent:*

(1) (u,σ) satisfy the testing condition

$$\mathcal{M}_\alpha = \sup_Q \sigma(Q)^{-1/p}\left(\int_Q M_\alpha(\chi_Q\sigma)(x)^q u(x)\,dx\right)^{\frac{1}{q}} < \infty;$$

(2) for every $f \in L^p(\sigma)$,

$$\left(\int_{\mathbb{R}^n} M_\alpha(f\sigma)(x)^q u(x)\,dx\right)^{\frac{1}{q}} \le C(n,p,\alpha)\mathcal{M}_\alpha\left(\int_{\mathbb{R}^n} |f(x)|^p\sigma(x)\,dx\right)^{\frac{1}{p}}.$$

To overcome the fact that the weight σ need not satisfy the A_∞ condition (which was central to the proof in the one weight case) we introduce a stopping time argument referred to as the corona decomposition. This technique was one of the tools introduced into the study of the A_2 conjecture by Lacey, Petermichl and Reguera [52]. The terminology goes back to David and Semmes [33, 34], but the construction itself seems to have first been used by Muckenhoupt and Wheeden [69] in one dimension, where they constructed "principal intervals." (See also Refs. [19, 90].)

Before proving Theorem 5.1 we first describe the corona construction in more general terms. Given a fixed dyadic cube Q_0 in a dyadic grid $\mathscr{D}$, a family of dyadic cubes $\mathscr{T} \subset \mathscr{D}$ all contained in Q_0, a non-negative, locally integrable function f, and a weight σ, we define a subfamily $\mathscr{F} \subset \mathscr{T}$ inductively. Let $\mathscr{F}_0 = \{Q_0\}$. For $k \ge 0$, given the collection of cubes $\mathscr{F}_k$ and $F \in \mathscr{F}_k$, let $\eta_{\mathscr{F}}(F)$ be the collection of maximal disjoint subcubes Q of F such that $\langle f\rangle_{Q,\sigma} > 2\langle f\rangle_{F,\sigma}$. (This collection could be empty; if it is the construction stops.) Then set

$$\mathscr{F}_{k+1} = \bigcup_{F\in\mathscr{F}_k} \eta_{\mathscr{F}}(F)$$

and define

$$\mathscr{F} = \bigcup_k \mathscr{F}_k.$$

We will refer to $\mathscr{F}$ as the corona cubes of f with respect to σ.

Given any cube $Q \in \mathscr{T}$, then by construction it is contained in some cube in $\mathscr{F}$. Let $\pi_{\mathscr{F}}(Q)$ be the smallest cube in $\mathscr{F}$ such that $Q \subset \pi_{\mathscr{F}}(Q)$. We will refer to the cubes $\eta_{\mathscr{F}}(F)$ as the *children* of F in $\mathscr{F}$, and $\pi_{\mathscr{F}}(Q)$ as the *parent* of Q in $\mathscr{F}$.[e]

[e] In the literature, the notation $ch_{\mathscr{F}}(F)$ is often used for the children of F. We wanted to use Greek letters to denote both sets. The letter η seemed appropriate since it is the Greek "h", and in Spanish the cubes in these collections are called *hijos* and *padres*.

The cubes in $\mathcal{F}$ have the critical property that they are sparse with respect to the measure $d\sigma$: given any $F \in \mathcal{F}$, if we compute the measure of the children of F we see that

$$\sum_{F' \in \eta_{\mathcal{F}}(F)} \sigma(F') \leq \frac{1}{2} \sum_{F' \in \eta_{\mathcal{F}}(F)} \frac{(f\sigma)(F')}{\langle f \rangle_{F,\sigma}} \leq \frac{1}{2} \frac{(f\sigma)(F)}{\langle f \rangle_{F,\sigma}} = \frac{1}{2}\sigma(F).$$

Therefore, if we define the set

$$E_{\mathcal{F}}(F) = F \setminus \bigcup_{F' \in \eta_{\mathcal{F}}(F)} F',$$

then

$$\sigma\big(E_{\mathcal{F}}(F)\big) \geq \frac{1}{2}\sigma(F).$$

We will refer to this as the A_∞ property of the cubes in $\mathcal{F}$.

Below we will perform this construction not just on a single cube Q_0 but on each cube in a fixed set of disjoint cubes. We will again refer to the collection of all the cubes that result from this construction applied to each cube in this set as $\mathcal{F}$.

Proof of Theorem 5.1. The necessity of the testing condition is immediate if we take $f = \chi_Q$.

To prove the sufficiency of the testing condition, first note that arguing as we did in the proof of Theorem 4.1 we may assume that f is non-negative, bounded and has compact support. Therefore, it will suffice to show that given any dyadic grid $\mathcal{D}$ and sparse set $\mathcal{S} \subset \mathcal{D}$, the strong type inequality holds for $L_\alpha^{\mathcal{S}}$ assuming the testing condition holds for $L_\alpha^{\mathcal{S}}$. Here we use the fact that given f there exists a sparse subset $\mathcal{S}$ such that $M_\alpha^{\mathcal{D}}(f\sigma)(x) \lesssim L_\alpha^{\mathcal{S}}(f\sigma)(x)$, and that for every such sparse set, $L_\alpha^{\mathcal{S}}(\chi_Q\sigma)(x) \leq M_\alpha^{\mathcal{D}}(\chi_Q\sigma)(x)$.

Fix $\mathcal{D}$, $\mathcal{S}$ and f. Then there exists a function $g \in L^{q'}(u)$, $\|g\|_{L^{q'}(u)} = 1$, such that

$$\begin{aligned}\|L_\alpha^{\mathcal{S}}(f\sigma)\|_{L^q(u)} &= \int_{\mathbb{R}^n} L_\alpha^{\mathcal{S}}(f\sigma)(x) g(x) u(x)\, dx \\ &= \sum_{Q \in \mathcal{S}} |Q|^{\frac{\alpha}{n}} \langle f\sigma \rangle_Q \int_{E(Q)} g(x) u(x)\, dx.\end{aligned}$$

To estimate the righthand side, fix $N \geq 0$ and let $\mathcal{S}_N$ be the set of cubes Q in $\mathcal{S}$ such that $\ell(Q) \leq 2^N$. Then by the monotone convergence theorem it will suffice to prove that

$$\sum_{Q \in \mathcal{S}_N} |Q|^{\frac{\alpha}{n}} \langle f\sigma \rangle_Q \int_{E(Q)} g(x) u(x)\, dx \leq C(n, p, \alpha) \mathcal{M}_\alpha \|f\|_{L^p(\sigma)}.$$

For each maximal cube $Q \in \mathscr{S}_N$, form the corona decomposition of f with respect to σ. Then we can rewrite the sum above as

$$\sum_{Q\in\mathscr{S}_N} |Q|^{\frac{\alpha}{n}} \langle f\sigma\rangle_Q \int_{E(Q)} g(x)u(x)\,dx = \sum_{F\in\mathscr{F}} \sum_{\substack{Q\in\mathscr{S}_N \\ \pi_{\mathscr{F}}(Q)=F}} |Q|^{\frac{\alpha}{n}} \langle f\sigma\rangle_Q \int_{E(Q)} g(x)u(x)\,dx.$$

Fix a cube F and Q such that $\pi_{\mathscr{F}}(Q) = F$. Then given any $F' \in \eta_{\mathscr{F}}(F)$, we must have that $F' \cap Q = \emptyset$ or $F' \subsetneq Q$. If the latter, then, since $\mathscr{S}$ is sparse, we must have that $F' \cap E(Q) = \emptyset$. Therefore,

$$\begin{aligned}\int_{E(Q)} g(x)u(x)\,dx &= \int_{E(Q)\cap E_{\mathscr{F}}(F)} g(x)u(x)\,dx + \sum_{F'\in\eta_{\mathscr{F}}(F)} \int_{E(Q)\cap F'} g(x)u(x)\,dx \\ &= \int_{E(Q)\cap E_{\mathscr{F}}(F)} g(x)u(x)\,dx.\end{aligned}$$

Let $g_F(x) = g(x)\chi_{E(F)}$ and argue as follows: by the definition of the corona cubes, the testing condition, and Hölder's inequality,

$$\begin{aligned}&\sum_{F\in\mathscr{F}} \sum_{\substack{Q\in\mathscr{S}_N \\ \pi_{\mathscr{F}}(Q)=F}} |Q|^{\frac{\alpha}{n}} \langle f\sigma\rangle_Q \int_{E(Q)} g(x)u(x)\,dx \\ &\quad= \sum_{F\in\mathscr{F}} \sum_{\substack{Q\in\mathscr{S}_N \\ \pi_{\mathscr{F}}(Q)=F}} |Q|^{\frac{\alpha}{n}} \langle f\rangle_{Q,\sigma} \langle \sigma\rangle_Q \int_{E(Q)} g_F(x)u(x)\,dx \\ &\quad\le 2\sum_{F\in\mathscr{F}} \langle f\rangle_{F,\sigma} \sum_{\substack{Q\in\mathscr{S}_N \\ \pi_{\mathscr{F}}(Q)=F}} |Q|^{\frac{\alpha}{n}} \langle \sigma\rangle_Q \int_{E(Q)} g_F(x)u(x)\,dx \\ &\quad\le 2\sum_{F\in\mathscr{F}} \langle f\rangle_{F,\sigma} \int_F L_\alpha^{\mathscr{S}}(\sigma\chi_F)(x) g_F(x)u(x)\,dx \\ &\quad\le 2\sum_{F\in\mathscr{F}} \langle f\rangle_{F,\sigma} \|L_\alpha^{\mathscr{S}}(\sigma\chi_F)\|_{L^q(u)} \|g_F\chi_{\mathscr{F}}\|_{L^{q'}(u)} \\ &\quad\le 2\mathscr{M}_\alpha \sum_{F\in\mathscr{F}} \langle f\rangle_{F,\sigma} \sigma(F)^{1/p} \|g_F\chi_{\mathscr{F}}\|_{L^{q'}(u)} \\ &\quad\le 2\mathscr{M}_\alpha \Big(\sum_{F\in\mathscr{F}} \langle f\rangle_{F,\sigma}^p \sigma(F)\Big)^{\frac{1}{p}} \Big(\sum_{F\in\mathscr{F}} \|g_F\chi_{\mathscr{F}}\|_{L^{q'}(u)}^{p'}\Big)^{\frac{1}{p'}}.\end{aligned}$$

We estimate each of these sums separately. For the first we use the A_∞ property of cubes in $\mathscr{F}$ and Lemma 4.1:

$$\Big(\sum_{F\in\mathscr{F}}\langle f\rangle_{F,\sigma}^p\sigma(F)\Big)^{\frac{1}{p}} \leq 2^{\frac{1}{p}}\Big(\sum_{F\in\mathscr{F}}\langle f\rangle_{F,\sigma}^p\sigma(E_{\mathscr{F}}(F))\Big)^{\frac{1}{p}}$$
$$\leq 2^{\frac{1}{p}}\Big(\sum_{F\in\mathscr{F}}\int_{E_{\mathscr{F}}(F)} M_\sigma^{\mathscr{D}}f(x)^p\,d\sigma\Big)^{\frac{1}{p}}$$
$$\leq 2^{\frac{1}{p}}\Big(\int_{\mathbb{R}^n} M_\sigma^{\mathscr{D}}f(x)^p\,d\sigma\Big)^{\frac{1}{p}} \leq C(n,p)\|f\|_{L^p(\sigma)}.$$

To estimate the second sum we use the fact that $q'\leq p'$:

$$\Big(\sum_{F\in\mathscr{F}}\|g_F\chi_{\mathscr{F}}\|_{L^{q'}(u)}^{p'}\Big)^{\frac{1}{p'}} \leq \Big(\sum_{F\in\mathscr{F}}\|g_F\chi_{\mathscr{F}}\|_{L^{q'}(u)}^{q'}\Big)^{\frac{1}{q'}}$$
$$= \Big(\int_{E_{\mathscr{F}}(F)} g(x)^{q'}u(x)\,dx\Big)^{\frac{1}{q'}}$$
$$\leq \Big(\int_{\mathbb{R}^n} g(x)^{q'}u(x)\,dx\Big)^{\frac{1}{q'}} = 1.$$

If we combine these two estimates we get the desired inequality. □

One consequence of this proof is that a weaker condition on the operator is actually sufficient. At the point we apply the testing condition, we could replace $L_\alpha^{\mathscr{S}}(\sigma\chi_F)$ with the smaller, localized operator

$$L_{\alpha,F}^{\mathscr{S},In}\sigma(x) = \sum_{\substack{Q\in\mathscr{S}\\ Q\subset F}}|Q|^{\frac{\alpha}{n}}\langle\sigma\rangle_Q\chi_{E(Q)}(x).$$

The discarded portion of the sum contains no additional information: for all $x\in F$,

$$\sum_{\substack{Q\in\mathscr{S}\\ F\subset Q}}|Q|^{\frac{\alpha}{n}}\langle\sigma\chi_F\rangle_Q\chi_{E(Q)}(x) \leq \sigma(F)\sum_{k=1}^{\infty}|F|^{\frac{\alpha}{n}-1}2^{\alpha-n}\chi_F(x)$$
$$\leq C(n,\alpha)|F|^{\frac{\alpha}{n}}\langle\sigma\rangle_F\chi_F(x).$$

The final characteristic function is over F instead of $E(F)$, but this yields a finite overlap and so does not substantially affect the rest of the estimate. We will consider such local testing conditions again below.

Testing conditions for fractional integral operators

We now prove a testing condition theorem for fractional integrals. If we try to modify the proof of Theorem 5.1 we quickly discover the main obstacle: since the sum defining $I_\alpha^{\mathscr{S}}$ is over the characteristic functions χ_Q and not $\chi_{E(Q)}$, the definition of the function g_F must change. There are additional terms in the sum and the estimate for the norm of g_F no longer works. Another condition is required to evaluate this sum.

The need for such a condition is natural: while a testing condition for I_α is clearly necessary, Sawyer [88] constructed a counter-example showing that by itself it is not sufficient. Motivated by work of Muckenhoupt and Wheeden [69] that suggested duality played a role, Sawyer [91] showed that the testing condition plus the testing condition derived from the dual inequality for I_α is necessary and sufficient. Necessity follows immediately: if $I_\alpha(\cdot\sigma) : L^p(\sigma) \to L^q(u)$, then, since I_α is a self-adjoint linear operator, we have that $I_\alpha(\cdot u) : L^{q'}(u) \to L^{p'}(\sigma)$. Moreover, it turns out that this "dual" testing condition is the right one for the weak type inequality.

Theorem 5.2. *Given* $0 \le \alpha < n$, $1 < p \le q < \infty$, *and a pair of weights* (u,σ), *then the following are equivalent:*

(1) The testing condition,

$$\mathscr{I}_\alpha = \sup_Q \sigma(Q)^{-\frac{1}{p}} \left(\int_Q I_\alpha(\chi_Q \sigma)(x)^q u(x)\, dx \right)^{\frac{1}{q}} < \infty,$$

and the dual testing condition,

$$\mathscr{I}_\alpha^* = \sup_Q u(Q)^{-\frac{1}{q'}} \left(\int_Q I_\alpha(\chi_Q u)(x)^{p'} \sigma(x)\, dx \right)^{\frac{1}{p'}} < \infty,$$

hold;

(2) For all $f \in L^p(\sigma)$,

$$\left(\int_{\mathbb{R}^n} |I_\alpha(f\sigma)(x)|^q u(x)\, dx \right)^{\frac{1}{q}} \le C(n,p,q)(\mathscr{I}_\alpha + \mathscr{I}_\alpha^*) \left(\int_{\mathbb{R}^n} |f(x)|^p \sigma(x)\, dx \right)^{\frac{1}{p}}.$$

Moreover, the dual testing condition is equivalent to the weak type inequality

$$\sup_{t>0} t\, u(\{x \in \mathbb{R}^n : |I_\alpha(f\sigma)(x)| > t\})^{\frac{1}{q}} \le C(n,p,q)\mathscr{I}_\alpha^* \left(\int_{\mathbb{R}^n} |f(x)|^p \sigma(x)\, dx \right)^{\frac{1}{p}}.$$

The equivalence between the dual testing condition and the weak type inequality has the following very deep corollary relating the weak and strong type inequalities.

Corollary 5.1. *Given* $0 < \alpha < n$ *and* $1 < p \le q < \infty$,

$$\|I_\alpha(\cdot\sigma)\|_{L^p(\sigma)\to L^q(u)} \approx \|I_\alpha(\cdot\sigma)\|_{L^p(\sigma)\to L^{q,\infty}(u)} + \|I_\alpha(\cdot u)\|_{L^{q'}(u)\to L^{p',\infty}(\sigma)}.$$

It is conjectured that a similar equivalence holds for singular integrals. However, this is a much more difficult problem and was only recently proved for the Hilbert transform on weighted L^2 by Lacey, *et al.* [53].

Theorem 5.2 was first proved by Sawyer [88, 91] (see also [93]). The proof of the weak type inequality is relatively straightforward and readily adapts to the case of dyadic operators (see [55]). We will omit this proof and refer the reader to these papers. The proof of the strong type inequality is more difficult and even for the dyadic fractional integral operator was initially quite complex: see Lacey, Sawyer and Uriarte-Tuero [55]. Recently, however, Hytönen has given a much simpler proof that relies on the corona decomposition and which is very similar to the proof given above for the fractional maximal operator. Besides its elegance, this proof has the advantage that it makes clear why two testing conditions are needed: it provides a means of evaluating a summation over non-disjoint cubes Q instead of over disjoint sets $E(Q)$ as we did for the fractional maximal operator. We give this proof below. Another proof that takes a somewhat different approach is due to Treil [99].

Proof of Theorem 5.2. As we already discussed, the necessity of the two testing conditions is immediate. To prove sufficiency, we will follow the outline of the proof of Theorem 5.1, highlighting the changes.

First, by arguing as we did in the proof of Theorem 4.3 we can assume that f is non-negative, bounded and has compact support. Further, it will suffice to prove the strong type inequality for the dyadic operator $I_\alpha^{\mathscr{D}}$, where $\mathscr{D}$ is any dyadic grid, assuming that the testing condition holds for this operator. (We could in fact pass to the sparse operator $I_\alpha^{\mathscr{S}}$, but unlike for the fractional maximal operator, sparseness with respect to Lebesgue measure does not simplify the proof.)

Fix a dyadic grid $\mathscr{D}$ and for each $N > 0$ let $\mathscr{D}_N$ be the collection of dyadic cubes Q in $\mathscr{D}$ such that $\ell(Q) \leq 2^N$. Then by duality and the monotone convergence theorem, it will suffice to prove that for any $g \in L^{q'}(u)$, $\|g\|_{L^{q'}(u)} = 1$,

$$\sum_{Q\in\mathscr{D}_N} |Q|^{\frac{\alpha}{n}} \langle f\sigma\rangle_Q \int_Q g(x)u(x)\,dx \leq C(n,p,q)(\mathscr{I}_\alpha + \mathscr{I}_\alpha^*)\|f\|_{L^p(\sigma)}.$$

We now form two "parallel" corona decompositions. For each cube in $\mathscr{D}_N$ of side-length 2^N form the corona decomposition of f with respect to σ; denote the union of all of these cubes by $\mathscr{F}$. (Since f has compact support we in fact only form a finite number of such decompositions.) Simultaneously, on the same cubes form the corona decomposition of g with respect to u; denote the union of these sets of cubes by $\mathscr{G}$.

We now decompose the sum above as follows:

$$\sum_{Q\in\mathscr{D}_N} |Q|^{\frac{\alpha}{n}} \langle f\sigma\rangle_Q \int_Q g(x)u(x)\,dx = \sum_{F\in\mathscr{F}} \sum_{\substack{G\in\mathscr{G}\\ G\subseteq F}} \sum_{\substack{Q\in\mathscr{D}_N\\ \pi_{\mathscr{F}}(Q)=F\\ \pi_{\mathscr{G}}(Q)=G}} + \sum_{G\in\mathscr{G}} \sum_{\substack{F\in\mathscr{F}\\ F\subsetneq G}} \sum_{\substack{Q\in\mathscr{D}_N\\ \pi_{\mathscr{F}}(Q)=F\\ \pi_{\mathscr{G}}(Q)=G}} .$$

Write the above decomposition as $\Sigma_1+\Sigma_2$. We first estimate Σ_1. Fix F, $G\subset F$ and Q such that $\pi_{\mathscr{F}}(Q)=F$ and $\pi_{\mathscr{G}}(Q)=G$. (If no such G or Q exists, then this term in the sum is vacuous and can be disregarded.) Let $F'\in\eta_{\mathscr{F}}(F)$ be such that $Q\cap F'\neq\emptyset$. We cannot have $Q\subseteq F'$, since this would imply that $\pi_{\mathscr{F}}(Q)\subseteq F'\subsetneq F$, a contradiction. Hence, $F'\subsetneq Q\subset G$. We now define the function g_F by

$$\begin{aligned}
\int_Q g(x)u(x)\,dx &= \int_{Q\cap E_{\mathscr{F}}(F)} g(x)u(x)\,dx + \sum_{F'\in\eta_{\mathscr{F}}(F)} \int_{Q\cap F'} g(x)u(x)\,dx\\
&= \int_Q \Big(g(x)\chi_{E_{\mathscr{F}}(F)} + \sum_{F'\in\eta_{\mathscr{F}}(F)} \langle g\rangle_{F',u}\chi_{F'}(x)\Big)u(x)\,dx\\
&= \int_Q g_F(x)u(x)\,dx.
\end{aligned}$$

Moreover, in the definition of g_F, the sum is over $F'\subsetneq Q\subset G$, so we can actually restrict the sum to be over F' in the set

$$\eta^*_{\mathscr{F}}(F)=\{F'\in\eta_{\mathscr{F}}(F):\pi_{\mathscr{G}}(F')\subseteq F\}.$$

We can now argue as follows: by the definition of corona cubes, the testing condition and Hölder's inequality,

$$\begin{aligned}
\Sigma_1 &\le 2\sum_{F\in\mathscr{F}} \langle f\sigma^{-1}\rangle_{F,\sigma} \sum_{\substack{G\in\mathscr{G}\\ G\subseteq F}} \sum_{\substack{Q\in\mathscr{D}_N\\ \pi_{\mathscr{F}}(Q)=F\\ \pi_{\mathscr{G}}(Q)=G}} |Q|^{\frac{\alpha}{n}}\langle\sigma\rangle_Q \int_Q g_F(x)u(x)\,dx\\
&\le 2\sum_{F\in\mathscr{F}} \langle f\sigma^{-1}\rangle_{F,\sigma} \sum_{\substack{Q\in\mathscr{D}\\ Q\subset F}} |Q|^{\frac{\alpha}{n}}\langle\sigma\rangle_Q \int_Q g_F(x)u(x)\,dx\\
&\le 2\sum_{F\in\mathscr{F}} \langle f\sigma^{-1}\rangle_{F,\sigma} \int_F I_\alpha^{\mathscr{D}}(\chi_F\sigma)(x)g_F(x)u(x)\,dx\\
&\le 2\sum_{F\in\mathscr{F}} \langle f\sigma^{-1}\rangle_{F,\sigma} \|I_\alpha^{\mathscr{D}}(\chi_F\sigma)\chi_F\|_{L^q(u)}\|g_F\|_{L^{q'}(u)}\\
&\le 2\mathscr{I}_\alpha \sum_{F\in\mathscr{F}} \langle f\sigma^{-1}\rangle_{F,\sigma}\sigma(F)^{1/p}\|g_F\|_{L^{q'}(u)}\\
&\le 2\mathscr{I}_\alpha \Big(\sum_{F\in\mathscr{F}} \langle f\sigma^{-1}\rangle_{F,\sigma}^p\sigma(F)\Big)^{\frac{1}{p}} \Big(\sum_{F\in\mathscr{F}} \|g_F\|_{L^{q'}(u)}^{p'}\Big)^{\frac{1}{p}}.
\end{aligned}$$

The first sum in the last term can be estimated exactly as in the proof of Theorem 5.1, getting that it is bounded by $C(n,p)\|f\|_{L^p(\sigma)}$. To estimate the second sum we use the fact that $q' \le p'$ and divide it into two parts to get

$$\begin{aligned}
\Big(\sum_{F\in\mathscr{F}} \|g_F\|_{L^{q'}(u)}^{p'}\Big)^{\frac{1}{p'}} &\le \Big(\sum_{F\in\mathscr{F}} \|g_F\|_{L^{q'}(u)}^{q'}\Big)^{\frac{1}{q'}} \\
&\le \Big(\sum_{F\in\mathscr{F}} \int_{E_{\mathscr{F}}(F)} g(x)^{q'} u(x)\,dx\Big)^{\frac{1}{q'}} + \Big(\sum_{F\in\mathscr{F}} \sum_{F'\in\eta^*_{\mathscr{F}}(F)} \langle g\rangle_{F',u}^{q'} u(F')\Big)^{\frac{1}{q'}}.
\end{aligned}$$

We again estimate the first sum as we did in the proof of Theorem 5.1, getting that it is bounded by 1. To bound the second sum, we use the properties of the corona cubes in $\mathscr{F}$ and $\mathscr{G}$, the definition of $\eta^*_{\mathscr{F}}(F)$, and Lemma 4.1:

$$\begin{aligned}
\sum_{F\in\mathscr{F}} \sum_{F'\in\eta^*_{\mathscr{F}}(F)} \langle g\rangle_{F',u}^{q'} u(F') &= \sum_{F\in\mathscr{F}} \sum_{\substack{G\in\mathscr{G}\\ G\subseteq F}} \sum_{\substack{F'\in\eta_{\mathscr{F}}(F)\\ \pi_{\mathscr{G}}(F')=G}} \langle g\rangle_{F',u}^{q'} u(F') \\
&\le 2^{q'} \sum_{F\in\mathscr{F}} \sum_{\substack{G\in\mathscr{G}\\ G\subseteq F}} \langle g\rangle_{G,u}^{q'} \sum_{\substack{F'\in\eta_{\mathscr{F}}(F)\\ \pi_{\mathscr{G}}(F')=G}} u(F') \\
&\le 2^{q'} \sum_{F\in\mathscr{F}} \sum_{\substack{G\in\mathscr{G}\\ G\subseteq F}} \langle g\rangle_{G,u}^{q'} u(G) \\
&\le 2^{q'+1} \sum_{F\in\mathscr{F}} \sum_{\substack{G\in\mathscr{G}\\ G\subseteq F}} \langle g\rangle_{G,u}^{q'} u(E_{\mathscr{G}}(G)) \\
&\le 2^{q'+1} \sum_{F\in\mathscr{F}} \sum_{\substack{G\in\mathscr{G}\\ G\subseteq F}} \int_{E_{\mathscr{G}}(G)} M_u^{\mathscr{D}} g(x)^{q'} u(x)\,dx \\
&\le 2^{q'+1} \int_{\mathbb{R}^n} M_u^{\mathscr{D}} g(x)^{q'} u(x)\,dx \\
&\le C(q) \int_{\mathbb{R}^n} g(x)^{q'} u(x)\,dx \\
&= C(q).
\end{aligned}$$

This completes the estimate of Σ_1.

The estimate for Σ_2 is exactly the same, exchanging the roles of (f,σ) and (g,u) and using the dual testing condition which yields the constant $\mathscr{I}^*_\alpha$. This completes the proof. □

Local and global testing conditions

An examination of the proof of Theorem 5.2 shows that we did not actually need the full testing conditions on the operator $I_\alpha^{\mathscr{D}}$; rather, we used the following localized testing conditions:

$$\mathcal{I}_{\mathcal{D},in} = \sup_Q \sigma(Q)^{-\frac{1}{p}} \left(\int_Q I_{\alpha,Q}^{\mathcal{D},in}(\chi_Q\sigma)(x)^q u(x)\, dx \right)^{\frac{1}{q}} < \infty,$$

$$\mathcal{I}_{\mathcal{D},in}^* = \sup_Q u(Q)^{-\frac{1}{q'}} \left(\int_Q I_{\alpha,Q}^{\mathcal{D},in}(\chi_Q u)(x)^{p'} \sigma(x)\, dx \right)^{\frac{1}{p'}} < \infty,$$

where for $x \in Q$,

$$I_{\alpha,Q}^{\mathcal{D},in}(\chi_Q\sigma)(x) = \sum_{\substack{P\in\mathcal{D}\\ P\subseteq Q}} |P|^{\frac{\alpha}{n}} \langle\sigma\rangle_P \chi_P(x).$$

Similarly, the weak type inequality is equivalent to the dual local testing condition (i.e., the condition that $\mathcal{I}_{\mathcal{D},in}^* < \infty$). This fact is not particular to the dyadic fractional integrals: it is a general property of positive dyadic operators and reflects the fact that they are, in some sense, local operators. See Lacey *et al.* [55].

Somewhat surprisingly, when $p < q$ the local testing conditions can be replaced with global testing conditions:

$$\mathcal{I}_{\mathcal{D},out} = \sup_Q \sigma(Q)^{-\frac{1}{p}} \left(\int_{\mathbb{R}^n} I_{\alpha,Q}^{\mathcal{D},out}(\chi_Q\sigma)(x)^q u(x)\, dx \right)^{\frac{1}{q}} < \infty,$$

$$\mathcal{I}_{\mathcal{D},out}^* = \sup_Q u(Q)^{-\frac{1}{q'}} \left(\int_{\mathbb{R}^n} I_{\alpha,Q}^{\mathcal{D},out}(\chi_Q u)(x)^{p'} \sigma(x)\, dx \right)^{\frac{1}{p'}} < \infty,$$

where for $x \in Q$,

$$I_{\alpha,Q}^{\mathcal{D},out}(\chi_Q u)(x) = \sum_{\substack{P\in\mathcal{D}\\ Q\subsetneq P}} |P|^{\frac{\alpha}{n}} \langle\sigma\chi_Q\rangle_P \chi_P(x).$$

We record this fact as theorem; we will discuss one of its consequences in Section 7. For a proof, see [55].

Theorem 5.3. *Given $0 \le \alpha < n$, $1 < p < q < \infty$, a dyadic grid $\mathcal{D}$, and a pair of weights (u,σ), then:*

(1) $\|I_\alpha^{\mathcal{D}}(\cdot\sigma)\|_{L^p(\sigma)\to L^q(u)} \approx I_{\mathcal{D},out} + \mathcal{I}_{\mathcal{D},out}^*$;

(2) $\|I_\alpha^{\mathcal{D}}(\cdot\sigma)\|_{L^p(\sigma)\to L^{q,\infty}(u)} \approx \mathcal{I}_{\mathcal{D},out}^*$.

Testing conditions for commutators

We conclude this section by considering the problem of testing conditions for commutators. This is completely open but we give some conjectures and also sketch some possible approaches and the obstacles which will be encountered.

In light of the testing conditions in Theorems 5.1 and 5.2, it seems reasonable to conjecture that for $1 < p \leq q < \infty$, $0 < \alpha < n$ and $b \in BMO$, the following two testing conditions,

$$\mathscr{C}_\alpha = \sup_Q \sigma(Q)^{-\frac{1}{p}} \left(\int_Q \left| [b, I_\alpha](\chi_Q \sigma)(x) \right|^q u(x)\, dx \right)^{\frac{1}{q}} < \infty,$$

$$\mathscr{C}_\alpha^* = \sup_Q u(Q)^{-\frac{1}{q'}} \left(\int_Q \left| [b, I_\alpha](\chi_Q u)(x) \right|^{p'} \sigma(x)\, dx \right)^{\frac{1}{p'}} < \infty,$$

are necessary and sufficient for the strong type inequality $[b, I_\alpha](\cdot \sigma) : L^p(\sigma) \to L^q(u)$, and that the dual testing condition (i.e., $\mathscr{C}_\alpha^* < \infty$) is necessary and sufficient for the weak type inequality. The necessity of both testing conditions for the strong type inequality is immediate. The necessity of the dual testing condition follows by duality: see, for instance, Sawyer [88] for the proof of necessity for I_α which adapts immediately to this case.

A significant obstacle for proving sufficiency is that we cannot pass directly to dyadic operators, such as the operator $C_b^{\mathscr{D}}$ defined in Proposition 3.4. The first problem is that since $[b, I_\alpha]$ is not a positive operator, we do not have an obvious pointwise equivalence between $[b, I_\alpha]$ and $C_b^{\mathscr{D}}$. Therefore, we cannot pass from a testing condition for the commutator to a dyadic testing condition as we did in the proof of Theorem 5.2. This means that we will be required to work directly with the non-dyadic testing conditions. This is very much the same situation as is encountered for the Hilbert transform, and we suspect that the same (sophisticated) techniques used there may be applicable to this problem. In addition, the recent work of Sawyer, *et al.* [92] on fractional singular integrals in higher dimensions should also be relevant.

An intermediate result would be to prove that testing conditions for the operator $C_b^{\mathscr{D}}$ are necessary and sufficient for that operator to be bounded, which would yield a sufficient condition for $[b, I_\alpha]$. In this case the parallel corona decomposition used in the proof of Theorem 5.2 should be applicable, but it is not clear how to use the fact that b is in BMO in a way which interacts well with the corona decomposition.

6. Bump conditions

In this section we discuss the second approach to two weight norm inequalities, the A_p-bump conditions. These were first introduced by Neugebauer [76], but they were systematically developed by Pérez [79, 81]. They are a generalization of the Muckenhoupt A_p and Muckenhoupt-Wheeden $A_{p,q}$ conditions. Compared to testing conditions they have several relative strengths and weaknesses.

They only provide sufficient conditions—they are not necessary, though examples show that they are in some sense sharp (see [29]). On the other hand, they are "universal" sufficient conditions: they give conditions that hold for families of operators and are not conditioned to individual operators. (This property is much more important in the study of singular integrals than it is for the study of fractional integrals.) The bump conditions are geometric conditions on the weights and do not involve the operator, so in practice it is easier to check whether a pair of weights satisfies a bump condition. In addition, there exists a very flexible technique for constructing pairs that satisfy a given condition: the method of factored weights which we will discuss below. Finally, since the bump conditions are defined with respect to cubes, they work well with the Calderón-Zygmund decomposition and with dyadic grids in general.

The $A^{\alpha}_{p,q}$ condition

We begin by defining the natural generalization of the one weight $A_{p,q}$ condition given in Definition 4.1. To state it we introduce the following notation for normalized, localized L^p norms: given $1 \le p < \infty$ and a cube Q,

$$\|f\|_{p,Q} = \left(\fint_Q |f(x)|^p \, dx\right)^{\frac{1}{p}}.$$

Definition 6.1. Given $1 < p \le q < \infty$ and $0 \le \alpha < n$, we say that a pair of weights (u,σ) is in the class $A^{\alpha}_{p,q}$ if

$$[u,\sigma]_{A^{\alpha}_{p,q}} = \sup_Q |Q|^{\frac{\alpha}{n}+\frac{1}{q}-\frac{1}{p}} \|u^{\frac{1}{q}}\|_{q,Q} \|\sigma^{\frac{1}{p'}}\|_{p',Q} < \infty.$$

We can extend this definition to the case $p = 1$ by using the L^∞ norm. However, in this case it makes more sense to express the endpoint weak type inequality in terms of pairs (u,v) as originally discussed in Section 5. We will consider these endpoint inequalities in Section 7.

The two weight $A^{\alpha}_{p,q}$ condition characterizes weak type inequalities for M_α. This result is well-known but a proof does not seem to have appeared in the literature. Since it is very similar to the proof of Theorem 4.1 we also omit the details. For a generalization to non-homogeneous spaces whose proof adapts well to dyadic grids, see García-Cuerva and Martell [37].

Theorem 6.1. *Given $1 < p \le q < \infty$, $0 < \alpha < n$, and a pair of weights (u,σ), the following are equivalent:*

(1) $(u,\sigma) \in A^{\alpha}_{p,q}$;

(2) for any $f \in L^p(\sigma)$,

$$\sup_{t>0} t\, u(\{x \in \mathbb{R}^n : M_\alpha(f\sigma)(x) > t\})^{\frac{1}{q}} \leq C(n,\alpha)[u,\sigma]_{A^\alpha_{p,q}} \left(\int_{\mathbb{R}^n} |f(x)|^p \sigma(x)\, dx\right)^{\frac{1}{p}}.$$

While the $A^\alpha_{p,q}$ condition characterizes the weak type inequality, it is not sufficient for the strong type inequality. This fact has been part of the folklore of the field, but a counter-example was not published until recently: see [25]. When $\alpha = 0$, a counter-example for the Hardy-Littlewood maximal operator was constructed by Muckenhoupt and Wheeden [68]. However, this example does not extend to the case $\alpha > 0$ and our construction is substantially different from theirs.

Example 6.1. Given $1 < p \leq q < \infty$ and $0 < \alpha < n$, there exists a pair of weights $(u,\sigma) \in A^\alpha_{p,q}$ and a function $f \in L^p(\sigma)$ such that $M_\alpha(f\sigma) \notin L^q(u)$.

To construct Example 6.1 we will make use of the technique of factored weights. Factored weights are generalization of the easier half of the Jones A_p factorization theorem: given $w_1, w_2 \in A_1$, then for $1 < p < \infty$, $w_1 w_2^{1-p} \in A_p$. (See Refs. [35, 38]; in [22] this was dubbed *reverse factorization.*) Precursors of this idea have been well-known since the 1970s (cf. the counter-example in [68]) but it was first systematically developed (in the case $p = q$) in [22], Chapter 6. The following result is from [25].

Lemma 6.1. *Given* $0 < \alpha < n$, *suppose* $1 < p \leq q < \infty$ *and* $\frac{1}{p} - \frac{1}{q} \leq \frac{\alpha}{n}$. *Let* w_1, w_2 *be locally integrable functions, and define*

$$u = w_1 (M_\gamma w_2)^{-\frac{q}{p'}}, \qquad \sigma = w_2 (M_\gamma w_1)^{-\frac{p'}{q}},$$

where

$$\gamma = \frac{\frac{\alpha}{n} + \frac{1}{q} - \frac{1}{p}}{\frac{1}{n}\left(1 + \frac{1}{q} - \frac{1}{p}\right)}.$$

Then $(u,\sigma) \in A^\alpha_{p,q}$ *and* $[u,\sigma]_{A^\alpha_{p,q}} \leq 1$.

Proof. By our assumptions on p, q and α, $0 \le \gamma \le \alpha$. Fix a cube Q. Then

$$\begin{aligned}
&|Q|^{\frac{\alpha}{n}+\frac{1}{q}-\frac{1}{p}}\left(\fint_Q w_1(x)(M_\gamma w_2(x)^{-\frac{q}{p'}}\,dx\right)^{\frac{1}{q}}\left(\fint w_2(x)\big(M_\gamma w_1(x)^{-\frac{p'}{q}}\,dx\right)^{\frac{1}{p'}} \\
&\quad \le |Q|^{\frac{\alpha}{n}+\frac{1}{q}-\frac{1}{p}}\left(\fint w_1(x)\,dx\right)^{\frac{1}{q'}}\left(|Q|^{\frac{\gamma}{n}}\left(\fint_Q w_2(x)\,dx\right)\right)^{-\frac{1}{p'}} \\
&\qquad\quad \times\left(\fint w_2(x)\,dx\right)^{\frac{1}{p'}}\left(|Q|^{\frac{\gamma}{n}}\left(\fint_Q w_1(x)\,dx\right)\right)^{-\frac{1}{q}} \\
&\quad = |Q|^{\frac{\alpha}{n}+\frac{1}{q}-\frac{1}{p}-\frac{\gamma}{n}\left(1+\frac{1}{q}-\frac{1}{p}\right)} \\
&\quad = 1.
\end{aligned}$$

□

Construction of Example 6.1. To construct the desired example, we need to consider two cases. In both cases we will work on the real line, so $n = 1$.

Suppose first that $\frac{1}{p} - \frac{1}{q} > \alpha$. Let $f = \sigma = \chi_{[-2,-1]}$ and let $u = x^t\chi_{[0,\infty)}$, where $t = q(1-\alpha)-1$. Given any $Q = (a,b)$, $Q \cap \operatorname{supp}(u) \cap \operatorname{supp}(\sigma) = \emptyset$ unless $a < -1$ and $b > 0$. In this case we have that

$$\begin{aligned}
&|Q|^{\alpha+\frac{1}{q}-\frac{1}{p}}\|u^{\frac{1}{q}}\|_{q,Q}\|\sigma^{\frac{1}{p'}}\|_{p',Q} \\
&\qquad \le b^{\alpha+\frac{1}{q}-\frac{1}{p}}\left(\frac{1}{b}\int_0^b x^t\,dx\right)^{\frac{1}{q}}\left(\frac{1}{b}\int_{-2}^{-1}dx\right)^{\frac{1}{p'}} \lesssim b^{\alpha+\frac{t+1}{q}-1} = 1.
\end{aligned}$$

Hence, $(u,\sigma) \in A^\alpha_{p,q}$. On the other hand, for all $x > 1$,

$$M_\alpha(f\sigma)(x) \approx x^{\alpha-1},$$

and so

$$\int_{\mathbb{R}} M_\alpha(f\sigma)(x)^q u(x)\,dx \gtrsim \int_1^\infty x^{q(\alpha-1)}x^{q(1-\alpha)-1}\,dx = \int_1^\infty \frac{dx}{x} = \infty.$$

Now suppose $\frac{1}{p} - \frac{1}{q} \le \alpha$. Fix γ as in Lemma 6.1. We first construct a set $E \subset [0,\infty)$ such that $M_\gamma(\chi_E)(x) \approx 1$ for $x > 0$. Let

$$E = \bigcup_{j\ge 0}[j, j+(j+1)^{-\gamma}).$$

Suppose $x \in [k, k+1)$; if $k = 0$, then it is immediate that if we take $Q = [0,2]$, then $M_\gamma(\chi_E) \ge 3\cdot 2^{\gamma-2} \approx 1$. If $k \ge 1$, let $Q = [0,x]$; then

$$\begin{aligned}
M_\gamma(\chi_E)(x) &\ge x^{\gamma-1}\sum_{0\le j\le\lfloor x\rfloor}(j+1)^{-\gamma} \\
&\ge (k+1)^{\gamma-1}\sum_{j=0}^{k}(j+1)^{-\gamma} \approx (k+1)^{\gamma-1}(k+1)^{1-\gamma} = 1.
\end{aligned}$$

To prove the reverse inequality we will show that $|Q|^{\gamma-1}|Q\cap E|\lesssim 1$ for every cube Q. If $|Q|\le 1$, then

$$|Q|^{\gamma-1}|Q\cap E|\le |Q|^{\gamma}\le 1,$$

so we only have to consider Q such that $|Q|\ge 1$. In this case, given Q let Q' be the smallest interval whose endpoints are integers that contains Q. Then $|Q'|\le |Q|+2\le 3|Q|$, and so $|Q|^{\gamma-1}|E\cap Q|\approx |Q'|^{\gamma-1}|E\cap Q'|$. Therefore, without loss of generality, we may assume that $Q=[a,a+h+1]$, where a, h are non-negative integers. Then

$$\begin{aligned}|Q|^{\gamma-1}|Q\cap E| &= (1+h)^{\gamma-1}\sum_{a\le j\le a+h}(j+1)^{-\gamma}\\ &\approx (1+h)^{\gamma-1}\int_a^{a+h}(t+1)^{-\gamma}\,dt\\ &\approx (1+h)^{\gamma-1}\big((a+h+1)^{1-\gamma}-(a+1)^{1-\gamma}\big).\end{aligned}$$

To estimate the last term suppose first that $h\le a$. Then by the mean value theorem the last term is dominated by

$$(1+h)^{\gamma-1}(1+h)(a+1)^{-\gamma}\le 1.$$

On the other hand, if $h>a$, then the last term is dominated by

$$(1+h)^{\gamma-1}(a+h+1)^{1-\gamma}\le 2^{1-\gamma}\approx 1.$$

This completes the proof that $M_\gamma(\chi_E)(x)\approx 1$.

We can now give our desired counter-example. Let $w_1=\chi_E$ and $w_2=\chi_{[0,1]}$. Then for all $x\ge 2$,

$$M_\gamma w_1(x)\approx 1,\qquad M_\gamma w_2(x)=\sup_Q |Q|^{\gamma-1}\int_Q w_2(y)\,dy\approx x^{\gamma-1}.$$

Define

$$u=w_1(M_\gamma w_2)^{-\frac{q}{p'}},\qquad \sigma=w_2(M_\gamma w_1)^{-\frac{p'}{q}};$$

then by Lemma 6.1, $(u,\sigma)\in A^{\alpha}_{p,q}$. Moreover, for $x\ge 2$, we have that

$$u(x)\approx x^{(1-\gamma)\frac{q}{p'}}\chi_E(x),\qquad \sigma(x)\approx\chi_{[0,1]}(x).$$

Fix $f\in L^p(\sigma)$: without loss of generality, we may assume $\operatorname{supp}(f)\subset[0,1]$. Then $f\sigma$ is locally integrable, and for $x\ge 2$ we have that

$$M_\alpha(f\sigma)(x)\ge x^{\alpha-1}\|f\sigma\|_1\approx x^{\alpha-1}.$$

Therefore, for $x\ge 2$,

$$M_\alpha(f\sigma)(x)^q u(x)\gtrsim x^{(\alpha-1)q}x^{(1-\gamma)\frac{q}{p'}}\chi_E(x).$$

By the definition of γ,

$$\gamma\Big(\frac{1}{q}+\frac{1}{p'}\Big)=\gamma\Big(1+\frac{1}{q}-\frac{1}{p}\Big)=\alpha+\frac{1}{q}-\frac{1}{p}=\alpha-1+\frac{1}{q}+\frac{1}{p'};$$

equivalently,

$$(\gamma-1)\Big(\frac{q}{p'}+1\Big)=q(\alpha-1),$$

and so

$$(\alpha-1)q+(1-\gamma)\frac{q}{p'}=\gamma-1.$$

Therefore, to show that $M_\alpha(f\sigma)\notin L^q(u)$, it will be enough to prove that

$$\int_2^\infty x^{\gamma-1}\chi_E(x)\,dx=\infty,$$

but this is straightforward:

$$\begin{aligned}\int_2^\infty x^{\gamma-1}\chi_E(x)\,dx&=\sum_{j=2}^\infty\int_j^{j+(j+1)^{-\gamma}}x^{\gamma-1}\,dx\\&\geq\sum_{j=2}^\infty(j+(j+1)^{-\gamma})^{\gamma-1}(j+1)^{-\gamma}\\&\geq\sum_{j=2}^\infty(j+1)^{\gamma-1}(j+1)^{-\gamma}=\sum_{j=2}^\infty(j+1)^{-1}=\infty.\end{aligned}$$

□

If we combine Example 6.1 with the pointwise inequalities in Section 3, we see that the $A^\alpha_{p,q}$ condition is also not sufficient for the fractional integral operator to satisfy the strong type inequality. This condition is also not sufficient for the weak (p,q) inequality. A counter-example when $p=q=n=2$ and $\alpha=\frac{1}{2}$ using measures was constructed by Kerman and Sawyer [46]. Here we construct a general counter-example that holds for all p, q and α. For simplicity we construct the example for $n=1$, but it can be modified to work in all dimensions. We want to thank E. Sawyer for useful comments on an earlier version of this construction.

Example 6.2. Let $n=1$. Given $1<p\leq q<\infty$ and $0<\alpha<1$, there exists a pair of weights $(u,\sigma)\in A^\alpha_{p,q}$ and a non-negative function $f\in L^p(\sigma)$ such that

$$\sup_{t>0}t\,u(\{x\in\mathbb{R}:I_\alpha(f\sigma)(x)>t\})^{\frac{1}{q}}=\infty.$$

Proof. Fix p, q and α and let $u = \chi_{[-1,1]}$. We will first construct a non-negative weight σ such that $[u,\sigma]_{A^\alpha_{p,q}} < \infty$. We will then find a non-negative function $f \in L^p(\sigma)$ such that $I_\alpha(f\sigma)(x) = \infty$ for all $x \in (0,1)$. Then we have that

$$\sup_{t>0} t\, u(\{x \in \mathbb{R} : I_\alpha(f\sigma)(x) > t\})^{\frac{1}{q}} \geq \sup_{t>0} t\, u([0,1])^{\frac{1}{q}} = \infty.$$

Let $\sigma = |x|^{-r}\chi_{\{|x|>1\}}$, where r is defined by

$$\alpha - \frac{1}{p} = \frac{r}{p'}.$$

Given that u and σ are symmetric around the origin and have disjoint supports, it is immediate that to check the $A^\alpha_{p,q}$ condition it suffices to consider intervals $Q = [0,t]$, $t > 1$. But in this case,

$$|Q|^{\alpha+\frac{1}{q}-\frac{1}{p}} \left(\fint_Q u(x)\,dx \right)^{\frac{1}{q}} \left(\fint_Q \sigma(x)\,dx \right)^{\frac{1}{p'}} = t^{\alpha+\frac{1}{q}-\frac{1}{p}} t^{-\frac{1}{q}} \left(\frac{1}{t} \int_1^t x^{-r}\,dx \right)^{\frac{1}{p'}}.$$

If $r < 1$, then x^{-r} is locally integrable at the origin, and so by our choice of r the righthand term is bounded by

$$t^{\alpha+\frac{1}{q}-\frac{1}{p}} t^{-\frac{1}{q}} \left(\frac{1}{t} \int_0^t x^{-r}\,dx \right)^{\frac{1}{p'}} \approx t^{\alpha+\frac{1}{q}-\frac{1}{p}-\frac{1}{q}-\frac{r}{p'}} = 1.$$

On the other hand, if $r > 1$, then $x^{-r} \in L^1(1,\infty)$, and so the righthand side is bounded by

$$t^{\alpha+\frac{1}{q}-\frac{1}{p}-\frac{1}{q}-\frac{1}{p'}} = t^{\alpha-1} \leq 1.$$

Hence, $[u,\sigma]_{A^\alpha_{p,q}} < \infty$.

We now construct f with the desired properties. Let

$$f(x) = \frac{x^{r-\alpha}}{\log(ex)} \chi_{(1,\infty)}(x);$$

then

$$f(x)^p \sigma(x) = \frac{x^{(r-\alpha)p-r}}{\log(ex)^p} \chi_{(1,\infty)}(x).$$

By our definition of r,

$$\alpha - \frac{1}{p} = r\left(1 - \frac{1}{p}\right),$$

or equivalently,

$$\frac{r}{p} - \frac{1}{p} = r - \alpha,$$

which in turn implies that $p(r-\alpha)=r-1$. Hence, since $p>1$,

$$f(x)^p\sigma(x)=\frac{1}{x\log(ex)^p}\chi_{(1,\infty)}(x)\in L^1(\mathbb{R}).$$

But for $x\in(0,1)$,

$$\begin{aligned}I_\alpha(f\sigma)(x)&=\int_1^\infty \frac{f(y)\sigma(y)}{|x-y|^{1-\alpha}}\,dy\\&=\int_1^\infty \frac{dy}{y^\alpha(y-x)^{1-\alpha}\log(ey)}\geq\int_1^\infty\frac{dy}{y\log(ey)}=+\infty.\end{aligned}$$

This completes the proof. □

Though it does not matter for our proof, we note in passing that in this example we actually have that $I_\alpha(f\sigma)(x)=\infty$ for all x.

Young functions and Orlicz norms

Given the failure of the $A^\alpha_{p,q}$ condition to be sufficient for strong type norm inequalities for fractional maximal and integral operators, our goal is to generalize this condition to get one that is sufficient, resembles the $A^\alpha_{p,q}$ condition and shares its key properties. In particular, the condition should be "geometric" in the sense that, unlike the testing conditions in Section 5, it does not involve the operator itself, and it should interact well with dyadic grids. Our approach will be to replace the L^q and $L^{p'}$ norms in the definition with larger norms. For A_p weights this was first done by Neugebauer, who replaced the L^p and $L^{p'}$ norms with L^{rp} and $L^{rp'}$ norms, $r>1$. Pérez [79, 81] greatly extended this idea by showing that Orlicz norms that lie between L^p and L^{rp} for any $r>1$ will also work.

To formulate his approach we first need to introduce some basic ideas about Young functions and Orlicz norms. For complete information see [47, 82]. A function $B:[0,\infty)\to[0,\infty)$ is a Young function if it is continuous, convex and strictly increasing, if $B(0)=0$, and if $B(t)/t\to\infty$ as $t\to\infty$. $B(t)=t$ is not properly a Young function, but in many instances what we say applies to this function as well. It is convenient, particularly when computing constants, to assume $B(1)=1$, but this normalization is not necessary. A Young function B is said to be doubling if there exists a positive constant C such that $B(2t)\leq CB(t)$ for all $t>0$.

Given a Young function B and a cube Q, we define the normalized Luxemburg norm of f on Q by

$$\|f\|_{B,Q}=\inf\left\{\lambda>0:\fint_Q B\left(\frac{|f(x)|}{\lambda}\right)dx\leq 1\right\}. \tag{6}$$

When $B(t) = t^p$, $1 \le p < \infty$, the Luxemburg norm coincides with the normalized L^p norm:

$$\|f\|_{B,Q} = \left(\fint_Q |f(x)|^p\,dx\right)^{1/p} = \|f\|_{p,Q}.$$

If $A(t) \le B(ct)$ for all $t \ge t_0 > 0$, then there exists a constant C, depending only on A and B, such that for all cubes Q and functions f, $\|f\|_{A,Q} \le C\|f\|_{B,Q}$.

Given a Young function B, the associate Young function $\bar{B}$ is defined by

$$\bar{B}(t) = \sup_{s>0}\{st - B(s)\}, \qquad t > 0;$$

B and $\bar{B}$ satisfy

$$t \le B^{-1}(t)\bar{B}^{-1}(t) \le 2t.$$

Note that the associate of $\bar{B}$ is again B. Using the associate Young function, Hölder's inequality can be generalized to the scale of Orlicz spaces: given any Young function B, then for all functions f and g and all cubes Q,

$$\fint_Q |f(x)g(x)|\,dx \le 2\|f\|_{B,Q}\|g\|_{\bar{B},Q}.$$

More generally, if A, B and C are Young functions such that for all $t \ge t_0 > 0$,

$$B^{-1}(t)C^{-1}(t) \le cA^{-1}(t),$$

then

$$\|fg\|_{A,Q} \le K\|f\|_{B,Q}\|g\|_{C,Q}.$$

Below we will need to impose a growth condition on Young functions that compares them to powers of t. This condition was first introduced by Pérez [81]. Given $1 < p < \infty$, we say that a Young function B satisfies the B_p condition if

$$\int_1^\infty \frac{B(t)}{t^p}\frac{dt}{t} < \infty.$$

Frequently, we will want to make an assumption of the form $\bar{B} \in B_p$. If both B and $\bar{B}$ are doubling, then this is equivalent to

$$\int_1^\infty \left(\frac{t^{p'}}{B(t)}\right)^{p-1}\frac{dt}{t} < \infty. \tag{7}$$

(See [22], Proposition 5.10.) There are two important examples of functions that satisfy the B_p condition. If $B(t) = t^{rp'}$, $r > 1$, or if $B(t) = t^p\log(e+t)^{p-1+\delta}$, $\delta > 0$, then $\bar{B} \in B_p$. For reasons that will be clear below, we will refer to these as power bumps and log bumps. One essential property of this condition is that if $\bar{B} \in B_p$, then $\bar{B} \lesssim t^p$ and $B \gtrsim t^{p'}$. Note, however, that if $B(t) = t^{p'}$, then $\bar{B}(t) = t^p$ is not in B_p.

The B_p condition was introduced by Pérez to characterize the boundedness of the Orlicz maximal operator. Given a Young function B and a measurable function f, define

$$M_B f(x) = \sup_Q \|f\|_{B,Q}\, \chi_Q(x).$$

Proposition 6.1. *Given a Young function B and $1 < p < \infty$, the following are equivalent:*

(1) $B \in B_p$;

(2) for all $f \in L^p$,

$$\left(\int_{\mathbb{R}^n} M_B f(x)^p\, dx\right)^{\frac{1}{p}} \le C(n,p)\left(\int_{\mathbb{R}^n} |f(x)|^p\, dx\right)^{\frac{1}{p}}.$$

As stated in [81], Proposition 6.1 included the assumption that B was doubling. In the proof of sufficiency this assumption was only required to use the B_p condition in the form of (7). This was correctly noted in [22], but we made the incorrect assertion that it was not needed for the proof of necessity in [81]. However, Liu and Luque [63] recently gave a proof that it is necessary without assuming doubling.

We can also define a fractional Orlicz maximal operator $M_{B,\alpha}$: see Section 7 below. There is also a corresponding B_p^α condition which is useful in determining sharp constant estimates for the fractional integral operator: see [25] for details.

The $A^\alpha_{p,q}$ bump conditions

Using the machinery introduced above, we can now state our generalizations of the $A^\alpha_{p,q}$ condition. Given $0 < \alpha < n$, $1 < p \le q < \infty$, Young functions A and B, $\bar{A} \in B_{q'}$ and $\bar{B} \in B_p$, and a pair of weights (u,σ), we define

$$[u,\sigma]_{A^\alpha_{p,q,B}} = \sup_Q |Q|^{\frac{\alpha}{n}+\frac{1}{q}-\frac{1}{p}} \|u^{\frac{1}{q}}\|_{q,Q} \|\sigma^{\frac{1}{p'}}\|_{B,Q} < \infty,$$

$$[u,\sigma]^*_{A^\alpha_{p,q,A}} = \sup_Q |Q|^{\frac{\alpha}{n}+\frac{1}{q}-\frac{1}{p}} \|u^{\frac{1}{q}}\|_{A,Q} \|\sigma^{\frac{1}{p'}}\|_{p',Q} < \infty.$$

By our hypotheses on A and B, both of these quantities are larger than $[u,\sigma]_{A^\alpha_{p,q}}$: we have "bumped up" one of the norms in the scale of Orlicz spaces. For this reason we refer to these as $A^\alpha_{p,q}$ bump conditions.

Note that the second condition is the "dual" of the first, in the sense that

$$[u,\sigma]^*_{A^\alpha_{p,q,A}} = [\sigma,u]_{A^\alpha_{q',p',A}}.$$

As we will see below, this condition will play a role analogous to that of the dual testing conditions discussed in Section 5. Informally, it is common to refer to the $[u,\sigma]_{A^\alpha_{p,q,B}}$ condition as having a bump on the right, and the $[u,\sigma]^*_{A^\alpha_{p,q,A}}$ as having a bump on the left, and collectively we refer to these as separated bump conditions.

We can combine these conditions by putting a bump on both norms simultaneously:

$$[u,\sigma]_{A^\alpha_{p,q,A,B}} = \sup_Q |Q|^{\frac{\alpha}{n}+\frac{1}{q}-\frac{1}{p}} \|u^{\frac{1}{q}}\|_{A,Q} \|\sigma^{\frac{1}{p'}}\|_{B,Q} < \infty.$$

We refer this as a conjoined bump condition. Clearly, it is larger than either of the separated bump conditions. In fact, assuming the conjoined bump condition is stronger than assuming both separated bump conditions. The following example (with $\alpha = 0$, $p = 2$) was constructed in [1].

Example 6.3. Given $0 < \alpha < n$ and $1 < p \le q < \infty$, there exists a pair of Young functions A and B, $\bar{A} \in B_{q'}$ and $\bar{B} \in B_p$, and a pair of weights (u,σ), such that $[u,\sigma]_{A^\alpha_{p,q,B}}, [u,\sigma]^*_{A^\alpha_{p,q,A}} < \infty$, but $[u,\sigma]_{A^\alpha_{p,q,A,B}} = \infty$.

Proof. We construct our example on the real line, so $0 < \alpha < 1$. Define the Young functions

$$A(t) = t^q \log(e+t)^q, \quad B(t) = t^{p'} \log(e+t)^{p'}.$$

Then $\bar{A} \in B_{q'}$ and $\bar{B} \in B_p$. By rescaling, if we let $\Psi(t) = t\log(e+t)^q, \Phi(t) = t\log(e+t)^{p'}$, then for any pair (u,σ),

$$\|u^{\frac{1}{q}}\|_{A,Q} \approx \|u\|^{\frac{1}{q}}_{\Psi,Q}, \qquad \|\sigma^{\frac{1}{p'}}\|_{B,Q} \approx \|\sigma\|^{\frac{1}{p'}}_{\Phi,Q}.$$

Therefore, it will suffice to estimate the norms of u and σ with respect to Ψ and Φ. Similarly, we can replace the localized L^q and $L^{p'}$ norms of $u^{\frac{1}{q}}$ and $\sigma^{\frac{1}{p'}}$ with the L^1 norms of u and σ.

Before we define u and σ we first construct a pair (u_0,σ_0) which will be the basic building block for our example. Fix an integer $k \ge 2$ and define $Q = (0,k)$, $\sigma_0 = \chi_{(0,1)}$ and $u_0 = K_k^q \chi_{(k-1,k)}$, where $K_k = k^{1-\alpha} \log(e+k)^{-\frac{3}{2}}$. Since $\Psi^{-1}(t) \approx t\log(e+t)^{-q}$ and $\Phi^{-1}(t) \approx t\log(e+t)^{-p'}$, by the definition of the Luxemburg norm,

$$\|u_0\|_{1,Q}^{\frac{1}{q}} = \frac{K_k}{k^{\frac{1}{q}}}, \qquad \|u_0\|_{\Psi,Q} \approx \frac{K_k \log(e+k)}{k^{\frac{1}{q}}},$$

$$\|\sigma_0\|_{1,Q}^{\frac{1}{p'}} = \frac{1}{k^{\frac{1}{p'}}}, \qquad \|\sigma_0\|_{\Phi,Q} \approx \frac{\log(e+k)}{k^{\frac{1}{p'}}}.$$

Therefore, we have that

$$|Q|^{\alpha+\frac{1}{q}-\frac{1}{p}}\|u_0\|_{1,Q}^{\frac{1}{q}}\|\sigma_0\|_{\Phi,Q}^{\frac{1}{p'}}, \quad |Q|^{\alpha+\frac{1}{q}-\frac{1}{p}}\|u_0\|_{\Phi,Q}^{\frac{1}{q}}\|\sigma_0\|_{1,Q}^{\frac{1}{p'}} \approx \frac{1}{\log(e+k)^{\frac{1}{2}}},$$

but

$$|Q|^{\alpha+\frac{1}{q}-\frac{1}{p}}\|u_0\|_{\Phi,Q}^{\frac{1}{q}}\|\sigma_0\|_{\Phi,Q}^{\frac{1}{p'}} \approx \log(e+k)^{\frac{1}{2}}.$$

We now define u and σ as follows:

$$u(x) = \sum_{k\geq 2} K_k^q \chi_{I_k}(x), \qquad \sigma(x) = \sum_{k\geq 2} \chi_{J_k}(x),$$

where $I_k = (e^k + k - 1, e^k + k)$ and $J_k = (e^k, e^k + 1)$. Since the above computations are translation invariant, we immediately get that if $Q_k = (e^k, e^k + k)$, then

$$|Q_k|^{\alpha+\frac{1}{q}-\frac{1}{p}}\|u\|_{\Phi,Q_k}^{\frac{1}{q}}\|\sigma\|_{\Phi,Q_k}^{\frac{1}{p'}} \approx \log(e+k)^{\frac{1}{2}},$$

and so $[u,\sigma]_{A^\alpha_{p,q,A,B}} = \infty$.

We will now prove that $[u,\sigma]_{A^\alpha_{p,q,B}}$ and $[\sigma,u]^*_{A^\alpha_{p,q,A}}$ are both finite. We will show $[u,\sigma]^*_{A^\alpha_{p,q,A}} < \infty$; the argument for the first condition is essentially the same. Fix an interval Q; we will show that $|Q|^{\alpha+\frac{1}{q}-\frac{1}{p}}\|u\|_{\Psi,Q}^{\frac{1}{q}}\|\sigma\|_{1,Q}^{\frac{1}{p'}}$ is uniformly bounded. Let N be an integer such that $N-1 \leq |Q| \leq N$. We need to consider those values of k such that Q intersects either I_k or J_k.

Suppose that for some $k \geq N+2$, Q intersects I_k. But in this case it cannot intersect J_j for any j and so $\|\sigma\|_{1,Q} = 0$. Similarly, if Q intersects J_k, then $\|u\|_{\Psi,Q} = 0$.

Now suppose that for some $k < N+2$, Q intersects one of I_k or J_k. If $\log(N) \lesssim k$ (more precisely, if $N < e^k - e^{k-1} - 1$), then for any $j \neq k$, Q cannot intersect I_j or J_j. In this case $|Q|^{\alpha+\frac{1}{q}-\frac{1}{p}}\|u\|_{\Psi,Q}^{\frac{1}{q}}\|\sigma\|_{1,Q}^{\frac{1}{p'}} \neq 0$ only if Q intersects both I_k and J_k, and will reach its maximum when $N \approx k$. But in this case we can replace Q by $(e^k, e^k + k)$ and the above computation shows that $|Q|^{\alpha+\frac{1}{q}-\frac{1}{p}}\|u\|_{\Psi,Q}^{\frac{1}{q}}\|\sigma\|_{1,Q}^{\frac{1}{p'}} \lesssim 1$.

Finally, suppose Q intersects one or more pairs I_k and J_k with $k \lesssim \log(N)$. Then $|\operatorname{supp}(u)\cap Q| \lesssim \log(N)$ and $\|u\|_{L^\infty(Q)} \approx K^q_{\lfloor \log(N)\rfloor} \lesssim \log(N)^{q(1-\alpha)}$. Therefore, for any $r>1$,

$$\|u\|^{\frac{1}{q}}_{\Psi,Q} \lesssim \|u\|^{\frac{1}{q}}_{r,Q} \leq \|u\|^{\frac{1}{q}}_{L^\infty(Q)}\left(\frac{|\operatorname{supp}(u)\cap Q|}{|Q|}\right)^{\frac{1}{rq}} \lesssim \frac{\log(N)^{1-\alpha+\frac{1}{rq}}}{N^{\frac{1}{rq}}}.$$

A similar calculation shows that

$$\|\sigma\|^{\frac{1}{p'}}_{1,Q} \lesssim \left(\frac{\log(N)}{N}\right)^{\frac{1}{p'}}.$$

Hence, we have that

$$|Q|^{\alpha+\frac{1}{q}-\frac{1}{p}}\|u\|^{\frac{1}{q}}_{\Phi,Q}\|\sigma\|^{\frac{1}{p'}}_{1,Q} \lesssim N^{\alpha+\frac{1}{q}-\frac{1}{p}-\frac{1}{rq}-\frac{1}{p'}}\log(N)^{1-\alpha+\frac{1}{rq}+\frac{1}{p'}}.$$

Since $\alpha<1$, if we fix $r>1$ sufficiently close to 1 we have that the exponent on N is negative, and so this quantity will be uniformly bounded for all N. We thus have that $[u,\sigma]^*_{A^\alpha_{p,q,A}} < \infty$ and our proof is complete. □

Bump conditions for fractional maximal operators

There is a parallel between bump conditions and the testing conditions described in Section 5. For maximal operators, only a single testing condition is needed for the strong type inequality; similarly, only a single bump (on the right) is required to get a sufficient condition. The following result is due to Pérez [79, 81] and our proof is based on his.

Theorem 6.2. *Given $0\leq\alpha<n$, $1<p\leq q<\infty$, and a Young function B such that $\bar{B}\in B_p$, suppose the pair of weights (u,σ) is such that $[u,\sigma]_{A^\alpha_{p,q,B}}<\infty$. Then for every $f\in L^p(\sigma)$,*

$$\left(\int_{\mathbb{R}^n} M_\alpha(f\sigma)(x)^q u(x)\,dx\right)^{\frac{1}{q}} \leq C(n,p,q)[u,\sigma]_{A^\alpha_{p,q,B}}\left(\int_{\mathbb{R}^n}|f(x)|^p\sigma(x)\,dx\right)^{\frac{1}{p}}.$$

Note that while our proof shows directly that the constant depends linearly on $[u,\sigma]_{A^\alpha_{p,q,B}}$, in fact this is always true in two weight inequalities. This observation is due to Sawyer: see [24], Remark 1.4.

Proof. Arguing as we did in the proof of Theorem 4.2, we may assume that f is non-negative, bounded and has compact support, and it will suffice to prove the desired inequality for $L^{\mathcal{S}}_\alpha$, where $\mathcal{S}$ is a sparse subset of a dyadic grid $\mathcal{D}$. Indeed, we begin as we did there, using the fact that the sets $E(Q)$ are disjoint. But instead of the A_∞ property we will use the generalized Hölder's inequality to introduce the Orlicz maximal operator. This allows us to sum over the cubes in $\mathscr{S}$ and apply Proposition 6.1 to get the desired estimate:

$$\begin{aligned}
\|L_\alpha^{\mathscr{S}}(f\sigma)\|_{L^q(u)}^q &= \sum_{Q\in\mathscr{S}} |Q|^{q\frac{\alpha}{n}} \langle f\sigma\rangle_Q^q u(E(Q)) \\
&= \sum_{Q\in\mathscr{S}} |Q|^{q\frac{\alpha}{n}+1-\frac{q}{p}} \langle f\sigma\rangle_Q^q \langle u\rangle_Q |Q|^{\frac{q}{p}} \\
&\le 2^{\frac{q}{p}+1} \sum_{Q\in\mathscr{S}} |Q|^{q\frac{\alpha}{n}+1-\frac{q}{p}} \langle u\rangle_Q \|\sigma^{\frac{1}{p'}}\|_{B,Q}^q \|f\sigma^{\frac{1}{p}}\|_{\bar{B},Q}^q |E(Q)|^{\frac{q}{p}} \\
&\le 2^{\frac{q}{p}+1} [u,\sigma]_{A_{p,q,B}^\alpha}^q \sum_{Q\in\mathscr{S}} \|f\sigma^{\frac{1}{p}}\|_{\bar{B},Q}^q |E(Q)|^{\frac{q}{p}} \\
&\le 2^{\frac{q}{p}+1} [u,\sigma]_{A_{p,q,B}^\alpha}^q \Big(\sum_{Q\in\mathscr{S}} \|f\sigma^{\frac{1}{p}}\|_{\bar{B},Q}^p |E(Q)|\Big)^{\frac{q}{p}} \\
&\le 2^{\frac{q}{p}+1} [u,\sigma]_{A_{p,q,B}^\alpha}^q \Big(\sum_{Q\in\mathscr{S}} \int_{E(Q)} M_{\bar{B}}(f\sigma^{\frac{1}{p}})(x)^p\,dx\Big)^{\frac{q}{p}} \\
&\le 2^{\frac{q}{p}+1} [u,\sigma]_{A_{p,q,B}^\alpha}^q \Big(\int_{\mathbb{R}^n} M_{\bar{B}}(f\sigma^{\frac{1}{p}})(x)^p\,dx\Big)^{\frac{q}{p}} \\
&\le C(n,p,q)[u,\sigma]_{A_{p,q,B}^\alpha}^q \Big(\int_{\mathbb{R}^n} |f(x)|^p\sigma(x)\,dx\Big)^{\frac{q}{p}}. \qquad \square
\end{aligned}$$

Bump conditions for fractional integral operators

We now consider bump conditions for the fractional integral operator. For the strong type condition, we need two bumps, analogous to the fact that you need two testing conditions. Our first result is for conjoined bumps; we will discuss separated bump conditions in Section 7 below. Theorem 6.3 was originally proved by Pérez [79] and our proof is modeled on his.

Theorem 6.3. *Given $0 < \alpha < n$, $1 < p \le q < \infty$, and Young functions A, B such that $\bar{A} \in B_{q'}$ and $\bar{B} \in B_p$, suppose the pair of weights (u,σ) is such that $[u,\sigma]_{A_{p,q,A,B}^\alpha} < \infty$. Then for every $f \in L^p(\sigma)$,*

$$\Big(\int_{\mathbb{R}^n} |I_\alpha(f\sigma)(x)|^q u(x)\,dx\Big)^{\frac{1}{q}} \le C(n,p)[u,\sigma]_{A_{p,q,A,B}^\alpha} \Big(\int_{\mathbb{R}^n} |f(x)|^p\sigma(x)\,dx\Big)^{\frac{1}{p}}.$$

We note that Theorem 6.3 was very influential in the study of two weight norm inequalities, and it led to the conjecture that an analogous result held for singular integral operators. This problem was solved recently by Lerner [58]; for prior results see Refs. [20, 21, 31].

Proof. Arguing as we did in the proof of Theorem 4.3, we may assume that f is non-negative, bounded and has compact support. Further, it will suffice to prove the desired inequality for $I_\alpha^{\mathcal{S}}$, where $\mathcal{S}$ is a sparse subset of a dyadic grid $\mathcal{D}$.

We begin as in the one weight case by applying duality. But here we use the generalized Hölder's inequality to introduce two Orlicz maximal operators and use these to sum over cubes in $\mathcal{S}$. We can then apply Proposition 6.1 twice. More precisely, by duality there exists $g \in L^{q'}(u)$, $\|g\|_{L^{q'}(u)} = 1$, such that

$$\begin{aligned}
\|I_\alpha^{\mathcal{S}}(f\sigma)\|_{L^q(u)} &= \int_{\mathbb{R}^n} I_\alpha^{\mathcal{S}}(f\sigma) g(x)u(x)\,dx = \sum_{Q\in\mathcal{S}} |Q|^{\frac{\alpha}{n}} \langle f\sigma\rangle_Q \langle gu\rangle_Q |Q| \\
&\leq 2^2 \sum_{Q\in\mathcal{S}} |Q|^{\frac{\alpha}{n}+\frac{1}{q}-\frac{1}{p}} \|u^{\frac{1}{q}}\|_{A,Q} \|\sigma^{\frac{1}{p'}}\|_{B,Q} \|f\sigma^{\frac{1}{p}}\|_{\bar{B},Q} \|gu^{\frac{1}{q'}}\|_{\bar{A},Q} |Q|^{\frac{1}{p}+\frac{1}{q'}} \\
&\leq 2^{\frac{1}{p}+\frac{1}{q'}+2} [u,\sigma]_{A^\alpha_{p,q,A,B}} \sum_{Q\in\mathcal{S}} \|f\sigma^{\frac{1}{p}}\|_{\bar{B},Q} |E(Q)|^{\frac{1}{p}} \|gu^{\frac{1}{q'}}\|_{\bar{A},Q} |E(Q)|^{\frac{1}{q'}}.
\end{aligned}$$

By Hölder's inequality and the fact that $q' \leq p'$, we have that

$$\begin{aligned}
&\leq 16[u,\sigma]_{A^\alpha_{p,q,A,B}} \Big(\sum_{Q\in\mathcal{S}} \|f\sigma^{\frac{1}{p}}\|^p_{\bar{B},Q} |E(Q)|\Big)^{\frac{1}{p}} \Big(\sum_{Q\in\mathcal{S}} \|gu^{\frac{1}{q'}}\|^{p'}_{\bar{A},Q} |E(Q)|^{\frac{p'}{q'}}\Big)^{\frac{1}{p'}} \\
&\leq 16[u,\sigma]_{A^\alpha_{p,q,A,B}} \Big(\sum_{Q\in\mathcal{S}} \|f\sigma^{\frac{1}{p}}\|^p_{\bar{B},Q} |E(Q)|\Big)^{\frac{1}{p}} \Big(\sum_{Q\in\mathcal{S}} \|gu^{\frac{1}{q'}}\|^{q'}_{\bar{A},Q} |E(Q)|\Big)^{\frac{1}{q'}} \\
&\leq 16[u,\sigma]_{A^\alpha_{p,q,A,B}} \Big(\sum_{Q\in\mathcal{S}} \int_{E(Q)} M_{\bar{B}}(f\sigma^{\frac{1}{p}})(x)^p\,dx\Big)^{\frac{1}{p}} \\
&\qquad \times \Big(\sum_{Q\in\mathcal{S}} \int_{E(Q)} M_{\bar{A}}(gu^{\frac{1}{q'}})(x)^{q'}\,dx\Big)^{\frac{1}{q'}} \\
&\leq 16[u,\sigma]_{A^\alpha_{p,q,A,B}} \Big(\int_{\mathbb{R}^n} M_{\bar{B}}(f\sigma^{\frac{1}{p}})(x)^p\,dx\Big)^{\frac{1}{p}} \Big(\int_{\mathbb{R}^n} M_{\bar{A}}(gu^{\frac{1}{q'}})(x)^{q'}\,dx\Big)^{\frac{1}{q'}} \\
&\leq C(n,p,q)[u,\sigma]_{A^\alpha_{p,q,A,B}} \Big(\int_{\mathbb{R}^n} f(x)^p\sigma(x)\,dx\Big)^{\frac{1}{p}}.
\end{aligned}$$

□

Bump conditions for commutators

Finally, we prove a conjoined bump condition for commutators. Because commutators are more singular, we need stronger bump conditions. To use the fact that b is a BMO function, it is most natural to state these in terms of log bumps. This result was originally proved in [24]; our proof is a simplification of the argument given there.

Theorem 6.4. *Given $0 < \alpha < n$, $1 < p \leq q < \infty$, and $b \in BMO$, suppose the pair* *of weights* (u,σ) *is such that* $[u,\sigma]_{A^{\alpha}_{p,q,A,B}} < \infty$, *where*

$$A(t) = t^q \log(e+t)^{2q-1+\delta}, \quad B(t) = t^{p'} \log(e+t)^{2p'-1+\delta}, \quad \delta > 0.$$

Then for every $f \in L^p(\sigma)$,

$$\left(\int_{\mathbb{R}^n} |[b, I_\alpha](f\sigma)(x)|^q u(x)\, dx\right)^{\frac{1}{q}} \leq C(n,p)\|b\|_{BMO}[u,\sigma]_{A^{\alpha}_{p,q,A,B}} \left(\int_{\mathbb{R}^n} |f(x)|^p \sigma(x)\, dx\right)^{\frac{1}{p}}.$$

The proof requires one lemma, which generalizes a result due to Sawyer and Wheeden [93], p. 829, and is proved in much the same way.

Lemma 6.2. *Fix* $0 < \alpha < n$, *a dyadic grid* $\mathscr{D}$, *and a Young function* Φ. *Then for any* $P \in \mathscr{D}$ *and any function* f,

$$\sum_{\substack{Q \in \mathscr{D} \\ Q \subset P}} |Q|^{\frac{\alpha}{n}} |Q| \|f\|_{\Phi,Q} \leq C(\alpha)|P|^{\frac{\alpha}{n}}|P|\|f\|_{\Phi,P}.$$

Proof. To prove this we need to replace the Luxemburg norm with the equivalent Amemiya norm (see [82], Section 3.3):

$$\|f\|_{\Phi,P} \leq \inf_{\lambda>0}\left\{\lambda \fint_P 1 + \Phi\left(\frac{|f(x)|}{\lambda}\right) dx\right\} \leq 2\|f\|_{\Phi,P}.$$

By the second inequality, we can fix $\lambda_0 > 0$ such that the middle quantity is less than $3\|f\|_{\Phi,P}$. Then by the first inequality,

$$\begin{aligned}
\sum_{\substack{Q \in \mathscr{D} \\ Q \subset P}} |Q|^{\frac{\alpha}{n}} |Q| \|f\|_{\Phi,Q} &= \sum_{k=0}^{\infty} \sum_{\substack{Q \subset P \\ \ell(Q) = 2^{-k}\ell(P)}} |Q|^{\frac{\alpha}{n}} |Q| \|f\|_{\Phi,Q} \\
&\leq |P|^{\frac{\alpha}{n}} \sum_{k=0}^{\infty} 2^{-k\alpha} \sum_{\substack{Q \subset P \\ \ell(Q) = 2^{-k}\ell(P)}} \lambda_0 \int_Q 1 + \Phi\left(\frac{|f(x)|}{\lambda_0}\right) dx \\
&= C(\alpha)|P|^{\frac{\alpha}{n}} \lambda_0 \int_P 1 + \Phi\left(\frac{|f(x)|}{\lambda_0}\right) dx \\
&\leq C(\alpha)|P|^{\frac{\alpha}{n}}|P|\|f\|_{\Phi,P}.
\end{aligned}$$

□

Proof of Theorem 6.4. Fix $b \in BMO$. We first make some reductions. Since $[b, I_\alpha]$ is linear, by splitting f into its positive and negative parts we may assume f is non-negative. By Fatou's lemma we may assume that f is bounded and has compact support. Finally, by Proposition 3.4 it will suffice to prove this result for the dyadic operator $C_b^{\mathscr{D}}$, where $\mathscr{D}$ is any dyadic grid.

We begin by applying duality: there exists $g \in L^{q'}(u)$, $\|g\|_{L^{q'}(u)} = 1$, such that

$$\begin{aligned}\|C_b^{\mathscr{D}}(f\sigma)\|_{L^q(u)} &= \int_{\mathbb{R}^n} C_b^{\mathscr{D}}(f\sigma)(x)g(x)u(x)\,dx \\ &= \sum_{Q\in\mathscr{D}} |Q|^{\frac{\alpha}{n}} \int_Q \fint_Q |b(x)-b(y)|f(y)\sigma(y)\,dy\,g(x)u(x)\,dx \\ &\le \sum_{Q\in\mathscr{D}} |Q|^{\frac{\alpha}{n}} \fint_Q |b(x)-\langle b\rangle_Q| g(x)u(x)\,dx\,\langle f\sigma\rangle_Q|Q| \\ &\quad + \sum_{Q\in\mathscr{D}} |Q|^{\frac{\alpha}{n}} \fint_Q |b(y)-\langle b\rangle_Q| f(y)\sigma(y)\,dy\,\langle gu\rangle_Q|Q|.\end{aligned}$$

We will estimate the first term; the estimate for the second is exactly the same, exchanging the roles of f, σ and g, u. Arguing as we did in the proof of Theorem 5.2, if we let $\mathscr{D}_N$ be the set of all dyadic cubes Q in $\mathscr{D}$ with $\ell(Q) = 2^N$, then it will suffice to bound this sum with $\mathscr{D}$ replaced by $\mathscr{D}_N$ and with a constant independent of N. Form the corona decomposition of $f\sigma$ with respect to Lebesgue measure for each cube in $\mathscr{D}_N$. Let $\mathscr{F}$ denote the union of all these cubes.

Let $\Phi(t) = t\log(e+t)$; then $\bar{\Phi}(t) \approx e^t - 1$, and so by the generalized Hölder's inequality and the John-Nirenberg inequality,

$$\fint_Q |b(x)-\langle b\rangle_Q| g(x)u(x)\,dx \le 2\|b-\langle b\rangle_Q\|_{\bar{\Phi},Q}\|gu\|_{\Phi,Q} \le C(n)\|b\|_{BMO}\|gu\|_{\Phi,Q}.$$

Furthermore, if we define $C(t) = t^{q'}/\log(e+t)^{1+(q'-1)\delta}$, then $C \in B_{q'}$ and $A^{-1}(t)C^{-1}(t) \lesssim \Phi^{-1}(t)$. (See [24], Lemma 2.12, for the details of this calculation.) We also have that $\bar{B} \in B_p$.

If we combine all of these facts and use Lemma 6.2 and the generalized Hölder's inequality twice, we get that

$$\begin{aligned}&\sum_{Q\in\mathscr{D}_N} |Q|^{\frac{\alpha}{n}} \fint_Q |b(x)-\langle b\rangle_Q| g(x)u(x)\,dx\,\langle f\sigma\rangle_Q|Q| \\ &\le C(n)\|b\|_{BMO} \sum_{F\in\mathscr{F}} \langle f\sigma\rangle_F \sum_{\pi_{\mathscr{F}}(Q)=F} |Q|^{\frac{\alpha}{n}}|Q|\|gu\|_{\Phi,Q} \\ &\le C(n,\alpha)\|b\|_{BMO} \sum_{F\in\mathscr{F}} \langle f\sigma\rangle_F |F|^{\frac{\alpha}{n}}|F|\|gu\|_{\Phi,F} \\ &\le C(n,\alpha)\|b\|_{BMO} \sum_{F\in\mathscr{F}} |F|^{\frac{\alpha}{n}}\|u^{\frac{1}{q}}\|_{A,F}\|\sigma^{\frac{1}{p'}}\|_{B,F}\|f\sigma^{\frac{1}{p}}\|_{\bar{B},F}\|gu^{\frac{1}{q'}}\|_{C,F}|E_{\mathscr{F}}(F)| \\ &\le C(n,\alpha)\|b\|_{BMO}[u,\sigma]_{A^{\alpha}_{p,q,A,B}} \sum_{F\in\mathscr{F}} \|f\sigma^{\frac{1}{p}}\|_{\bar{B},F}\|gu^{\frac{1}{q'}}\|_{C,F}|E_{\mathscr{F}}(F)|^{\frac{1}{p}+\frac{1}{q'}}.\end{aligned}$$

We can now apply Hölder's inequality and use the fact that $\bar{B} \in B_p$ and $C \in B_{q'}$ to finish the argument exactly as we did in the proof of Theorem 6.3. □

7. Separated bump conditions

We conclude with a discussion of some very recent work and some additional open problems for fractional integral operators and their commutators. To put these into context, we will first review the Muckenhoupt-Wheeden conjectures for singular integral operators and their relation to bump conditions. For a more detailed overview of these conjectures, see Refs. [21, 22].

The Muckenhoupt-Wheeden conjectures

In the late 1970s, while studying two weight norm inequalities for the Hilbert transform, Muckenhoupt and Wheeden made a series of conjectures relating this problem to two weight norm inequalities for the maximal operator.([f]) These conjectures were quickly extended to general singular integral operators. Restated in terms of weights (u,σ) instead of weights (u,v) as they were originally framed, they conjectured that for $1<p<\infty$, a sufficient condition for a singular integral operator to satisfy $T(\cdot\sigma): L^p(\sigma)\to L^p(u)$ is that the maximal operator satisfy

$$M(\cdot\sigma): L^p(\sigma)\to L^p(u),$$

and the dual inequality,

$$M(\cdot u): L^{p'}(u)\to L^{p'}(\sigma).$$

They further conjectured that the weak type inequality $T(\cdot\sigma): L^p(\sigma)\to L^{p,\infty}(u)$ holds if the maximal operator only satisfies the dual inequality. (Note the parallels between these conjectures and the testing conditions described in Section 5.) Finally, they conjectured that the following weak $(1,1)$ inequality holds:

$$\sup_{t>0} t\,u(\{x\in\mathbb{R}^n : |Tf(x)|>t\}) \le C\int_{\mathbb{R}^n} |f(x)|Mu(x)\,dx.$$

In the one weight case (i.e., with Muckenhoupt A_p weights) all of these conjectures are true, and with additional assumptions on the weights (e.g., $u,v\in A_\infty$) they are true in the two weight case. However, all three conjectures were recently shown to be false in general. The weak $(1,1)$ conjecture was disproved by Reguera and Thiele [84]; the strong (p,p) conjecture by Reguera and Scurry [83]; and building on this the weak (p,p) conjecture was disproved in [32].

---

[f] I first learned these conjectures from Pérez, and later learned some of their history directly from Muckenhoupt. However, they do not appear to have ever been published until they appeared in [22]. The weak $(1,1)$ conjecture appeared shortly before this in [61].

On the other hand, an "off-diagonal" version of this conjecture is true (see [23]): if $1 < p < q < \infty$, and the maximal operator satisfies

$$M(\cdot\sigma) : L^p(\sigma) \to L^q(u),$$

and the dual inequality,

$$M(\cdot u) : L^{q'}(u) \to L^{p'}(\sigma),$$

then $T(\cdot\sigma) : L^p(\sigma) \to L^q(u)$. If the dual inequality holds, then the weak (p,q) inequality $T(\cdot\sigma) : L^p(\sigma) \to L^{q,\infty}(u)$ holds as well. Examples of such weights can be easily constructed using the Sawyer testing condition (Theorem 5.1 with $\alpha = 0$). For instance, $u = \chi_{[0,1]}$ and $\sigma = \chi_{[2,3]}$ work for all $p > 1$.

It follows from Theorem 6.2 (with $\alpha = 0$) that these two off-diagonal inequalities for the maximal operator are implied by a pair of separated bump conditions, $[u,\sigma]_{A^0_{p,q,B}}, [u,\sigma]^*_{A^0_{p,q,A}} < \infty$. When $p = q$ this leads to the separated bump conjectures for singular integrals: if $[u,\sigma]_{A^0_{p,p,B}}, [u,\sigma]^*_{A^0_{p,p,A}} < \infty$, then a singular integral satisfies the strong (p,p) inequality, and if the dual condition holds, it satisfies the weak (p,p) inequality. This conjecture is due to Pérez: his study of bump conditions was partly motivated by the Muckenhoupt-Wheeden conjectures. It was first published, however, in [32], where it was proved for log bumps: $A(t) = t^p \log(e+t)^{p-1+\delta}$, $B(t) = t^{p'} \log(e+t)^{p'-1+\delta}$, $\delta > 0$, and some closely related bump conditions (the so called "loglog" bumps). The proof was quite technical, relying on a "freezing" argument and a version of the corona decomposition. For another, simpler proof that also holds in spaces of homogeneous type, see [1]. It is not clear if the separated bump conjecture is true for singular integrals only assuming bumps that satisfy the B_p condition. For very recent work that suggests it may be false, see Lacey [48] and Treil and Volberg [100].

Separated bump conditions for fractional integral operators

Though never addressed by Muckenhoupt and Wheeden, their conjectures for singular integrals extend naturally to fractional integrals as well. Such a generalization was first considered by Carro, *et al.* [4], who showed that the analog of the Muckenhoupt weak $(1,1)$ conjecture,

$$\sup_{t>0} t\, u(\{x \in \mathbb{R}^n : |I_\alpha f(x)| > t\}) \le C \int_{\mathbb{R}^n} |f(x)| M_\alpha u(x)\, dx, \tag{8}$$

is false.

In [25] we made the following conjectures: given $0 < \alpha < n$ and $1 < p \le q < \infty$, suppose the fractional maximal operator satisfies

$$M_\alpha(\cdot\sigma) : L^p(\sigma) \to L^q(u), \tag{9}$$

and the dual inequality,

$$M_\alpha(\cdot u) : L^{q'}(u) \to L^{p'}(\sigma). \tag{10}$$

Then the strong (p,q) inequality for I_α holds, and if the dual inequality holds, the weak (p,q) inequality holds. Analogous to the case of singular integrals, both of these conjectures are true when $p < q$: this was proved in [25]. Earlier, in [26] we proved a weaker version of this conjecture for separated bump conditions when the difference $\frac{1}{p} - \frac{1}{q}$ is very close to $\frac{\alpha}{n}$.

Theorem 7.1. *Given $0 < \alpha < n$ and $1 < p < q < \infty$, suppose the pair of weights (u,σ) are such that* (9) *and* (10) *hold. Then $I_\alpha(\cdot\sigma) : L^p(\sigma) \to L^q(u)$. If* (10) *holds, then $I_\alpha(\cdot\sigma) : L^p(\sigma) \to L^{q,\infty}(u)$.*

Proof. It will suffice to prove this for the dyadic fractional integral operator $I_\alpha^{\mathscr{D}}$, where $\mathscr{D}$ is any dyadic grid. We will show that the desired inequalities follow immediately from Theorem 5.3. To see this we will first show that the testing condition,

$$\mathscr{I}_{\mathscr{D},out} = \sup_Q \sigma(Q)^{-\frac{1}{p}} \left(\int_{\mathbb{R}^n} I_{\alpha,Q}^{\mathscr{D},out}(\sigma\chi_Q)(x)^q u(x)\, dx \right)^{\frac{1}{q}} < \infty,$$

holds. Fix a cube Q and $x \in \mathbb{R}^n$ such that there exists a dyadic cube $P \in \mathscr{D}$ with $x \in P$ and $Q \subsetneq P$. (If no such cube exists then $I_{\alpha,Q}^{\mathscr{D},out}(\sigma\chi_Q)(x) = 0$.) Let Q_0 be the smallest such cube, and for $k \ge 1$ let Q_k be the unique dyadic cube such that $Q_0 \subset Q_k$ and $\ell(Q_k) = 2^k \ell(Q_0)$. Then

$$\begin{aligned} I_{\alpha,Q}^{\mathscr{D},out}(\sigma\chi_Q)(x) &= \sum_{k=0}^{\infty} |Q_k|^{\frac{\alpha}{n}} \langle \sigma\chi_Q \rangle_{Q_k} \chi_{Q_k}(x) \\ &= |Q_0|^{\frac{\alpha}{n}} \langle \sigma\chi_Q \rangle_{Q_0} \sum_{k=0}^{\infty} 2^{k(\alpha-n)} \\ &\le C(n,\alpha) |Q_0|^{\frac{\alpha}{n}} \langle \sigma\chi_Q \rangle_{Q_0} \le C(n,\alpha) M_\alpha(\sigma\chi_Q)(x). \end{aligned}$$

Therefore, we can replace $I_{\alpha,Q}^{\mathscr{D},out}$ by M_α in the testing condition, and if (9) holds, then we immediately get that $\mathscr{I}_{\mathscr{D},out} < \infty$. Similarly, if we assume (10), then we get the dual testing condition: i.e., that $\mathscr{I}_{\mathscr{D},out}^* < \infty$. The strong and weak type inequalities then follow from Theorem 5.3. □

We do not know whether Theorem 7.1 is true when $p = q$, though the failure of the Muckenhoupt-Wheeden conjectures for singular integrals suggests that it is false. However, it is not clear where to look for a counter-example. One possibility is to modify the example of Reguera and Scurry [83]. However, this example depends strongly on the cancellation in the Hilbert transform, which is not present in the fractional integral, and it is not certain how this would affect the example. An alternative would be to consider the counter-example to (8) in [4].

When $p = q$ there is a weaker conjecture that we believe is true. As we noted above, by Theorem 6.2 we have that (9) holds if $[u,\sigma]_{A^\alpha_{p,p,B}} < \infty$, $\bar{B} \in B_p$, and (10) holds if $[u,\sigma]^*_{A^\alpha_{p,p,A}} < \infty$, $\bar{A} \in B_{p'}$. We therefore conjecture that if $[u,\sigma]_{A^\alpha_{p,p,B}}, [u,\sigma]^*_{A^\alpha_{p,p,A}} < \infty$, then $I_\alpha(\cdot\sigma) : L^p(\sigma) \to L^p(u)$, and if $[u,\sigma]^*_{A^\alpha_{p,p,A}} < \infty$, then $I_\alpha(\cdot\sigma) : L^p(\sigma) \to L^{p,\infty}(u)$.

This conjecture is the analog of the separated bump conjecture for singular integrals. For fractional integrals, this conjecture is only known for "double" log bumps: i.e., $A(t) = t^p \log(e+t)^{2p-1+\delta}$, $B(t) = t^{p'} \log(e+t)^{2p'-1+\delta}$. In [22], Theorem 9.42, it was shown that for this choice of A the weak (p,p) inequality is true if $[u,\sigma]^*_{A^\alpha_{p,p,A}} < \infty$. We therefore also have that the weak (p',p') inequality is true if $[u,\sigma]_{A^\alpha_{p,p,B}} < \infty$. Then by Corollary 5.1 we have that the two bump conditions together imply the strong (p,p) inequality.

The proof that the bump condition implies the weak type inequality has two steps. First, using a sharp function estimate and two weight extrapolation, we prove a weak $(1,1)$ inequality similar to (8):

$$\sup_{t>0} t\, u(\{x \in \mathbb{R}^n : |I_\alpha f(x)| > t\}) \leq C \int_{\mathbb{R}^n} |f(x)| M_{\Phi,\alpha} u(x)\, dx,$$

where $\Phi(t) = t\log(e+t)^{1+\epsilon}$, $\epsilon > 0$, and $M_{\Phi,\alpha}$ is the Orlicz fractional maximal operator,

$$M_{\Phi,\alpha} u(x) = \sup_Q |Q|^{\frac{\alpha}{n}} \|u\|_{\Phi,Q}\, \chi_Q(x).$$

The weak (p,p) inequality then follows by again applying two weight extrapolation and a two weight norm inequality for $M_{\Phi,\alpha}$.

We conjecture that the weak $(1,1)$ inequality is true if we replace Φ with $\Psi(t) = t\log(e+t)^\epsilon$. If this were the case, then the same extrapolation argument would yield the weak (p,p) inequality for log bumps, and the strong type inequality would follow as before. The analogous weak $(1,1)$ inequality is true for singular integrals: this was proved by Pérez [80]. (See also [1].) Unfortunately, every attempt to adapt these proofs to fractional integrals has failed.

An alternate approach would be to prove the weak (p, p) inequality directly using the testing conditions in Theorem 5.2. One way to do this would be to adapt the corona decomposition argument used in [32] to fractional integrals. We tried to do this, but our proof in [26] only worked if $\frac{1}{p}-\frac{1}{q}\approx\frac{\alpha}{n}$. More recently, we have shown in [27] that it can be modified to work provided $p<q$; but again the argument fails when $p=q$. We strongly believe that the separated bump conjecture is true for log bumps, and suspect that it is true in general. However, it is clear that either new ideas or a non-trivial adaptation of existing ones will be needed to prove it.

Two conjectures for commutators

We conclude with two conjectures and a question for commutators of fractional integrals. The first is a weak separated bump conjecture. A close examination of the proof of Theorem 6.4 shows that we actually proved something stronger: we showed that for $0<\alpha<n$ and $1<p\le q<\infty$, if a pair of weights (u,σ) satisfies

$$\sup_Q |Q|^{\frac{\alpha}{n}+\frac{1}{q}-\frac{1}{p}}\|u^{\frac{1}{q}}\|_{A,Q}\|\sigma^{\frac{1}{p'}}\|_{B,Q}<\infty, \tag{11}$$

with $A(t)=t^q\log(e+t)^{2q-1+\delta}$, and $B(t)=t^{p'}\log(e+t)^{p'-1+\delta}$, and

$$\sup_Q |Q|^{\frac{\alpha}{n}+\frac{1}{q}-\frac{1}{p}}\|u^{\frac{1}{q}}\|_{C,Q}\|\sigma^{\frac{1}{p'}}\|_{D,Q}<\infty, \tag{12}$$

with $C(t)=t^q\log(e+t)^{q-1+\delta}$, and $D(t)=t^{p'}\log(e+t)^{2p'-1+\delta}$, then the strong (p,q) inequality $[b,I_\alpha](\cdot\sigma):L^p(\sigma)\to L^q(u)$ holds.

There is no comparable result known for the weak (p,q) inequality. However, in [30] two weight weak type inequalities were proved for singular integral operators and we believe that the proofs there could be adapted to prove that $[b,I_\alpha](\cdot\sigma):L^p(\sigma)\to L^{q,\infty}(u)$ provided that (12) holds with $C(t)=t^{rq}$, $r>1$, and $D(t)=t^{p'}\log(e+t)^{p'}$. Further, using ideas from [15], we could in fact take $C(t)$ to be from a family of smaller Young functions called exponential log bumps.

Moreover, we also conjecture that the following separated bump conditions are sufficient: the strong (p,q) inequality holds if (u,σ) satisfy (11) and (12) but with $B(t)=t^{p'}$ and $C(t)=t^{q'}$. Similarly, the weak (p,q) inequality holds if (11) holds with $B(t)=t^{p'}$. To prove these conjectures it would suffice to prove them for the dyadic operator $C_b^{\mathscr{D}}$ in Proposition 3.4. It will probably be easier to prove these conjectures in the off-diagonal case when $p<q$. One approach in this case would be to prove a "global" version of the testing condition conjectures for commutators given at the end of Section 5. This might be done by adapting the arguments in [55]. Further, though it would probably not yield the

full conjecture, it would be interesting to see if the proof in [22], Theorem 9.42, for fractional integrals could be modified to prove a non-optimal weak type inequality for commutators. It seems possible that this approach would yield the weak type inequality with $A(t) = t^q \log(e+t)^{3q-1+\delta}$.

We finish with a question concerning the necessity of BMO for commutators to be bounded. Chanillo [7] showed that if $[b, I_\alpha] : L^p \to L^q$, $\frac{1}{p} - \frac{1}{q} = \frac{\alpha}{n}$, and $n-\alpha$ is an even integer, then $b \in BMO$. Very recently, this restriction on $n-\alpha$ was removed by Chaffee [5]. At the end of the meeting in Antequera, J. L. Torrea asked if anything could be said about b if there exists a pair of weights (u,σ) (or perhaps a family of such pairs) such that $[b, I_\alpha](\cdot\sigma) : L^p(\sigma) \to L^q(u)$. Nothing is known about this question, but it merits further investigation.

References

[1] T. C. Anderson, D. Cruz-Uribe and K. Moen, Logarithmic bump conditions for Calderón-Zygmund operators on spaces of homogeneous type, *Publ. Mat.* **59**, pp. 17–43 (2015).

[2] A. P. Calderón, Inequalities for the maximal function relative to a metric, *Studia Math.* **57**, pp. 297–306 (1976).

[3] A. P. Calderón and A. Zygmund, On the existence of certain singular integrals, *Acta Math.* **88**, pp. 85–139 (1952).

[4] M. J. Carro, C. Pérez, F. Soria and J. Soria, Maximal functions and the control of weighted inequalities for the fractional integral operator, *Indiana Univ. Math. J.* **54**, pp. 627–644 (2005).

[5] L. Chaffee, Characterizations of BMO through commutators of bilinear singular integral operators, *preprint* (2014), ArXiv 1410.4587.

[6] S.-Y. A. Chang, J. M. Wilson and T. H. Wolff, Some weighted norm inequalities concerning the Schrödinger operators, *Comment. Math. Helv.* **60**, pp. 217–246 (1985).

[7] S. Chanillo, A note on commutators, *Indiana Univ. Math. J.* **31**, pp. 7–16 (1982).

[8] F. Chiarenza and M. Franciosi, A generalization of a theorem by C. Miranda, *Ann. Mat. Pura Appl. (4)* **161**, pp. 285–297 (1992).

[9] M. Christ and R. Fefferman, A note on weighted norm inequalities for the Hardy-Littlewood maximal operator, *Proc. Amer. Math. Soc.* **87**, pp. 447–448 (1983).

[10] D. Chung, C. Pereyra and C. Pérez, Sharp bounds for general commutators on weighted Lebesgue spaces, *Trans. Amer. Math. Soc.* **364**, pp. 1163–1177 (2012).

[11] R. R. Coifman and C. Fefferman, Weighted norm inequalities for maximal functions and singular integrals, *Studia Math.* **51**, pp. 241–250 (1974).

[12] J. M. Conde, A note on dyadic coverings and nondoubling Calderón-Zygmund theory, *J. Math. Anal. Appl.* **397**, pp. 785–790 (2013).

[13] J. M. Conde-Alonso and G. Rey, A pointwise estimate for positive dyadic shifts and some applications, *Math. Ann.*, pp. 1–25 (2015). ArXiv 1409.4351.

[14] D. Cruz-Uribe, Elementary proofs of one weight norm inequalities for fractional integral operators and commutators, in *Harmonic Analysis, Partial Differential Equations, Banach Spaces and Operator Theory. Celebrating Cora Sadosky's*

Life, Vol. 2, M. C. Pereyra et al. Eds. (Springer AWM Series). (To appear). ArXiv 1507.02559.

[15] D. Cruz-Uribe and A. Fiorenza, The A_∞ property for Young functions and weighted norm inequalities, *Houston J. Math.* **28**, pp. 169–182 (2002).

[16] D. Cruz-Uribe and A. Fiorenza, Endpoint estimates and weighted norm inequalities for commutators of fractional integrals, *Publ. Mat.* **47**, pp. 103–131 (2003).

[17] D. Cruz-Uribe and A. Fiorenza, *Variable Lebesgue Spaces: Foundations and Harmonic Analysis* (Birkhäuser, Basel, 2013).

[18] D. Cruz-Uribe, J. M. Martell and C. Pérez, Extrapolation from A_∞ weights and applications, *J. Funct. Anal.* **213**, pp. 412–439 (2004).

[19] D. Cruz-Uribe, J. M. Martell and C. Pérez, Weighted weak-type inequalities and a conjecture of Sawyer, *Int. Math. Res. Not.* , pp. 1849–1871 (2005).

[20] D. Cruz-Uribe, J. M. Martell and C. Pérez, Sharp two-weight inequalities for singular integrals, with applications to the Hilbert transform and the Sarason conjecture, *Adv. Math.* **216**, pp. 647–676 (2007).

[21] D. Cruz-Uribe, J. M. Martell and C. Pérez, Sharp weighted estimates for classical operators, *Adv. Math.* **229**, pp. 408–441 (2011).

[22] D. Cruz-Uribe, J. M. Martell and C. Pérez, *Weights, extrapolation and the theory of Rubio de Francia*, Operator Theory: Advances and Applications, Vol. 215, pp. xiv+280 (Birkhäuser/Springer Basel AG, Basel, 2011).

[23] D. Cruz-Uribe, J. M. Martell and C. Pérez, A note on the off-diagonal Muckenhoupt-Wheeden conjecture, in *Advanced Courses in Mathematical Analysis V*, (World Scientific, to appear). ArXiv 1303.3424.

[24] D. Cruz-Uribe and K. Moen, Sharp norm inequalities for commutators of classical operators, *Publ. Mat.* **56**, pp. 147–190 (2012).

[25] D. Cruz-Uribe and K. Moen, A fractional Muckenhoupt-Wheeden theorem and its consequences, *Integr. Equ. Oper. Theory* **76**, pp. 421–446 (2013).

[26] D. Cruz-Uribe and K. Moen, One and two weight norm inequalities for Riesz potentials, *Illinois J. Math.* **57**, pp. 295–323 (2013).

[27] D. Cruz-Uribe and K. Moen, Super sparse families with applications to the separated bump conjecture for fractional integral operators, *preprint* (2014).

[28] D. Cruz-Uribe, K. Moen and S. Rodney, Regularity results for weak solutions of elliptic PDEs below the natural exponent, *Ann. Mat. Pura Appl.* (to appear), ArXiv 1408.6759.

[29] D. Cruz-Uribe and C. Pérez, Sharp two-weight, weak-type norm inequalities for singular integral operators, *Math. Res. Lett.* **6**, pp. 417–427 (1999).

[30] D. Cruz-Uribe and C. Pérez, Two-weight, weak-type norm inequalities for fractional integrals, Calderón-Zygmund operators and commutators, *Indiana Univ. Math. J.* **49**, pp. 697–721 (2000).

[31] D. Cruz-Uribe and C. Pérez, On the two-weight problem for singular integral operators, *Ann. Sc. Norm. Super. Pisa Cl. Sci. (5)* **1**, pp. 821–849 (2002).

[32] D. Cruz-Uribe, A. Reznikov and A. Volberg, Logarithmic bump conditions and the two-weight boundedness of Calderón-Zygmund operators, *Adv. Math.* **255**, pp. 706–729 (2014).

[33] G. David and S. Semmes, Singular integrals and rectifiable sets in $\mathbf{R}^n$: Beyond Lipschitz graphs, *Astérisque*, pp. 152 (1991).

[34] G. David and S. Semmes, *Analysis of and on uniformly rectifiable sets*, Mathematical Surveys and Monographs, Vol. 38, pp. xii+356 (American Mathematical Society, Providence, RI, 1993).

[35] J. Duoandikoetxea, *Fourier analysis*, Graduate Studies in Mathematics, Vol. 29, pp. xviii+222 (American Mathematical Society, Providence, RI, 2001).

[36] C. Fefferman, The uncertainty principle, *Bull. Amer. Math. Soc. (N.S.)* **9**, pp. 129–206 (1983).

[37] J. García-Cuerva and J. M. Martell, Two-weight norm inequalities for maximal operators and fractional integrals on non-homogeneous spaces, *Indiana Univ. Math. J.* **50**, pp. 1241–1280 (2001).

[38] J. García-Cuerva and J. L. Rubio de Francia, *Weighted norm inequalities and related topics*, North-Holland Mathematics Studies, Vol. 116, pp. x+604 (North-Holland Publishing Co., Amsterdam, 1985).

[39] J. B. Garnett and P. W. Jones, BMO from dyadic BMO, *Pacific J. Math.* **99**, pp. 351–371 (1982).

[40] L. Grafakos, *Classical Fourier Analysis*, Graduate Texts in Mathematics, Vol. 249, 2nd edn. (Springer, New York, 2008).

[41] L. Grafakos, *Modern Fourier Analysis*, Graduate Texts in Mathematics, Vol. 250, 2nd edn. (Springer, New York, 2008).

[42] T. Hytönen, The sharp weighted bound for general Calderón-Zygmund operators, *Ann. of Math. (2)* **175**, pp. 1473–1506 (2012).

[43] T. Hytönen, The A_2 theorem: remarks and complements, in *Harmonic analysis and partial differential equations*, Contemp. Math., Vol. 612, pp. 91–106 (Amer. Math. Soc., Providence, RI, 2014).

[44] T. Hytönen and C. Pérez, Sharp weighted bounds involving A_∞, *Anal. PDE* **6**, pp. 777–818 (2013).

[45] A. Kairema, Two-weight norm inequalities for potential type and maximal operators in a metric space, *Publ. Mat.* **57**, pp. 3–56 (2013).

[46] R. Kerman and E. Sawyer, The trace inequality and eigenvalue estimates for Schrödinger operators, *Ann. Inst. Fourier (Grenoble)* **36**, pp. 207–228 (1986).

[47] M. A. Krasnosel'skiĭ and J. B. Rutickiĭ, *Convex functions and Orlicz spaces*, pp. xi+249 (P. Noordhoff Ltd., Groningen, 1961), Translated from the first Russian edition by Leo F. Boron.

[48] M. T. Lacey, On the separated bumps conjecture for Calderón-Zygmund operators, *Hokkaido Math. J.* (to appear). ArXiv 1310.3507.

[49] M. T. Lacey, The two weight inequality for the Hilbert transform: a primer, *preprint* (2013), ArXiv 1304.5004.

[50] M. T. Lacey, Two weight inequality for the Hilbert transform: a real variable characterization, II, *Duke Math. J.* **163**, no. 15, pp. 2821–2840 (2014).

[51] M. T. Lacey, K. Moen, C. Pérez and R. H. Torres, Sharp weighted bounds for fractional integral operators, *J. Funct. Anal.* **259**, pp. 1073–1097 (2010).

[52] M. T. Lacey, S. Petermichl and M. C. Reguera, Sharp A_2 inequality for Haar shift operators, *Math. Ann.* **348**, pp. 127–141 (2010).

[53] M. T. Lacey, E. T. Sawyer, C.-Y. Shen and I. Uriarte-Tuero, Two-weight inequality for the Hilbert transform: a real variable characterization, I, *Duke Math. J.* **163**, pp. 2795–2820 (2014).

[54] M. T. Lacey, E. T. Sawyer and I. Uriarte-Tuero, A characterization of two weight norm inequalities for maximal singular integrals with one doubling measure, *Anal. PDE* **5**, pp. 1–60 (2012).

[55] M. T. Lacey, E. T. Sawyer and I. Uriarte-Tuero, Two weight inequalities for discrete positive operators, *preprint* (2012), ArXiv 0911.3437.

[56] R. Lechner, The one-third-trick and shift operators, *Bull. Pol. Acad. Sci. Math.* **61**, pp. 219–238 (2013).

[57] A. K. Lerner, An elementary approach to several results on the Hardy-Littlewood maximal operator, *Proc. Amer. Math. Soc.* **136**, pp. 2829–2833 (2008).

[58] A. K. Lerner, On an estimate of Calderón-Zygmund operators by dyadic positive operators, *J. Anal. Math.* (2012).

[59] A. K. Lerner, A simple proof of the A_2 conjecture, *Int. Math. Res. Not. IMRN*, pp. 3159–3170 (2013).

[60] A. K. Lerner and F. Nazarov, Intuitive dyadic calculus: the basics, *preprint* (2014), www.math.kent.edu/~zvavitch/Lerner_Nazarov_Book.pdf.

[61] A. K. Lerner, S. Ombrosi and C. Pérez, Sharp A_1 bounds for Calderón-Zygmund operators and the relationship with a problem of Muckenhoupt and Wheeden, *Int. Math. Res. Not. IMRN*, pp. Art. ID rnm161, 11 (2008).

[62] J. Li, J. Pipher and L. Ward, Dyadic structure theorems for multiparameter function spaces, *Rev. Mat. Iberoam.* (to appear), http://www.math.brown.edu/~jpipher/LiPipherWard_RMI_2013.pdf.

[63] L. Liu and T. Luque, A B_p condition for the strong maximal function, *Trans. Amer. Math. Soc.* **366**, pp. 5707–5726 (2014).

[64] R. L. Long and F. S. Nie, Weighted Sobolev inequality and eigenvalue estimates of Schrödinger operators, in *Harmonic analysis (Tianjin, 1988)*, Lecture Notes in Math., Vol. 1494, pp. 131–141 (Springer, Berlin, 1991).

[65] V. G. Maz'ya, *Sobolev spaces*, pp. xix+486 (Springer-Verlag, Berlin, 1985), Translated from the Russian by T. O. Shaposhnikova.

[66] T. Mei, BMO is the intersection of two translates of dyadic BMO, *C. R. Math. Acad. Sci. Paris* **336**, pp. 1003–1006 (2003).

[67] B. Muckenhoupt and R. L. Wheeden, Weighted norm inequalities for fractional integrals, *Trans. Amer. Math. Soc.* **192**, pp. 261–274 (1974).

[68] B. Muckenhoupt and R. L. Wheeden, Two weight function norm inequalities for the Hardy-Littlewood maximal function and the Hilbert transform, *Studia Math.* **55**, pp. 279–294 (1976).

[69] B. Muckenhoupt and R. L. Wheeden, Some weighted weak-type inequalities for the Hardy-Littlewood maximal function and the Hilbert transform, *Indiana Univ. Math. J.* **26**, pp. 801–816 (1977).

[70] C. Muscalu, T. Tao and C. Thiele, Multi-linear operators given by singular multipliers, *J. Amer. Math. Soc.* **15**, pp. 469–496 (2002).

[71] F. Nazarov and S. Treil, The hunt for a Bellman function: applications to estimates for singular integral operators and to other classical problems of harmonic analysis, *Algebra i Analiz* **8**, pp. 32–162 (1996).

[72] F. Nazarov, S. Treil and A. Volberg, Cauchy integral and Calderón-Zygmund operators on nonhomogeneous spaces, *Internat. Math. Res. Notices*, pp. 703–726 (1997).

[73] F. Nazarov, S. Treil and A. Volberg, The Bellman functions and two-weight inequalities for Haar multipliers, *J. Amer. Math. Soc.* **12**, pp. 909–928 (1999).

[74] F. Nazarov, S. Treil and A. Volberg, The Tb-theorem on non-homogeneous spaces, *Acta Math.* **190**, pp. 151–239 (2003).

[75] F. Nazarov, S. Treil and A. Volberg, Two weight inequalities for individual Haar multipliers and other well localized operators, *Math. Res. Lett.* **15**, pp. 583–597 (2008).

[76] C. J. Neugebauer, Inserting A_p-weights, *Proc. Amer. Math. Soc.* **87**, pp. 644–648 (1983).

[77] K. Okikiolu, Characterization of subsets of rectifiable curves in $\mathbf{R}^n$, *J. London Math. Soc. (2)* **46**, pp. 336–348 (1992).

[78] M. C. Pereyra, Lecture notes on dyadic harmonic analysis, in *Second Summer School in Analysis and Mathematical Physics (Cuernavaca, 2000)*, Contemp. Math., Vol. 289, pp. 1–60 (Amer. Math. Soc., Providence, RI, 2001).

[79] C. Pérez, Two weighted inequalities for potential and fractional type maximal operators, *Indiana Univ. Math. J.* **43**, pp. 663–683 (1994).

[80] C. Pérez, Weighted norm inequalities for singular integral operators, *J. London Math. Soc. (2)* **49**, pp. 296–308 (1994).

[81] C. Pérez, On sufficient conditions for the boundedness of the Hardy-Littlewood maximal operator between weighted L^p-spaces with different weights, *Proc. London Math. Soc. (3)* **71**, pp. 135–157 (1995).

[82] M. M. Rao and Z. D. Ren, *Theory of Orlicz spaces*, Monographs and Textbooks in Pure and Applied Mathematics, Vol. 146, pp. xii+449 (Marcel Dekker Inc., New York, 1991).

[83] M. Reguera and J. Scurry, On joint estimates for maximal functions and singular integrals on weighted spaces, *Proc. Amer. Math. Soc.* **141**, pp. 1705–1717 (2013).

[84] M. Reguera and C. Thiele, The Hilbert transform does not map $L^1(Mw)$ to $L^{1,\infty}(w)$, *Math. Res. Let.* **19**, pp. 1–7 (2012).

[85] M. Riesz, L'intégrale de Riemann-Liouville et le problème de Cauchy, *Acta Math.* **81**, pp. 1–223 (1949).

[86] H. L. Royden, *Real analysis*, third edn., pp. xx+444 (Macmillan Publishing Company, New York, 1988).

[87] E. T. Sawyer, A characterization of a two-weight norm inequality for maximal operators, *Studia Math.* **75**, pp. 1–11 (1982).

[88] E. T. Sawyer, A two weight weak type inequality for fractional integrals, *Trans. Amer. Math. Soc.* **281**, pp. 339–345 (1984).

[89] E. T. Sawyer, Weighted inequalities for the two-dimensional Hardy operator, *Studia Math.* **82**, pp. 1–16 (1985).

[90] E. T. Sawyer, A weighted weak type inequality for the maximal function, *Proc. Amer. Math. Soc.* **93**, pp. 610–614 (1985).

[91] E. T. Sawyer, A characterization of two weight norm inequalities for fractional and Poisson integrals, *Trans. Amer. Math. Soc.* **308**, pp. 533–545 (1988).

[92] E. T. Sawyer, C.-Y. Shen and I. Uriarte-Tuero, A two weight theorem for fractional singular integrals with an energy side condition, *preprint* (2013), ArXiv 1302.5093.

[93] E. T. Sawyer and R. L. Wheeden, Weighted inequalities for fractional integrals on Euclidean and homogeneous spaces, *Amer. J. Math.* **114**, pp. 813–874 (1992).

[94] E. M. Stein, *Singular integrals and differentiability properties of functions*, Princeton Mathematical Series, Vol. 30, pp. xiv+290 (Princeton University Press, Princeton, N.J., 1970).

[95] E. M. Stein, *Harmonic analysis: real-variable methods, orthogonality, and oscillatory integrals*, Princeton Mathematical Series, Vol. 43, pp. xiv+695 (Princeton University Press, Princeton, NJ, 1993), With the assistance of Timothy S. Murphy, Monographs in Harmonic Analysis, III.

[96] E. M. Stein and G. Weiss, *Introduction to Fourier analysis on Euclidean spaces*, pp. x+297 (Princeton University Press, Princeton, N.J., 1971), Princeton Mathematical Series, No. 32.

[97] H. Tanaka, A characterization of two-weight trace inequalities for positive dyadic operators in the upper triangle case, *Potential Anal.* **41**, pp. 487–499 (2014).

[98] A. de la Torre, On the adjoint of the maximal function, in *Function spaces, differential operators and nonlinear analysis (Paseky nad Jizerou, 1995)*, pp. 189–194 (Prometheus, Prague, 1996).

[99] S. Treil, A remark on two weight estimates for positive dyadic operators, *preprint* (2012), ArXiv 1201.1455.

[100] S. Treil and A. Volberg, Entropy conditions in two weight inequalities for singular integral operators, *preprint* (2014), ArXiv 1408.0385.

[101] I. E. Verbitsky, Weighted norm inequalities for maximal operators and Pisier's theorem on factorization through $L^{p,\infty}$, *Integr. Equ. Oper. Theory* **15**, pp. 124–153 (1992).

[102] A. Volberg, *Calderón-Zygmund capacities and operators on nonhomogeneous spaces*, CBMS Regional Conference Series in Mathematics, Vol. 100, pp. iv+167 (Published for the Conference Board of the Mathematical Sciences, Washington, DC; by the American Mathematical Society, Providence, RI, 2003).

[103] J. M. Wilson, Weighted norm inequalities for the continuous square function, *Trans. Amer. Math. Soc.* **314**, pp. 661–692 (1989).

[104] T. H. Wolff, Two algebras of bounded functions, *Duke Math. J.* **49**, pp. 321–328 (1982).

[105] W. P. Ziemer, *Weakly Differentiable Functions*, Graduate Texts in Mathematics, Vol. 120, pp. xvi+308 (Springer-Verlag, New York, 1989).

Composition operators on Hardy spaces

Pascal Lefèvre

U-Artois, Laboratoire de Mathématiques de Lens EA 2462,
Fédération CNRS Nord-Pas-de-Calais FR 2956,
F-62 300 Lens, France
E-mail: pascal.lefevre@univ-artois.fr
http://lefevre.perso.math.cnrs.fr/

The topic of this short survey is composition operators, i.e., maps of the shape $f \longmapsto f \circ \varphi$, where the symbol $\varphi : \mathbb{D} \to \mathbb{D}$ is holomorphic. We shall give a (non-exhaustive) overview of -more or less recent- results on such operators, acting on Hardy spaces. The story involves some classical tools of complex analysis, as Nevanlinna counting function and Carleson measures. We shall illustrate this presentation with miscellaneous examples and questions. We shall pay attention to their possible membership to various classes of operator ideals, and in particular, in the membership of the class of absolutely summing operators.

Keywords: Absolutely summing operators, Carleson measure, compactness, composition operator, Hardy space, Hardy-Orlicz space, Nevanlinna counting function.

Acknowledgments

This survey is based on a series of lectures given during the "VI Curso Internacional de Análisis Matemático en Andalucía", held in Antequera in September 2014. The author wishes to thank warmly the organizers for the very nice atmosphere during this conference. He wishes to thank too both the referee and the editors for their help to improve the quality of this paper.

1. Introduction

We shall give an overview on composition operators on various kind of Hardy spaces, but, obviously, it is far from being exhaustive. Here are the topics developed in the paper:

- Classical Hardy spaces on $\mathbb{D}$.
- Composition operators: general framework and examples, boundedness.
- Compactness.
- The case of H^∞.

- The case of Hardy-Orlicz spaces.
- The Carleson function. Carleson versus Nevanlinna.
- Schatten classes.
- Absolutely summing composition operators.
- Some open problems...

1.1. *Classical Hardy spaces on the unit disk*

In the present paper, we are interested in Banach spaces of holomorphic functions on the unit disk of the complex plane: $\mathbb{D} = \{z \in \mathbb{C} : |z| < 1\}$. Its boundary, the torus, is denoted: $\mathbb{T} = \{z \in \mathbb{C} : |z| = 1\} = \partial\mathbb{D}$ (and can be viewed also as $\mathbb{R}/\mathbb{Z}$). We denote by $\mathcal{H}(\mathbb{D})$ the class of holomorphic functions on the unit disk.

More specifically, we shall focus on Hardy spaces. First let us recall quickly how classical Hardy spaces are defined (see [6] for more details).

When $p \in [1, +\infty)$:

$$H^p = \Big\{f \in \mathcal{H}(\mathbb{D}) : \sup_{r<1} \int_{\mathbb{T}} |f(rz)|^p \, d\lambda < \infty\Big\},$$

and

$$\|f\|_p = \sup_{r<1} \Big(\int_{\mathbb{T}} |f(rz)|^p \, d\lambda\Big)^{1/p} = \sup_{r<1} \|f_r\|_{L^p(\mathbb{T})}.$$

Here λ stands for the Lebesgue measure on the torus, and $f_r(z) = f(rz)$ with $r \in (0,1)$ and $z \in \overline{\mathbb{D}}$.

In the particular case $p = 2$ (the hilbertian case), thanks to the Parseval formula, we can express the norm with the Taylor coefficients of the function. Indeed, let $f(z) = \sum_{n=0}^{+\infty} a_n z^n$ be analytic on $\mathbb{D}$:

$$\|f\|_2 = \Big(\sum_{n=0}^{+\infty} |a_n|^2\Big)^{1/2}.$$

The case $p = +\infty$ defines the space of bounded analytic functions on $\mathbb{D}$:

$$H^\infty = \{f \in \mathcal{H}(\mathbb{D}) : \|f\|_\infty = \sup_{z\in\mathbb{D}} |f(z)| < \infty\}.$$

It is easy to check that all these spaces are Banach spaces. Let us mention some (very useful) properties of analytic functions belonging to the Hardy spaces. Every $f \in H^p$ admits almost everywhere radial limits on the torus, i.e., $f^*(e^{it}) = \lim_{r\to 1^-} f(re^{it})$ is defined for almost every e^{it} of the torus. It is known that $f^* \in L^p(\mathbb{T})$ and $\|f\|_{H^p} = \|f^*\|_{L^p(\mathbb{T})}$. In fact, f^* belongs to the space $\{g \in L^p(\mathbb{T}) \mid \hat{g}(m) = 0 \text{ for every } m < 0\}$. Here $\hat{g}(m)$ stands for the usual m^{th} Fourier coefficient of g.

Conversely, if we consider any $g \in L^p(\mathbb{T})$, with $\hat{g}(m) = 0$ for every $m < 0$, the Poisson integral of g at the point $z = re^{2i\pi\theta}$,

$$P[g](z) = P_r * g(\theta) = \int_0^1 P_r(\theta - t) g(e^{2i\pi t})\, dt,$$

belongs to H^p. Moreover $(P[g])^* = g$.

Hence, H^p is naturally identified with the (closed) subspace of $L^p(\mathbb{T})$: $\{g \in L^p(\mathbb{T}) \mid \hat{g}(m) = 0 \text{ for every } m < 0\}$.

These remarks allow to adopt a double viewpoint on Hardy spaces: we can have the viewpoint of functions on the open unit disk, and the viewpoint of functions on the unit circle. Gathering all this, $f \in H^p$ is defined, not only on $\mathbb{D}$, but on the whole $\overline{\mathbb{D}} = \mathbb{D} \cup \mathbb{T}$. In the sequel, we shall sometimes omit the star, i.e., identifying f and f^*.

An inner function is an H^∞ function I with $|I^*| = 1$ a.e. on $\mathbb{T}$. By the Riesz factorization theorem (see [6], Th 2.8), every $f \in H^p$ can be written as $f = Bg$, where B is inner and g does not vanish on $\mathbb{D}$. Since $|f^*| = |g^*|$ (a.e. on $\mathbb{T}$), we have $\|f\|_{H^p} = \|g\|_{H^p}$.

For every $z \in \mathbb{D}$, the point evaluation at $z \in \mathbb{D}$ is defined on H^p by $\delta_z(f) = f(z)$. The linear functional δ_z is a continuous functional and we can easily compute its norm.

Proposition 1.1. *We have*

- *On the Hilbert space H^2, the functional δ_z is associated to the reproducing kernel $w \in \overline{\mathbb{D}} \longmapsto \dfrac{1}{1 - \bar{z}w}$.*
- $\|\delta_z\|_{(H^p)^*} = \left(\dfrac{1}{1 - |z|^2}\right)^{1/p} \approx \dfrac{1}{(1 - |z|)^{1/p}}$ *(when $|z| \to 1^-$).*

Proof. Indeed, we have the Taylor expansion

$$\delta_z(f) = f(z) = \sum_{n \geq 0} a_n z^n = \langle f, D_z \rangle,$$

where $D_z(w) = \frac{1}{1 - \bar{z}w} = \sum_{n \geq 0} \bar{z}^n w^n$. In particular,

$$\|\delta_z\|^2_{(H^2)^*} = \|D_z\|^2_{H^2} = \delta_z(D_z) = \frac{1}{1 - |z|^2}.$$

Now, by the Riesz factorization theorem again and the fact that there exist holomorphic branches of the logarithm of nonvanishing holomorphic functions on $\mathbb{D}$, we may write $f \in H^p$ as $f = Bg = BG^{2/p}$ on $\mathbb{D}$, with G holomorphic on $\mathbb{D}$ and B inner. Thus,

$$|\delta_z(f)|^p = |B(z)|^p |G(z)|^2 \leq |G(z)|^2 \leq \|\delta_z\|^2_{(H^2)^*} \|G\|^2_{H^2} = \frac{1}{1 - |z|^2} \|f\|^p_{H^p}.$$

So

$$\|\delta_z\|_{(H^p)^*} \leq \left(\frac{1}{1-|z|^2}\right)^{1/p}.$$

Testing $D_z^{2/p}$, we get the reverse inequality. □

Let us emphasize that, on the unit ball of H^p, the weak-star topology (which is the weak topology when $p > 1$) coincides with the topology of uniform convergence on compact subsets of $\mathbb{D}$. We do not prove it: it is an easy exercise left to the reader.

1.2. *Composition operators*

Let X and Y be two spaces of holomorphic functions on the unit disk. Composition operators are the operators of kind:

$$C_\varphi : f \in X \longmapsto f \circ \varphi \in Y,$$

where $\varphi : \mathbb{D} \to \mathbb{D}$ is analytic.

There are immediately several natural questions: when is it defined, bounded? When is it compact? When is it very compact?

More generally, we wish to understand the link

$$\text{“Operator } C_\varphi\text{”} \stackrel{??}{\longleftrightarrow} \text{“symbol } \varphi\text{”}.$$

The interesting part of this subject is to understand the link between the properties of the *operator* C_φ (which involves theory of operators) and the properties of the *symbol* φ (which involves theory of functions).

2. Boundedness

The first remarkable result is

Theorem 2.1. *$C_\varphi : H^p \longrightarrow H^p$ are always bounded.*

There are several ways to see or understand this statement! There is a nice purely hilbertian proof originally due to Littlewood (we shall not detail it and it is well explained in Shapiro's monograph [24]). We shall give below several proofs of Theorem 2.1: maybe the shortest and most general argument uses the (so-called) Littlewood subordination principle. Another proof uses a well known object of the one complex variable theory: the Nevanlinna counting function N_φ, which turns out to play a key role in the theory of composition operators.

On the other hand, thanks to the results of Carleson on (now so called) Carleson's embeddings, we can rewrite the boundedness of a composition operator in terms of Carleson's measures.

2.1. *Automorphism of the disk*

Let us first treat the special (but important) case of Moebius transformations, which are automorphisms of the disk.

Consider the Moebius transformation $q_a(z) = \frac{a-z}{1-\bar{a}z}$, where $a \in \mathbb{D}$. We can use a true change of variable: for every polynomial f, we have

$$\|f \circ q_a\|_{H^p}^p = \int_{\mathbb{T}} |f(z)|^p \frac{1-|a|^2}{|1-\bar{a}z|^2}\, d\lambda.$$

The norm of C_{q_a} is clearly the norm of the multiplier operator on H^p, associated to the weight $M(z) = \left(\frac{1-|a|^2}{(1-\bar{a}z)^2}\right)^{1/p}$. We get

$$\|C_{q_a}\| = \|M\|_\infty = \left(\frac{1+|a|}{1-|a|}\right)^{1/p}.$$

For a general automorphism of the disk θ, which can be written $\theta = cq_a$ (for some $a \in \mathbb{D}$ and some $c \in \mathbb{T}$), we get

$$\|C_\theta\| = \left(\frac{1+|\theta(0)|}{1-|\theta(0)|}\right)^{1/p}.$$

2.2. *Reduction to the case* $\varphi(0) = 0$

We can now reduce the problem to composition operators C_ϕ with a symbol verifying $\phi(0) = 0$.

Indeed, write $a = \varphi(0)$ and consider $\phi = q_a \circ \varphi \iff q_a \circ \phi = \varphi$. We have $\phi(0) = 0$.

Hence once we prove that C_ϕ is bounded (and we shall prove it below with the estimate $\|C_\phi\| = 1$), we get that $C_\varphi = C_\phi \circ C_{q_a}$ is bounded as well! Thus

$$\|C_\varphi\| \leq \left(\frac{1+|\varphi(0)|}{1-|\varphi(0)|}\right)^{1/p}.$$

2.3. *When* $\varphi(0) = 0$

Let us give several proofs of the fact C_φ is bounded with $\|C_\varphi\| = 1$.

2.3.1. *Boundedness via the subordination principle*

Let $\varphi\colon \mathbb{D} \to \mathbb{D}$ be analytic with $\varphi(0) = 0$, and let $g\colon \mathbb{D} \to [0, +\infty)$ be a subharmonic function. For every $r \in (0,1)$, we claim that

$$\int_0^{2\pi} g\big(\varphi(re^{it})\big)\, dt \leq \int_0^{2\pi} g(re^{it})\, dt.$$

Indeed, let G be the harmonic function on $r\mathbb{D}$, continuous on $\overline{r\mathbb{D}}$, such that $G = g$ on $r\mathbb{T}$. Thus, $g \le G$ on $r\mathbb{D}$. Thanks to the Schwarz lemma, $\varphi(re^{it})$ belongs to $r\mathbb{D}$. So, not forgetting that $G \circ \varphi$ is harmonic, we have

$$\int_0^{2\pi} g\big(\varphi(re^{it})\big)\,dt \le \int_0^{2\pi} G\big(\varphi(re^{it})\big)\,dt = G\circ\varphi(0)$$
$$= G(0) = \int_0^{2\pi} G(re^{it})\,dt = \int_0^{2\pi} g(re^{it})\,dt,$$

which was our claim.

Now we apply this inequality to the function $g(z) = |f(z)|^p$, where $f \in H^p$. Letting $r \nearrow 1^-$, we get

$$\|f\circ\varphi\|_{H^p}^p \le \|f\|_{H^p}^p, \qquad \text{i.e.,} \qquad \|C_\varphi\| = 1.$$

The boundedness is proved. □

2.3.2. *Boundedness in the hilbertian case via the Nevanlinna function*

Let us first recall how this function is defined

$$N_\varphi(w) = \begin{cases} \sum_{\varphi(z)=w} \log\frac{1}{|z|} & \text{if } w \ne \varphi(0), \text{ and } w \in \varphi(\mathbb{D}), \\ 0 & \text{otherwise.} \end{cases}$$

(In the sum, every z occurs as many times as its multiplicity.)

We have for every $w \ne \varphi(0)$:

$$N_\varphi(w) \le \log\left|\frac{1-\overline{\varphi(0)}\,w}{\varphi(0)-w}\right| = O\big((1-|w|)\big) \quad \text{when } |w| \to 1^-.$$

Let us mention that this very nice inequality is a "super Schwarz lemma": it means, when $\varphi(0) = 0$,

$$\big|\varphi(z)\big| \le \prod_{\varphi(a)=\varphi(z)} |a| \quad \big(\le |z|\big).$$

Now, we want to establish the boundedness of C_φ. We shall use the Littlewood-Paley formula ($p = 2$),

$$\|f\|_2^2 = \big|f(0)\big|^2 + 2\int_{\mathbb{D}} |f'|^2 \log\frac{1}{|z|}\,d\mathscr{A}.$$

Here, $d\mathscr{A}$ stands for the normalized Lebesgue area measure on $\mathbb{D}$. This relation is easy to prove through a simple straightforward computation (use Parseval formula).

Let us see how it implies the boundedness of C_φ on H^2:

$$\begin{aligned}\|f\circ\varphi\|_2^2 &= \left|f\circ\varphi(0)\right|^2 + 2\int_{\mathbb{D}} |(f\circ\varphi)'|^2 \log\frac{1}{|z|}\,d\mathscr{A}\\ &= \left|f\circ\varphi(0)\right|^2 + 2\int_{\mathbb{D}} \left(|f'|^2\circ\varphi\right)\times\left|\varphi'\right|^2 \log\frac{1}{|z|}\,d\mathscr{A}\\ &= \left|f\circ\varphi(0)\right|^2 + 2\int_{\mathbb{D}} |f'|^2 N_\varphi\,d\mathscr{A},\end{aligned}$$

using the generalized (non injective) change of variable formula.

Now, since we assumed that $\varphi(0) = 0$, using the majorization $N_\varphi(w) \le \log\left|\frac{1}{w}\right|$, we get

$$\|f\circ\varphi\|_2^2 \le \left|f(0)\right|^2 + 2\int_{\mathbb{D}} |f'|^2 \log\left|\frac{1}{w}\right| d\mathscr{A} = \|f\|_2^2,$$

which proves (again) the boundedness of C_φ on H^2. □

2.3.3. *Boundedness from the Carleson embedding point of view*

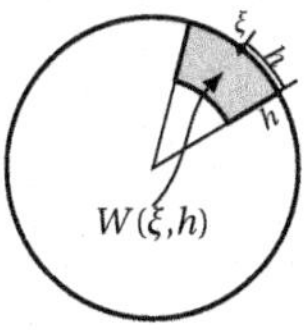

Recall that a (classical) Carleson measure is a (finite) positive Borel measure μ, defined on $\mathbb{D}$ (or $\overline{\mathbb{D}}$), for which there exists a constant $K > 0$ such that

$$\mu\big(W(\xi,h)\big) \le Kh, \quad \text{for all } \xi\in\mathbb{T}, \text{ and all } h\in(0,1).$$

Here, $W(\xi,h) = \{z\in\overline{\mathbb{D}} : 1-h < |z| \le 1, \ \ |\arg(z\overline{\xi})| < h\}$ is the typical Carleson window based at $\xi\in\mathbb{T}$ with "side length" $h\in(0,1)$.

Carleson's embedding Theorem (see [6, Th. 9.3], or [3, Th. 2.33]) asserts that, for $0 < p < \infty$, the linear embedding from H^p into $L^p(\mu)$ is continuous if and only if μ is a Carleson measure.

Now, defining the pullback measure λ_φ (on $\overline{\mathbb{D}}$) as $\lambda_\varphi(E) = \lambda\big({\varphi^*}^{-1}(E)\big)$, for $E\subset\overline{\mathbb{D}}$ Borel set, we point out that

$$\left\|f\circ\varphi\right\|_p^p = \int_{\overline{\mathbb{D}}} |f|^p\,d\lambda_\varphi,$$

and thus, the boundedness of C_φ on H^p is equivalent to λ_φ being a Carleson measure.

Hence, as a consequence of the boundedness of C_φ on H^p, we obtain

Proposition 2.1. *$\rho_\varphi(h) := \sup_{\xi\in\mathbb{T}} \lambda_\varphi\big(W(\xi,h)\big) = O(h)$, as $h\to 0$.*

3. Compactness

An operator $T\colon X \to Y$ is *compact* if $T(B_X)$ is relatively compact in Y. The first results on compactness are due to H. Schwartz:

Theorem 3.1 ([21]). *(1) The operator $C_\varphi\colon H^p \to H^p$ is compact if and only if for every bounded sequence $\{f_n\}_n$ in H^p converging to 0 uniformly on compact subsets of $\mathbb{D}$, we have $f_n \circ \varphi \to 0$ in H^p.*

(2) If C_φ is compact on H^p, then $\lambda_\varphi(\mathbb{T}) = 0$.

(3) If C_φ is compact on H^p, then $\displaystyle\lim_{|z|\to 1^-} \frac{1-|z|}{1-|\varphi(z)|} = 0$.

Proof. (1) It is an easy exercise, with the help of Montel's theorem (or use weak-star compactness).

(2) The sequence (z^n) converges uniformly to 0 on compact subsets of $\mathbb{D}$, so

$$\left\|C_\varphi(z^n)\right\|_p = \left\|\varphi^n\right\|_p \longrightarrow 0,$$

but

$$\|\varphi^n\|_p^p = \int_{\mathbb{T}} |\varphi^*|^{np}\, d\lambda \longrightarrow \lambda_\varphi(\mathbb{T}).$$

(3) Remember that the functional δ_z has norm $(1-|z|^2)^{-1/p}$ and point out that $C_\varphi^*(\delta_z) = \delta_{\varphi(z)}$. Then for any sequence $z_n \in \mathbb{D}$ such that $|z_n| \longrightarrow 1^-$, the sequence $\mu_n = (1-|z_n|^2)^{1/p}\delta_{z_n}$ lies in the unit sphere of the dual of H^p. Since C_φ^* is compact on $(H^p)^*$ and μ_n is weak-star convergent to 0, we have

$$\|C_\varphi^*(\mu_n)\|_{(H^p)^*} \longrightarrow 0.$$

But

$$\|C_\varphi^*(\mu_n)\|_{(H^p)^*} = (1-|z_n|^2)^{1/p}\left\|\delta_{\varphi(z_n)}\right\|_{(H^p)^*} = \frac{(1-|z_n|^2)^{1/p}}{\left(1-|\varphi(z_n)|^2\right)^{1/p}}. \qquad \square$$

The study of the compactness reduces to the hilbertian case:

Theorem 3.2 ([22]). *Let $p \geq 1$. C_φ is compact on H^p if and only if C_φ is compact on H^2.*

Proof. We only have to prove: if $p, q \geq 1$ and C_φ is compact on H^q then C_φ is compact on H^p.

Take $\{f_n\}_n \in B_{H^p}$ uniformly converging to 0 on compact subsets of $\mathbb{D}$. Write $f_n = B_n{\cdot}g_n$ with $|B_n^*| = 1$ *a.e.* and $g_n \in H^p$ without zeros in $\mathbb{D}$. The sequence $G_n = g_n^{p/q}$ is defined and lies in the unit ball of H^q.

Up to (enough) subsequences, we may assume that $G_n \to G$ and $B_n \to B$, uniformly on compact subsets of $\mathbb{D}$, and that the sequence $(G_n \circ \varphi)$ converges to $G\circ\varphi$ in H^q (since $C_\varphi\colon H^q \to H^q$ is compact). For every $z \in \mathbb{D}$, we get that $|f_n(z)|^p = |G_n(z)|^q |B_n(z)|^p \longrightarrow 0$, hence $|G(z)|^q|B(z)|^p = 0$.

Now, we can compute:

$$\int_{\mathbb{T}} |f_n\circ\varphi^*|^p\,d\lambda \lesssim \int_{\mathbb{T}} |(G_n - G)\circ\varphi^*|^q |B_n\circ\varphi^*|^p\,d\lambda + \int_{\mathbb{T}} |G\circ\varphi^*|^q |B_n\circ\varphi^*|^p\,d\lambda,$$

so

$$\int_{\mathbb{T}} |f_n\circ\varphi^*|^p\,d\lambda \lesssim \int_{\mathbb{T}} |(G_n - G)\circ\varphi^*|^q\,d\lambda + \int_{\mathbb{T}} |G\circ\varphi^*|^q |B_n\circ\varphi^*|^p\,d\lambda.$$

Remembering that $\lambda_\varphi(\mathbb{T}) = 0$, we have $\varphi^*(z) \in \mathbb{D}$ for almost every $z \in \mathbb{T}$. Therefore the dominated convergence theorem gives that the second term converges to 0 and the compactness of C_φ on H^p is proved. □

3.1. *Hilbert-Schmidt operators*

An operator $T\colon H \to H$ is *Hilbert-Schmidt* if for an (any) orthonormal basis (b_n) of the Hilbert space H, we have

$$\|T\|_{HS}^2 = \sum \|T(b_n)\|_H^2 < +\infty.$$

It is easy to check that Hilbert-Schmidt operators are compact. One of the main interest here is that the membership to the class of Hilbert-Schmidt operators is very easy to check:

Theorem 3.3 ([22]). *C_φ is Hilbert-Schmidt if and only if*

$$\|C_\varphi\|_{HS}^2 = \int_{\mathbb{T}} \frac{1}{1-|\varphi^*|^2}\,d\lambda < \infty.$$

Proof. Indeed, the sequence $b_n(z) = z^n$ (where $n \in \mathbb{N}$) is an orthonormal basis of H^2. Now, let us compute

$$\|C_\varphi\|_{HS}^2 = \sum_{n=0}^{\infty} \|\varphi^n\|_{H^2}^2 = \sum_{n=0}^{\infty} \int_{\mathbb{T}} |\varphi^*|^{2n}\,d\lambda = \int_{\mathbb{T}} \frac{1}{1-|\varphi^*|^2}\,d\lambda,$$

and the theorem is proved. □

It can be also written

$$\int_{\overline{\mathbb{D}}} \frac{1}{1-|z|^2}\,d\lambda_\varphi.$$

3.2. *Some characterizations of compactness*

Recall that, for $0 < h < 1$, $\rho_\varphi(h) = \sup_{\xi\in\mathbb{T}} \lambda_\varphi\big(W(\xi,h)\big)$.

Theorem 3.4 ([18, 20]). *C_φ is compact if and only if λ_φ is a vanishing Carleson measure, i.e.,*

$$\rho_\varphi(h) = o(h), \qquad \textit{when } h \to 0.$$

The next result gives the essential norm $\|C_\varphi\|_e$ of C_φ, i.e., its distance, in the operator norm, from the space of compact operators on H^2.

Theorem 3.5 ([23]). *C_φ is compact if and only if*

$$v_\varphi(h) = \sup_{|w|\geq 1-h} N_\varphi(w) = o(h), \quad \textit{when } h \to 0.$$

Actually,

$$\|C_\varphi\|_e = \limsup_{|w|\to 1^-}\Big(\frac{N_\varphi(w)}{1-|w|}\Big)^{1/2} = \limsup_{h\to 0}\Big(\frac{v_\varphi(h)}{h}\Big)^{1/2},$$

and, ([1]),

$$\|C_\varphi\|_e = \limsup_{|a|\to 1^-}\Big\| C_\varphi\Big(\frac{k_a}{\|k_a\|_2}\Big)\Big\|_{H^2}.$$

Proof. We shall not prove the full theorem (we refer to [23] and [1]) but let us prove that C_φ is compact when $\sup\limits_{|w|\geq 1-h} N_\varphi(w) = o(h)$ when $h \to 0$.

Consider $\{f_n\}_n \in B_{H^2}$ uniformly converging to 0 on compact subsets of $\mathbb{D}$, and remember the Littlewood-Paley formula

$$\begin{aligned}\|f_n\circ\varphi\|_2^2 &= \big|f_n\circ\varphi(0)\big|^2 + 2\int_{\mathbb{D}} |f_n'|^2 N_\varphi(z)\,d\mathscr{A}\\ &= \big|f_n\circ\varphi(0)\big|^2 + 2\int_{r\mathbb{D}} |f_n'|^2 N_\varphi(z)\,d\mathscr{A} + 2\int_{\mathbb{D}\setminus r\mathbb{D}} |f_n'|^2 N_\varphi(z)\,d\mathscr{A},\end{aligned}$$

for any $r \in (0,1)$.

But, fixing $\varepsilon > 0$, we may consider some $r \in (0,1)$ (now fixed) such that

$$\forall z \notin r\mathbb{D}, \qquad N_\varphi(z) \leq \varepsilon \log(1/|z|).$$

On the other hand, both $f_n\circ\varphi(0) \longrightarrow 0$ and $\int_{r\mathbb{D}} |f_n'|^2 N_\varphi(z)\,d\mathscr{A} \longrightarrow 0$. Hence, for n large enough

$$\|f_n\circ\varphi\|_2^2 \leq \varepsilon + 2\varepsilon\int_{\mathbb{D}\setminus r\mathbb{D}} |f_n'|^2 \log(1/|z|)\,d\mathscr{A} \leq \varepsilon + 2\varepsilon\|f_n\|_2^2 = 3\varepsilon. \qquad \square$$

As a corollary, we get that

Corollary 3.1. *C_φ is compact on H^p if and only if $\lim\limits_{|z|\to 1^-} \dfrac{1-|\varphi(z)|}{1-|z|} = \infty$ when φ univalent (or finitely valent).*

Proof. Indeed, if φ is p-valent:

$$\frac{N_\varphi(w)}{1-|w|} \leq p \frac{\max\{\log(1/|z|) \mid \varphi(z) = w\}}{1-|w|}.$$

But, when $|w| \to 1$,

$$\frac{\max\{\log(1/|z|) \mid \varphi(z) = w\}}{1-|w|} \sim \max\left\{\frac{1-|z|}{1-|\varphi(z)|} : \varphi(z) = w\right\} \longrightarrow 0. \qquad \square$$

It is worth mentioning that the converse is false in general without any assumption: MacCluer and Shapiro [19] constructed inner functions φ admitting no angular derivatives at any point of the circle (see the definition below).

3.3. *Angular derivative*

We shall say that φ satisfies (NC) if $\lim_{|z|\to 1^-} \frac{1-|\varphi(z)|}{1-|z|} = \infty$.
We say that φ has an angular derivative at $\xi \in \mathbb{T}$, if for some $a \in \mathbb{T}$ the following non-tangential limit exists in $\mathbb{C}$:

$$\angle \lim_{z\to\xi} \frac{\varphi(z) - a}{z - \xi}. \qquad (AD)$$

Theorem 3.6 (Julia-Carathéodory). *φ satisfies (NC) if and only φ has angular derivative at no point $\xi \in \mathbb{T}$.*

Observe that if φ has angular derivative at ξ and $a \in \mathbb{T}$ is like in (AD) then

$$\angle \lim_{z\to\xi} \varphi(z) = a.$$

This allowed MacCluer and Shapiro [19] to construct an example of a (finitely valent) symbol $\varphi\colon \mathbb{D} \to \mathbb{D}$ such that C_φ is compact, and φ is onto: $\varphi(\mathbb{D}) = \mathbb{D}$.

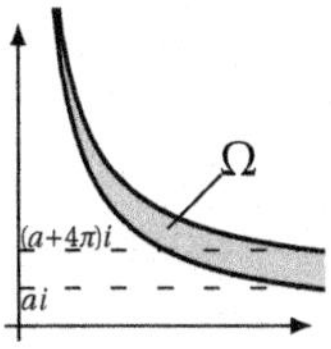

Let us give the idea of the construction of the MacCluer-Shapiro's example. Let $g\colon (0, +\infty) \to \mathbb{R}$ be a continuous decreasing function such that $\lim_{x\to 0^+} g(x) = +\infty$ (for instance $g(x) = 1/x$). And consider the domain

$$\Omega = \{x + iy : g(x) < y < g(x) + 4\pi\}.$$

Let $f\colon \mathbb{D} \to \Omega$ be a Riemann mapping (a conformal representation) and define, for $z \in \mathbb{D}$,

$$\varphi_1(z) = \exp\big(-f(z)\big).$$

Keep in mind that $|\varphi_1(z)| \to 1 \iff \operatorname{Re} f(z) \to 0$. Also φ_1 is 2-valent and has no point in $\mathbb{T}$ as radial limit. The only way to approach $\mathbb{T}$ is turning and turning inside the disk $\mathbb{D}$. Further, φ_1 is almost onto: $\varphi_1(\mathbb{D}) = \mathbb{D} \setminus \{0\}$. Take $a \in \mathbb{D} \setminus \{0\}$ and consider $\varphi = Q_a \circ \varphi_1$, where $Q_a(z) = \big(\frac{a-z}{1-\bar{a}z}\big)^2$. The symbol φ is now onto and $C_\varphi = C_{\varphi_1} \circ C_{Q_a}$ is compact.

4. The case of H^∞

The boundedness of $C_\varphi \colon H^\infty \longrightarrow H^\infty$ is obvious and $\|C_\varphi\| = 1$.

Theorem 4.1 ([21]). *C_φ is compact on H^∞ if and only if $\|\varphi\|_\infty < 1$.*

When $\|\varphi\|_\infty < 1$, C_φ is a nuclear operator, i.e., a (absolutely convergent) sum of rank one operators:

$$f \circ \varphi = \sum_{n=0}^{\infty} \hat{f}(n)\varphi^n \quad \text{where} \quad \sum_{n=0}^{\infty} \|\varphi^n\|_\infty < +\infty \text{ and } f \mapsto \hat{f}(n) \text{ has norm } 1.$$

The converse is easy to prove, but we actually have a stronger result due to Aron, Galindo and Lindström.

Theorem 4.2 ([2]). *If C_φ is a weakly compact operator on H^∞ then $\|\varphi\|_\infty < 1$.*

Hence the following assertions are equivalent:

- $\|\varphi\|_\infty < 1$.
- C_φ is a compact operator.
- C_φ is a nuclear operator.
- C_φ is a weakly compact operator.

The same holds replacing H^∞ by $A(\mathbb{D})$ (once $\varphi \in A(\mathbb{D})$).

Proof of Theorem 4.2. Let us assume that $\|\varphi\|_\infty = 1$. There are $a \in \mathbb{T}$ and $(z_j) \in \mathbb{D}$ s.t. $\varphi(z_j) \longrightarrow a$. Consider $f_n(z) = \Big(\dfrac{1+\bar{a}z}{2}\Big)^n$. Clearly

$$f_n \in A(\mathbb{D}), \qquad \|f_n\|_\infty = 1, \qquad f_n(a) = 1, \qquad f_n(z) \longrightarrow 0 \ \forall z \in \mathbb{D}.$$

Since C_φ is weaky compact, $f_{n_j} \circ \varphi \overset{\omega}{\longrightarrow} \sigma \in H^\infty$. In particular, for every $z \in \mathbb{D}$, $f_{n_k}\big(\varphi(z)\big)$ converges both to $\sigma(z)$ and 0.

By the Banach-Mazur theorem, we have for some convex combination

$$\sum_{I_m} c_k f_{n_k} \circ \varphi \overset{\|\cdot\|_\infty}{\longrightarrow} \sigma = 0.$$

But, for every m,

$$\Big\|\sum_{I_m} c_k f_{n_k}\circ\varphi\Big\|_\infty \geq \lim_{j\to+\infty}\Big|\sum_{I_m} c_k f_{n_k}\big(\varphi(z_j)\big)\Big| = \sum_{I_m} c_k = 1.$$

We get the desired contradiction. □

We even know the distance from C_φ to many operator ideals:

Theorem 4.3 ([25]). $\|C_\varphi\|_e = 1$ *if* $\|\varphi\|_\infty = 1$, *and* $\|C_\varphi\|_e = 0$ *if* $\|\varphi\|_\infty < 1$.

Actually,

Theorem 4.4 ([7]). *Assuming that* $\mathscr{K}(H^\infty)\subset\mathscr{I}\subset\mathscr{W}(H^\infty)$, *then* $\|C_\varphi\|_{e,\mathscr{I}} = 1$ *if* $\|\varphi\|_\infty = 1$, *and* $\|C_\varphi\|_{e,\mathscr{I}} = 0$ *if* $\|\varphi\|_\infty < 1$, *where* $\|.\|_{e,\mathscr{I}} = d\big(.,\mathscr{I}\big)$.

5. Hardy-Orlicz spaces

From the compactness point of view, the behavior of C_φ on H^p (the same as H^2) is different from the one on H^∞, for any $p \geq 1$. We aim to understand this "discontinuity", replacing the "L^p" framework by a new scale, the Orlicz framework. Shall we emphasize new phenomena? The behavior of C_φ will be there closer to the one on H^2 or on H^∞?

The results of this section come mainly from [9].

5.1. *Orlicz spaces*

Let $\Psi\colon [0,+\infty) \to [0,+\infty)$ be an Orlicz function: Ψ is continuous, convex, strictly increasing and $\Psi(0) = 0$. For instance, think of $\Psi(x) = x^p$, where $p \geq 1$; or $\Psi(x) = \Psi_2(x) = e^{x^2} - 1$; or $\Psi(x) = x\log(1+x)$,... We define the Orlicz space $L^\Psi(\mathbb{T})$ as formed by the (classes of) measurable functions $f\colon \mathbb{T}\to\mathbb{C}$ such that

$$\text{there exists } A > 0 \text{ with } \quad \int_{\mathbb{T}} \Psi(|f|/A)\,d\lambda < +\infty.$$

We have that $L^\Psi(\mathbb{T})$ is a Banach space, when equipped with the norm

$$\|f\|_\Psi = \inf\Big\{A > 0 : \int_{\mathbb{T}} \Psi(|f|/A)\,d\lambda \leq 1\Big\}.$$

Point out that a measurable f belongs to the unit ball of L^Ψ if and only if $\int_{\mathbb{T}} \Psi(|f|)\,d\lambda \leq 1$.

Let $M^\Psi(\mathbb{T})$ be the closure of $L^\infty(\mathbb{T})$ in $L^\Psi(\mathbb{T})$, usually called the Morse-Transue space. So, a measurable function $f\colon \mathbb{T}\to\mathbb{C}$ belongs to $M^\Psi(\mathbb{T})$ if and only if

$$\text{for every } C > 0, \text{ we have } \quad \int_{\mathbb{T}} \Psi(C|f|)\,d\lambda < +\infty.$$

Examples:

- When $\Psi(x) = x^p$, we have $L^\Psi = M^\Psi = L^p$.
- When $\Psi(x) = \Psi_2(x) = e^{x^2} - 1$, we have $L^\Psi \neq M^\Psi$.

Actually, $L^\Psi(\mathbb{T}) = M^\Psi(\mathbb{T})$ if and only if Ψ satisfies the so-called Δ_2 condition, that is:

$$\limsup_{x\to+\infty} \frac{\Psi(2x)}{\Psi(x)} < +\infty.$$

In M^Ψ there exists a Dominated Convergence Theorem, which clearly is not true in L^Ψ.

Proposition 5.1 (Dominated Convergence Theorem in M^Ψ). *Let $\{f_n\}_n$ be a sequence of measurable functions converging pointwise a.e. to f. If there exists $g \in M^\Psi(\mathbb{T})$ such that $|f_n| \leq g$ a.e. for every n, then $f_n \xrightarrow{\|\cdot\|_\Psi} f$.*

5.2. *Hardy-Orlicz spaces*

Hardy spaces H^p are defined in the framework of the Lebesgue spaces L^p. In the same way, we may define the Hardy–Orlicz spaces H^Ψ:

H^Ψ is formed by the analytic functions $f : \mathbb{D} \to \mathbb{C}$ such that

$$\|f\|_{H^\Psi} = \sup_{0\leq r<1} \|f_r\|_{L^\Psi(\mathbb{T})} < +\infty, \qquad (\bullet)$$

where $f_r(e^{it}) = f(re^{it})$. The $\sup_{0\leq r<1}$ in $(\bullet)$ is in fact equal to $\lim_{r\to1^-}$. Since, $H^\Psi \subset H^1$, every $f \in H^\Psi$ has radial limit a.e. f^*. We have

$$f^* \in L^\Psi(\mathbb{T}) \quad \text{and} \quad \|f\|_{H^\Psi} = \|f^*\|_{L^\Psi}.$$

In fact,

$$H^\Psi = \{f \in H^1 : f^* \in L^\Psi(\mathbb{T})\}.$$

We say that Ψ satisfies the condition Δ^2 if there exists $A > 1$ and $x_0 > 0$, such that

$$\Psi(Ax) \geq \left(\Psi(x)\right)^2, \quad \text{for all } x \geq x_0.$$

For example, think of $\Psi(x) = \Psi_q(x) = e^{x^q} - 1$.

When $1 < p < r < +\infty$, and Ψ satisfies the condition Δ^2, then

$$H^1 \supset H^p \supset H^r \supset H^\Psi \supset H^\infty,$$

and

$$\|\cdot\|_1 \leq \|\cdot\|_p \leq \|\cdot\|_r \lesssim \|\cdot\|_\Psi \lesssim \|\cdot\|_\infty.$$

We shall often use the functions $u_{a,r}$, $a \in \mathbb{T}$ and $r \in [0,1)$, defined by

$$u_{a,r}(z) = \Big(\frac{1-r}{1-\overline{a}rz}\Big)^2.$$

We have

$$\|u_{a,r}\|_{H^\infty} = 1, \qquad \|u_{a,r}\|_{H^1} = \frac{1-r}{1+r} \le 1-r,$$

and

$$\Big|u_{a,r}(z)\Big| \ge 1/4, \quad \text{when } |z-a| \le 1-r. \tag{$\clubsuit$}$$

For every Orlicz function Ψ, and any $A > 0$ we have

$$\int_{\mathbb{T}} \Psi(|u^*_{a,r}|/A)\, d\lambda \le \int_{\mathbb{T}} |u^*_{a,r}| \Psi(1/A)\, d\lambda \le (1-r)\Psi(1/A).$$

So $\|u_{a,r}\|_{H^\Psi} \le 1/\Psi^{-1}(\frac{1}{1-r})$.
Remembering ($\clubsuit$), we actually have

$$\|u_{a,r}\|_{H^\Psi} \approx \frac{1}{\Psi^{-1}\big(1/(1-r)\big)}.$$

The point evaluation δ_z, where $z \in \mathbb{D}$, is a linear functional on H^Ψ and its norm can be estimated:

$$\|\delta_z\|_{(H^\Psi)^*} \approx \Psi^{-1}\Big(\frac{1}{1-|z|}\Big).$$

In order to justify one direction one tests δ_z on $u_{a,r}$, for $r = |z|$, and $z = ra$. For the other direction, we use the Poisson kernel P_z. We know $P_z \ge 0$, $\|P_z\|_1 = 1$, $\|P_z\|_\infty = \dfrac{1+|z|}{1-|z|}$, and, for $f \in B_{H^\Psi}$,

$$f(z) = \int_{\mathbb{T}} P_z f^*\, d\lambda, \qquad \text{and}$$

$$\Psi\big(|f(z)|\big) \le \int_{\mathbb{T}} \Psi\big(|f^*|\big) P_z\, d\lambda \le \|P_z\|_\infty \le \frac{2}{1-r}.$$

Actually, for many Ψ, we have: $\quad \|\delta_z\|_{(H^\Psi)^*} \le \Psi^{-1}\Big(\dfrac{1}{1-|z|^2}\Big)$.

5.3. *Boundedness on Hardy-Orlicz spaces*

For every symbol $\varphi\colon \mathbb{D} \to \mathbb{D}$ and any Orlicz function Ψ, the operator $C_\varphi\colon H^\Psi \to H^\Psi$ is (well defined and) bounded.

Assuming $\varphi(0) = 0$, like in the case of Hardy spaces, we can use the Littlewood's Subordination Principle: for every $f \in B_{H^\Psi}$, applying this principle with $g(z) = \Psi(|f(z)|)$, we get

$$\|C_\varphi f\|_{H^\Psi} \le 1.$$

For general φ, once again we write $C_\varphi = C_\phi \circ C_{q_a}$ where $\phi(0) = 0$.

5.4. *Compactness on Hardy–Orlicz spaces*

The Schwartz's criterium for compactness is still valid:

Theorem 5.1. *The composition operator $C_\varphi\colon H^\Psi \to H^\Psi$ is compact if and only if for every bounded sequence $\{f_n\}_n$ in H^Ψ converging to 0 uniformly on compact subsets of $\mathbb{D}$, we have $f_n \circ \varphi \to 0$ in H^Ψ.*

And we have the same first consequence as for Theorem 3.1 when we apply it to $f_n(z) = z^n$:

Corollary 5.1. *If $C_\varphi\colon H^\Psi \to H^\Psi$ is compact, then $\lambda_\varphi(\mathbb{T}) = 0$, i.e., $|\varphi^*| < 1$ almost everywhere on $\mathbb{T}$.*

A second consequence goes as follows. Remember

$$u_{a,r}(z) = \Big(\frac{1-r}{1-\bar{a}rz}\Big)^2, \quad \text{and} \quad \|u_{a,r}\|_{H^\Psi} \approx \frac{1}{\Psi^{-1}\big(1/(1-r)\big)}.$$

For every sequence $\{a_n\}_n$ in $\mathbb{T}$, if $r_n \to 1^-$, we have

$$\Psi^{-1}\Big(\frac{1}{1-r_n}\Big) u_{a_n,r_n} \longrightarrow 0 \qquad \text{uniformly on compact subsets of } \mathbb{D}.$$

Corollary 5.2. *If $C_\varphi\colon H^\Psi \to H^\Psi$ is compact, then*

$$\lim_{r\to 1^-} \sup_{a\in\mathbb{T}} \Psi^{-1}\Big(\frac{1}{1-r}\Big) \Big\| C_\varphi u_{a,r} \Big\|_{H^\Psi} = 0. \tag{U}$$

5.5. *Order bounded composition operators*

Theorem 5.2. *If Ψ satisfies (Δ^2), then TFAE:*

(1) $C_\varphi\colon H^\Psi \to H^\Psi$ is weakly compact.
(2) $C_\varphi\colon H^\Psi \to H^\Psi$ is compact.
(3) $C_\varphi\colon H^\Psi \to H^\Psi$ is order bounded in M^Ψ: there exists $g \in M^\Psi(\mathbb{T})$, with $|(C_\varphi f)^*| \le g$ a.e., for all $f \in B_{H^\Psi}$.
(4) We have $\Psi^{-1}\big(\frac{1}{1-|\varphi^*|}\big) \in M^\Psi(\mathbb{T})$.

It is easy to check that order bounded operators on L^2 are exactly the Hilbert–Schmidt operators.

Proof. Observe that the best g we can choose to majorize $|(C_\varphi f)^*|$, for every f in the unit ball is

$$g(e^{it}) = \big\|\delta_{\varphi^*(e^{it})}\big\|_{(H^\Psi)^*} \approx \Psi^{-1}\Big(\frac{1}{1-|\varphi^*(e^{it})|}\Big).$$

So clearly, the last two statements in the previous theorem are equivalent.

If C_φ is order bounded in M^Ψ (by g), and $\{f_n\}_n \in B_{H^\Psi}$, converging to 0 uniformly on compact subsets of $\mathbb{D}$, then, (point out $|\varphi^*| < 1$ a.e.) $\{f_n \circ \varphi^*\}_n \xrightarrow{\text{a.e.}} 0$ and the convergence is still dominated by $g \in M^\Psi$. Thus $f_n \circ \varphi^* \xrightarrow{\|\cdot\|_\Psi} 0$. Therefore C_φ is compact on H^Ψ.

To finish the proof of the theorem, it remains to prove that, under condition (Δ^2), weak compactness implies order boundedness.

By Theorem 3.20 in [9], weak compactness, together with a condition weaker than (Δ^2), implies

$$\lim_{r\to 1^-} \sup_{a\in\mathbb{T}} \Psi^{-1}\Big(\frac{1}{1-r}\Big)\Big\|C_\varphi(u_{a,r})\Big\|_{H^\Psi} = 0. \tag{U}$$

This yields, for every $\varepsilon > 0$, the existence of $r_0 < 1$ such that, for every $a \in \mathbb{T}$, and all $r_0 < r < 1$:

$$\int_{\overline{\mathbb{D}}} \Psi\Big(\frac{1}{\varepsilon}\Psi^{-1}\Big(\frac{1}{1-r}\Big)|u_{a,r}(z)|\Big)\,d\lambda_\varphi = \int_{\mathbb{T}} \Psi\Big(\frac{1}{\varepsilon}\Psi^{-1}\Big(\frac{1}{1-r}\Big)|u_{a,r}\circ\varphi^*|\Big)\,d\lambda \le 1.$$

Using Markov's inequality and the fact that $|u_{a,r}(z)| \ge 1/4$, whenever $|z-a| \le 1-r$,

$$1 \ge \lambda(\{|\varphi^* - a| \le 1-r\})\Psi\Big(\frac{1}{4\varepsilon}\Psi^{-1}\Big(\frac{1}{1-r}\Big)\Big), \quad \text{for every } a \in \mathbb{T}.$$

Observe that the annulus $\{z \in \mathbb{C} : 1-h < |z| < 1\}$, for h small enough, can be covered by less than C/h balls of radii $2h$ and centers in $\mathbb{T}$ and therefore, taking $2h = 1-r$,

$$\begin{aligned}\frac{C}{h} &\ge m(\{|\varphi^*| > 1-h\})\Psi\Big(\frac{1}{4\varepsilon}\Psi^{-1}\Big(\frac{1}{2h}\Big)\Big)\\ &\ge \lambda\Big(\Big\{\frac{1}{1-|\varphi^*|} > \frac{1}{h}\Big\}\Big)\Psi\Big(\frac{1}{8\varepsilon}\Psi^{-1}\Big(\frac{1}{h}\Big)\Big).\end{aligned}$$

Putting $g = \Psi^{-1}\big(\frac{1}{1-|\varphi^*|}\big)$, and $x = \Psi^{-1}(1/h)$, we see that, for x big enough,

$$C\Psi(x) \ge \lambda(\{g > x\}) \cdot \Psi(x/8\varepsilon) \ge \lambda(\{g > x\})\big[\Psi(x/8A^2\varepsilon)\big]^4, \tag{$*$}$$

using condition (Δ^2) $(\Psi(Ax) \ge (\Psi(x))^2)$ twice.
If we would have been given any $B > 1$, we could have chosen ε to have $B = \frac{1}{8A^2\varepsilon}$. Then $(*)$ would yield, for x large enough:

$$\lambda(\{\Psi(Bg) > \Psi(Bx)\})\big[\Psi(Bx)\big]^4 \le C\Psi(x) \le C\Psi(Bx).$$

That is, for t large enough, $\lambda(\{\Psi(Bg) > t\}) \le \dfrac{C}{t^3}$, and $\Psi(Bg)$ is integrable, for every $B > 1$. Namely $g \in M^\Psi(\mathbb{T})$ and C_φ is order bounded in M^Ψ. □

5.6. *Back to pullback measures*

We already mentioned that, when $f \in H^p$, we have

$$\|C_\varphi f\|_{H^p}^p = \|(f\circ\varphi)^*\|_{L^p(\mathbb{T})}^p = \int_{\mathbb{T}} |f|^p \circ \varphi^* \, d\lambda = \|f\|_{L^p(\lambda_\varphi)}^p,$$

where λ_φ to the pullback measure of λ by the map φ^*: $\lambda_\varphi(B) = \lambda(\{\varphi^* \in B\})$, for every Borel set $B \subset \overline{\mathbb{D}}$. So properties like boundedness or compactness of the operator C_φ are the same than the properties of the inclusion (embedding) operator

$$j_{\lambda_\varphi} : H^p \hookrightarrow L^p(\lambda_\varphi).$$

Clearly, the same argument works in the Orlicz framework. We focus on this point of view in the sequel.

5.7. *Compactness on H^Ψ*

The compactness of composition operator $C_\varphi : H^\Psi \to H^\Psi$ is equivalent to the compactness of the inclusion operator of H^Ψ in $L^\Psi(\lambda_\varphi)$. So we could try to characterize for which finite measure μ on $\mathbb{D}$, is the inclusion operator $H^\Psi \hookrightarrow L^\Psi(\mu)$ compact. In this setting there is something similar to Schwartz's criterium:

Proposition 5.2. *Let μ be a finite measure on $\mathbb{D}$. The following assertions are equivalent:*

(1) The inclusion operator $H^\Psi \hookrightarrow L^\Psi(\mu)$ is compact.
(2) For every bounded sequence $\{f_n\}$ in H^Ψ converging to 0 uniformly on compact sets, we have $\|f_n\|_{L^\Psi(\mu)} \longrightarrow 0$.
(3) $H^\Psi(\mathbb{D})$ is included in $L^\Psi(\mu)$ and, putting $I_r(f) = f\mathbb{1}_{\mathbb{D}\setminus r\mathbb{D}}$, we have $\lim_{r\to 1^-} \|I_r\|_{H^\Psi \to L^\Psi(\mu)} = 0$.

Let μ be a finite measure on $\overline{\mathbb{D}}$, $h \in (0,1)$ and $A > 0$. We denote:

$$\rho_\mu(h) = \sup_{\xi\in\mathbb{T}} \mu\big(W(\xi,h)\big), \qquad K_\mu(h) = \sup_{0<t\le h} \frac{\rho_\mu(t)}{t},$$

$$\text{and} \qquad \gamma_A(h) = \frac{1}{\Psi\big(A\Psi^{-1}(1/h)\big)}.$$

Theorem 5.3. *Consider the following conditions:*

(R_0) *For every* $A > 0$, $\quad \rho_\mu(h) = o\big(\gamma_A(h)\big), \quad h \to 0^+.$
(K_0) *For every* $A > 0$, $\quad K_\mu(h) = o\Big(\frac{\gamma_A(h)}{h}\Big), \quad h \to 0^+.$

(C_0) *The inclusion of $H^\Psi(\mathbb{D})$ in $L^\Psi(\mu)$ is a compact operator.*

Then we have $\quad (K_0) \Longrightarrow (C_0) \Longrightarrow (R_0).$

For general measures we do not have a complete characterization of compactness of the inclusion for every Ψ. But in the study of composition operators C_φ we are interested in pullback measures λ_φ associated to *analytic* functions. This particular additional property is crucial to make things go round:

Theorem 5.4 (Regularity of the pullback measure). *There exists a constant $k_1 > 0$ so that, for every holomorphic map $\varphi\colon \mathbb{D} \to \mathbb{D}$, and for every $\xi \in \mathbb{T}$, we have*

$$\lambda_\varphi\big(W(\xi,\varepsilon h)\big) \le k_1\, \varepsilon\, \lambda_\varphi\big(W(\xi,h)\big),$$

whenever $0 < \varepsilon < 1$, and $0 < h < 1 - |\varphi(0)|$.

As a consequence: for $\mu = \lambda_\varphi$ we have

$$K_\mu(h) = \sup_{0<t\le h} \frac{\rho_\mu(t)}{t} \approx \frac{\rho_\mu(h)}{h},$$

therefore (R_0) and (K_0) are equivalent; hence

Theorem 5.5 (Characterization of compactness). *The composition operator $C_\varphi\colon H^\Psi \to H^\Psi$ is compact if and only if*

$$\lim_{h\to 0^+} \frac{\rho_{\lambda_\varphi}(h)}{\gamma_A(h)} = 0, \quad \textit{for every } A > 0,$$

if and only if

$$\lim_{h\to 0^+} \frac{\Psi^{-1}(1/h)}{\Psi^{-1}\big(1/\rho_{\lambda_\varphi}(h)\big)} = 0.$$

Let us mention some consequences of this theorem:

- If C_φ is compact on H^Ψ, then it is compact on H^2.
- When Ψ satisfies the Δ_2 condition, C_φ is compact on H^Ψ if and only if C_φ is compact on H^2.
- Conversely, if Ψ does not satisfy the Δ_2 condition, there exists a symbol φ such that C_φ is compact on H^2, without being compact on H^Ψ.

6. Carleson versus Nevanlinna

We just saw the efficiency of the Carleson's measures in characterizing compactness. On the other hand, remember that there is also a nice characterization via the Nevanlinna function. So we have

$$C_\varphi \text{ is compact on } H^2 \text{ if and only if } \sup_{|w|\geq 1-h} N_\varphi(w) = o(h),$$
$$\text{if and only if } \rho_\varphi(h) = \sup_{\xi\in\mathbb{T}} \lambda_\varphi\big(W(\xi,h)\big) = o(h).$$

Hence there should be a link between these two notions. It is indeed possible to prove that the Nevanlinna counting function is "equivalent" (in some sense) to the λ_φ-measure of the Carleson's windows. The following theorem gives the precise statement of this equivalence:

Theorem 6.1 ([10]). *There exist $c, C > 0$ (numerical) s.t.*

- $N_\varphi(w) \leq C\lambda_\varphi\Big(W\big(\frac{w}{|w|}, c(1-|w|)\big)\Big)$.
- $\lambda_\varphi\big(W(\xi,h)\big) \leq \dfrac{C}{\mathscr{A}\big(W(\xi,ch)\big)} \displaystyle\int_{W(\xi,ch)} N_\varphi(w)\, d\mathscr{A} \leq C \sup_{w\in W(\xi,ch)} N_\varphi(w)$.

An immediate consequence:

Theorem 6.2 ([11]). • *C_φ is compact on H^Ψ if and only if*

$$\lim_{|w|\to 1^-} \frac{\Psi^{-1}\big(1/(1-|w|)\big)}{\Psi^{-1}\big(1/N_\varphi(w)\big)} = 0.$$

- *When φ is finitely-valent, $C_\varphi\colon H^\Psi \to H^\Psi$ is compact if and only if*

$$\lim_{|z|\to 1^-} \frac{\Psi^{-1}\Big(\dfrac{1}{1-|\varphi(z)|}\Big)}{\Psi^{-1}\Big(\dfrac{1}{1-|z|}\Big)} = 0.$$

7. Schatten classes

Definition 7.1. Let H be a (separable) Hilbert space, and T a bounded operator on H. For $p \geq 1$, define the Schatten p-norm of T as

$$\|T\|_{\mathscr{S}^p} := \Big(\sum_{n\geq 1} \lambda_n^p(|T|)\Big)^{1/p} = \big(tr(|T|^p)\big)^{1/p},$$

where $\lambda_1(|T|) \geq \lambda_2(|T|) \geq \cdots \geq \lambda_n(|T|) \geq \cdots$ are the eigenvalues of the operator $|T| = \sqrt{(T^*T)}$. The operator T belongs to the Schatten class $\mathscr{S}^p$ if its Schatten p-norm is finite.

Remark 7.1. T belongs to $\mathscr{S}^2$ if and only if T is Hilbert-Schmidt.

The characterization of composition operators belonging to $\mathscr{S}_2$ was already settled in this paper (since it coincides with Hilbert-Schmidt operators) and the general case was solved by Luecking.

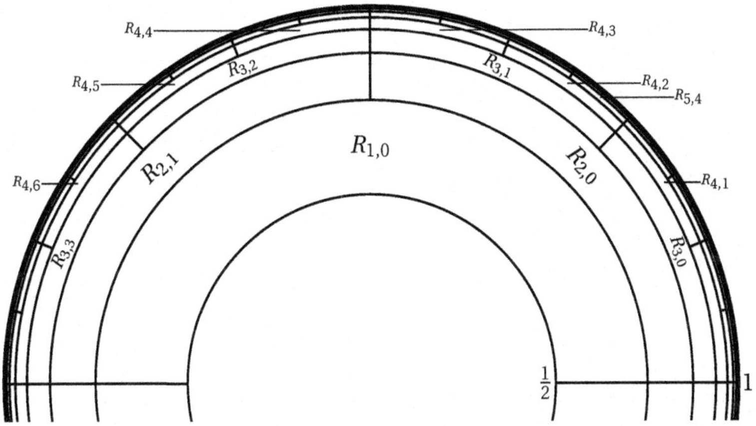

Its characterization uses Luecking windows $R_{n,j}$, $n \geq 1, 0 \leq j \leq 2^n - 1$,

$$R_{n,j} = \left\{ z \in \mathbb{D} : 1 - 2^{-n} \leq |z| < 1 - 2^{-n-1} \text{ and } \frac{2j\pi}{2^n} \leq \arg z < \frac{2(j+1)\pi}{2^n} \right\}.$$

These are half dyadic Carleson's windows, $W_{n,j}$:

$$W_{n,j} = \left\{ z \in \mathbb{D} : 1 - 2^{-n} \leq |z| \leq 1 \text{ and } \frac{2j\pi}{2^n} \leq \arg z < \frac{2(j+1)\pi}{2^n} \right\}.$$

The result states

Theorem 7.1 ([16]). *We assume that* $\lambda_\varphi(\mathbb{T}) = 0$.

$$C_\varphi \in \mathscr{S}_p \quad \textit{if and only if} \quad \sum_{n \geq 0} \sum_{j=0}^{2^n - 1} \left[2^n \lambda_\varphi(R_{n,j}) \right]^{p/2} < +\infty.$$

Actually, it is worth pointing out that

Proposition 7.1 ([11]).

$$C_\varphi \in \mathscr{S}_p \quad \textit{if and only if} \quad \sum_{n \geq 0} \sum_{j=0}^{2^n - 1} \left[2^n \lambda_\varphi(W_{n,j}) \right]^{p/2} < +\infty.$$

Another characterization involves the Nevanlinna counting function

Theorem 7.2 ([17]).

$$C_\varphi \in \mathscr{S}_p \quad \textit{if and only if} \quad \int_{\mathbb{D}} \Big(\frac{N_\varphi(z)}{\log(1/|z|)}\Big)^{p/2} \frac{d\mathscr{A}}{(1-|z|^2)^2} < +\infty.$$

Once the "equivalence" between the Nevanlinna counting function and the pullback measure is known, it is easy to deduce the two previous theorems one from the other (see Theorem 6.1. [11]).

8. Absolutely summing composition operators

Suppose $1 \le q < +\infty$ and let $T\colon X \to Y$ be a (bounded) operator between Banach spaces. Recall that T is a q-summing operator if there exists $C > 0$ such that

$$\Big(\sum_{j=1}^{n} \|Tx_j\|^q\Big)^{1/q} \le C \sup_{x^* \in B_{X^*}} \Big(\sum_{j=1}^{n} |\langle x^*, x_j\rangle|^q\Big)^{1/q} = C \sup_{a \in B_{\ell^{q'}}} \Big\|\sum_{j=1}^{n} a_j x_j\Big\|,$$

for every finite sequence $x_1, x_2, \dots, x_n$ in X.

The q-summing norm of T, denoted by $\pi_q(T)$, is the least suitable constant $C > 0$. The class of summing operators forms an operator ideal (for instance see [4] for more details). The 1-summing operators are also called absolutely summing operators.

We are interested in solving the following problem: when is a composition operator $C_\varphi\colon H^p \to H^p$ q-summing?

This question is equivalent to the following: when is the identity from H^p to $L^p(\overline{Dd}, \lambda_\varphi)$ q-summing?

That is why, we are naturally interested in the following more general problem: assume from now on that μ is concentrated in the open disk $\mathbb{D}$ and that μ is a Carleson measure; and find a characterization of the q-summingness of the Carleson embedding $j_\mu\colon H^p \hookrightarrow L^p(\mu)$.

Let us first recall some known facts. There were actually very few:

Theorem 8.1 ([22]). *Let $p \ge 2$. The composition operator $C_\varphi\colon H^p \to H^p$ is p-summing if and only if*

$$\int_{\mathbb{T}} \frac{1}{1-|\varphi^*|}\, d\lambda < +\infty.$$

Point out that in the Carleson embedding framework, the condition is

$$\int_{\mathbb{D}} \frac{1}{1-|z|}\, d\mu(z) < +\infty.$$

Actually, when $p \geq 1$, it is easy to check that

$$\int_{\mathbb{T}} \frac{1}{1-|z|}\, d\mu < +\infty \quad \text{if and only if} \quad j_\mu \colon H^p \hookrightarrow L^p(\mu) \text{ is order bounded.}$$

In particular, the condition implies that j_μ is p-summing for every $p \geq 1$. But the converse is false for $p \in [1,2)$.

The following result shows that it is not true anymore when $p < 2$.

Theorem 8.2 ([5]). *Let $p \in [1,2)$. There exist p-summing composition operators on H^p which are not order bounded.*

Very recently, Rodríguez-Piazza and the author were able to give some answers. The results stated below come from [12]. In the case of a measure concentrated on an annulus, we are able to solve the problem. Let us fix a finite measure μ on $\mathbb{D}$ and an integer n. We denote by μ_n the restriction of μ to the annulus $\Gamma_n = \{z \in \mathbb{D} : 1-2^{-n} \leq |z| < 1-2^{-n-1}\}$ and by j_n the inclusion of H^p into $L^p(\mu_n)$. Let Γ_n be the union of the 2^n Luecking windows $R_{n,j}$. Now consider, for $n \in \mathbb{N}$, the 2^n-dimensional subspace H_n^p of H^p generated by the monomials z^k, with $2^n \leq k < 2^{n+1}$. We have the decomposition

$$\{f \in H^p : f(0) = 0\} = \bigoplus_{n \geq 0} H_n^p,$$

which is an orthogonal decomposition when $p = 2$ (i.e., for H^2). Moreover $H_n^p \sim \ell_{2^n}^p$. Let α_n be the restriction of j_n to H_n^p.

Proposition 8.1. *For $1 < p < +\infty$, the following quantities are equivalent:*

(1) $\pi_q\big(j_n \colon H^p \to L^p(\mu_n)\big)$,
(2) $\pi_q\big(\alpha_n \colon H_n^p \to L^p(\mu_n)\big)$,
(3) $\pi_q(D_a)$, *where $D_a \colon \ell_{2^n}^p \to \ell_{2^n}^p$ is the diagonal operator whose multipliers are $a_j = \big(2^n \mu(R_{n,j})\big)^{1/p}$ (where $j = 1,2,\dots,2^n$).*

Since the summing norms of multipliers on sequence spaces are known, we deduce:

Theorem 8.3.

(1) *When $1 < p \leq 2$:* $\displaystyle \pi_q(j_n) \approx \Big(\sum_{j=1}^{2^n} [2^n \mu(R_{n,j})]^{2/p}\Big)^{1/2}.$

(2) *When $p > 2$:*

(a) *if $1 \leq q \leq p'$,* $\displaystyle \pi_q(j_n) \approx \Big(\sum_{j=1}^{2^n} [2^n \mu(R_{n,j})]^{p'/p}\Big)^{1/p'},$

$$\text{(b) if } p' \le q \le p, \quad \pi_q(j_n) \approx \Big(\sum_{j=1}^{2^n} [2^n \mu(R_{n,j})]^{q/p} \Big)^{1/q},$$

$$\text{(c) if } p \le q, \quad \pi_q(j_n) \approx \Big(\sum_{j=1}^{2^n} [2^n \mu(R_{n,j})] \Big)^{1/p},$$

It remains to glue the pieces. In some cases, we succeed in doing it:

Theorem 8.4. *In the case $q \ge p \ge 2$ we have:*

$$\pi_q(j_\mu) \approx \Big(\sum_n [\pi_q(j_n)]^p \Big)^{1/p} \approx \Big(\sum_{n,j} [2^n \mu(R_{n,j})] \Big)^{1/p} \approx \Big(\int_{\mathbb{D}} \frac{1}{1-|z|}\, d\mu(z) \Big)^{1/p}.$$

In the case $2 \le q \le p$ we have:

$$\pi_q(j_\mu) \approx \Big(\sum_n [\pi_q(j_n)]^q \Big)^{1/q} \approx \Big(\sum_{n,j} [2^n \mu(R_{n,j})]^{q/p} \Big)^{1/p}.$$

For $p > 2$, the case $1 \le q < 2$ is still open. When $p \le 2$, another approach is involved and we have

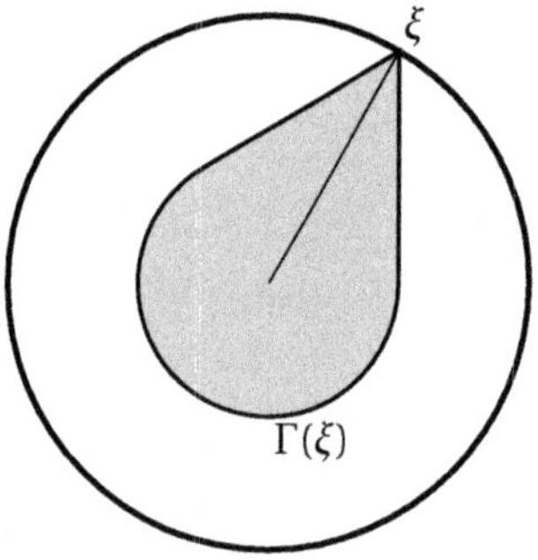

Theorem 8.5. *Let $1 < p \le 2$. The Carleson embedding $j_\mu \colon H^p \to L^p(\mu)$ is absolutely summing if and only if*

$$\int_{\mathbb{T}} \Big(\int_{\Gamma(\xi)} \frac{d\mu(z)}{(1-|z|)^{1+p/2}} \Big)^{2/p} d\lambda(\xi) < +\infty,$$

where $\Gamma(\xi)$ is a Stolz domain in ξ.

9. Some open problems

Let us mention here some open problems

- Compute the norm of any composition operator acting on H^2. This *a priori* simple question remains completely open. Of course, the same holds on H^p.
- Compute the value of the approximation numbers of any composition operator acting on H^2. Despite many recent progress by Li–Queffélec–Rodríguez-Piazza (see for instance [13], [14] and [15]), it is still open.
- Finish the characterization of summing composition operators. The same relatively to other operator ideals.

There are many other questions of course...

References

[1] J. R. Akeroyd, On Shapiro's compactness criterion for composition operators, *J. Math. Anal. Appl.* **379**, pp. 1–7 (2011).

[2] R. Aron, P. Galindo and M. Lindström, Compact homomorphisms between algebras of analytic functions, *Studia Math.* **123**, pp. 235–247 (1997).

[3] C. C. Cowen and B. D. MacCluer, *Composition operators on spaces of analytic functions*, Studies in Advanced Mathematics pp. xii+388, (CRC Press, Boca Raton, FL, 1995).

[4] J. Diestel, H. Jarchow and A. Tonge, *Absolutely summing operators*, Cambridge Studies in Advanced Mathematics, Vol. 43 pp. xvi+474, (Cambridge University Press, Cambridge, 1995).

[5] T. Domenig, Composition operators belonging to operator ideals, *J. Math. Anal. Appl.* **237**, pp. 327–349 (1999).

[6] P. L. Duren, *Theory of H^p spaces*, Pure and Applied Mathematics, Vol. 38 pp. xii+258, (Academic Press, New York-London, 1970).

[7] P. Lefèvre, Essential norms of weighted composition operators on the space H^∞ of Dirichlet series, *Studia Math.* **191**, pp. 57–66 (2009).

[8] P. Lefèvre, D. Li, H. Queffélec and L. Rodríguez-Piazza, Some examples of compact composition operators on H^2, *J. Funct. Anal.* **255**, pp. 3098–3124 (2008).

[9] P. Lefèvre, D. Li, H. Queffélec and L. Rodríguez-Piazza, Composition operators on Hardy-Orlicz spaces, *Mem. Amer. Math. Soc.* **207**, pp. vi+74 (2010).

[10] P. Lefèvre, D. Li, H. Queffélec and L. Rodríguez-Piazza, Nevanlinna counting function and Carleson function of analytic maps, *Math. Ann.* **351**, pp. 305–326 (2011).

[11] P. Lefèvre, D. Li, H. Queffélec and L. Rodríguez-Piazza, Some revisited results about composition operators on Hardy spaces, *Rev. Mat. Iberoam.* **28**, pp. 57–76 (2012).

[12] P. Lefèvre and L. Rodríguez-Piazza, Absolutely summing Carleson embeddings. Preprint.

[13] D. Li, H. Queffélec and L. Rodríguez-Piazza, On approximation numbers of composition operators, *J. Approx. Theory* **164**, pp. 431–459 (2012).

[14] D. Li, H. Queffélec and L. Rodríguez-Piazza, Estimates for approximation numbers of some classes of composition operators on the Hardy space, *Ann. Acad. Sci. Fenn. Math.* **38**, pp. 547–564 (2013).

[15] D. Li, H. Queffélec and L. Rodríguez-Piazza, A spectral radius type formula for approximation numbers of composition operators, *J. Funct. Anal.* **267**, pp. 4753–4774 (2014).

[16] D. H. Luecking, Embedding derivatives of Hardy spaces into Lebesgue spaces, *Proc. London Math. Soc. (3)* **63**, pp. 595–619 (1991).

[17] D. H. Luecking and K. H. Zhu, Composition operators belonging to the Schatten ideals, *Amer. J. Math.* **114**, pp. 1127–1145 (1992).

[18] B. D. MacCluer, Compact composition operators on $H^p(B_N)$, *Michigan Math. J.* **32**, pp. 237–248 (1985).

[19] B. D. MacCluer and J. H. Shapiro, Angular derivatives and compact composition operators on the Hardy and Bergman spaces, *Canad. J. Math.* **38**, pp. 878–906 (1986).

[20] S. C. Power, Vanishing Carleson measures, *Bull. London Math. Soc.* **12**, pp. 207–210

(1980).

[21] H. J. Schwartz, *Composition operators on H^p*, pp. 84, (ProQuest LLC, Ann Arbor, MI, 1969), Thesis (Ph.D.)–The University of Toledo.

[22] J. H. Shapiro and P. D. Taylor, Compact, nuclear, and Hilbert-Schmidt composition operators on H^2, *Indiana Univ. Math. J.* **23**, pp. 471–496 (1973/74).

[23] J. H. Shapiro, The essential norm of a composition operator, *Ann. of Math. (2)* **125**, pp. 375–404 (1987).

[24] J. H. Shapiro, *Composition operators and classical function theory*, Universitext: Tracts in Mathematics pp. xvi+223, (Springer-Verlag, New York, 1993).

[25] L. Zheng, The essential norms and spectra of composition operators on H^∞, *Pacific J. Math.* **203**, pp. 503–510 (2002).

On the boundedness of Bergman projection*

J.A. Peláez
Departamento de Análisis Matemático Estadística e I.O., y Matemática Aplicada,
Universidad de Málaga,
Campus de Teatinos, 29071 Málaga, Spain
E-mail: japelaez@uma.es
http://webpersonal.uma.es/~japelaez/

J. Rättyä
Department of Physics and Mathematics,
University of Eastern Finland,
P.O. Box 111, 80101 Joensuu, Finland
E-mail: jouni.rattya@uef.fi

The main purpose of this survey is to gather results on the boundedness of the Bergman projection. First, we shall go over some equivalent norms on weighted Bergman spaces A^p_ω which are useful in the study of this question. In particular, we shall focus on a decomposition norm theorem for radial weights ω with the doubling property $\int_r^1 \omega(s)\,ds \le C\int_{\frac{1+r}{2}}^1 \omega(s)\,ds$.

Keywords: Bergman space, decomposition norm theorem, Bergman projection.

1. Introduction

Let $\mathscr{H}(\mathbb{D})$ be the space of all analytic functions in the unit disc $\mathbb{D} = \{z : |z| < 1\}$. If $0 < r < 1$ and $f \in \mathscr{H}(\mathbb{D})$, set

$$M_p(r,f) = \left(\frac{1}{2\pi}\int_0^{2\pi} |f(re^{it})|^p\,dt\right)^{1/p}, \quad 0 < p < \infty,$$
$$M_\infty(r,f) = \sup_{|z|=r} |f(z)|.$$

For $0 < p \le \infty$, the Hardy space H^p consists of functions $f \in \mathscr{H}(\mathbb{D})$ such that $\|f\|_{H^p} = \sup_{0<r<1} M_p(r,f) < \infty$. A nonnegative integrable function ω on the

*This research was supported in part by the Ramón y Cajal program of MICINN (Spain); by Ministerio de Economía y Competitivivad, Spain, project MTM2014-52865-P; by La Junta de Andalucía, (FQM210) and (P09-FQM-4468); by Academy of Finland project no. 268009, by Väisälä Foundation of Finnish Academy of Science and Letters, and by Faculty of Science and Forestry of University of Eastern Finland project no. 930349.

unit disc $\mathbb{D}$ is called a weight. It is radial if $\omega(z) = \omega(|z|)$ for all $z \in \mathbb{D}$. For $0 < p < \infty$ and a weight ω, the weighted Bergman space A^p_ω is the space of $f \in \mathcal{H}(\mathbb{D})$ for which

$$\|f\|^p_{A^p_\omega} = \int_{\mathbb{D}} |f(z)|^p \omega(z)\, dA(z) < \infty,$$

where $dA(z) = \frac{dx\,dy}{\pi}$ is the normalized Lebesgue area measure on $\mathbb{D}$. That is, $A^p_\omega = L^p_\omega \cap \mathcal{H}(\mathbb{D})$ where L^p_ω is the corresponding weighted Lebesgue space. As usual, we write A^p_α for the standard weighted Bergman space induced by the radial weight $(1-|z|^2)^\alpha$, where $-1 < \alpha < \infty$ [12, 17, 33]. We denote $dA_\alpha = (\alpha + 1)(1-|z|^2)^\alpha\, dA(z)$ and $\omega(E) = \int_E \omega(z)\, dA(z)$ for short. The Carleson square $S(I)$ based on an interval $I \subset \mathbb{T}$ is the set $S(I) = \{re^{it} \in \mathbb{D} : e^{it} \in I, 1 - |I| \le r < 1\}$, where $|E|$ denotes the Lebesgue measure of $E \subset \mathbb{T}$. We associate to each $a \in \mathbb{D} \setminus \{0\}$ the interval $I_a = \{e^{i\theta} : |\arg(ae^{-i\theta})| \le \frac{1-|a|}{2}\}$, and denote $S(a) = S(I_a)$.

If the norm convergence in the Bergman space A^2_ω implies the uniform convergence on compact subsets, then the point evaluations L_z are bounded linear functionals on A^2_ω. Therefore, there are reproducing kernels $B^\omega_z \in A^2_\omega$ with $\|L_z\| = \|B^\omega_z\|_{A^2_\omega}$ such that

$$L_z f = f(z) = \langle f, B^{\omega}_z \rangle_{A^2_\omega} = \int_{\mathbb{D}} f(\zeta)\, \overline{B^\omega_z(\zeta)}\, \omega(\zeta)\, dA(\zeta), \quad f \in A^2_\omega.$$

Since A^2_ω is a closed subspace of L^2_ω, we may consider the orthogonal Bergman projection P_ω from L^2_ω to A^2_ω, that is usually called the Bergman projection. It is the integral operator

$$P_\omega(f)(z) = \int_{\mathbb{D}} f(\zeta) \overline{B^\omega_z(\zeta)}\, \omega(\zeta) dA(\zeta).$$

The main purpose of these lectures is to gather results on the inequality

$$\|P_\omega(f)\|_{L^p_v} \le C \|f\|_{L^p_v}. \tag{1}$$

We shall also provide a collection of equivalent norms on A^p_ω which have been used to study this problem. A solution for (1) is known for the class of standard weights $\omega(z) = (1-|z|^2)^\alpha$ and $1 < p < \infty$;

$$P_\alpha(f)(z) = (\alpha+1) \int_{\mathbb{D}} \frac{f(\zeta)(1-|\zeta|^2)^\alpha}{(1-z\overline{\zeta})^{2+\alpha}}\, dA(\zeta), \quad \alpha > -1,$$

is bounded on L^p_v if and only if $\frac{v(z)}{(1-|z|^2)^\alpha}$ belongs to the Bekollé-Bonami class $B_p(\alpha)$ [6, 8]. We remind the reader that $v \in B_p(\alpha)$ if

$$B_{p,\alpha}(v) = \sup_{I \subset \mathbb{T}} \frac{\left(\int_{S(I)} v(z)\, dA_\alpha(z)\right)\left(\int_{S(I)} v(z)^{\frac{-p'}{p}}\, dA_\alpha(z)\right)^{\frac{p}{p'}}}{A_\alpha\big(S(I)\big)^p} < \infty. \tag{2}$$

It is worth mentioning that the above result remains true replacing P_α by its sublinear positive counterpart

$$P_\alpha^+(f)(z) = (\alpha+1)\int_{\mathbb{D}} \frac{|f(\zeta)|(1-|\zeta|^2)^\alpha}{|1-z\overline{\zeta}|^{2+\alpha}}\,dA(\zeta).$$

Roughly speaking, this means that cancellation does not play an essential role in this question.

The situation is completely different when ω is not a standard weight, because of the lack of explicit expressions for the Bergman reproducing kernels B_z^ω. If ω is a radial weight, then the normalized monomials $\frac{z^n}{\sqrt{2\int_0^1 r^{2n+1}\omega(r)\,dr}}$, $n \in \mathbb{N}\cup\{0\}$, form the standard orthonormal basis of A_ω^2 and then [33, Theorem 4.19] yields

$$B_z^\omega(\zeta) = \sum_{n=0}^\infty \frac{(\zeta\bar{z})^n}{2\int_0^1 r^{2n+1}\omega(r)\,dr}, \quad z,\zeta\in\mathbb{D}. \tag{3}$$

This formula and a decomposition norm theorem has been used recently in order to obtain precise estimates for the L_v^p-integral of B_z^ω, see Theorem 4.2 below. This is a key to tackle the two weight inequality (1) when ω and v belong to a certain class of radial weights [28].

If ω is not necessarily radial, the theory of weighted Bergman spaces is at its early stages, and plenty of essential properties such as the density of polynomials (polynomials may not be dense in A_ω^p if ω is not radial, [26, Section 1.5] or [12, p. 138]) have not been described yet. Because of this fact, from now on we shall be mainly focused on Bergman spaces induced by radial weights.

Throughout the paper $\frac{1}{p}+\frac{1}{p'}=1$. Further, the letter $C=C(\cdot)$ will denote an absolute constant whose value depends on the parameters indicated in the parenthesis, and may change from one occurrence to another. We will use the notation $a\lesssim b$ if there exists a constant $C=C(\cdot)>0$ such that $a\le Cb$, and $a\gtrsim b$ is understood in an analogous manner. In particular, if $a\lesssim b$ and $a\gtrsim b$, then we will write $a\asymp b$.

2. Background on radial weights

We shall write $\widehat{\mathcal{D}}$ for the class of radial weights such that $\widehat{\omega}(z)=\int_{|z|}^1\omega(s)\,ds$ is doubling, that is, there exists $C=C(\omega)\ge 1$ such that $\widehat{\omega}(r)\le C\widehat{\omega}(\frac{1+r}{2})$ for all $0\le r<1$. We call a radial weight ω regular, denoted by $\omega\in\mathcal{R}$, if $\omega\in\widehat{\mathcal{D}}$ and $\omega(r)$ behaves as its integral average over $(r,1)$, that is,

$$\omega(r)\asymp\frac{1}{1-r}\int_r^1\omega(s)\,ds, \quad 0\le r<1.$$

As to concrete examples, we mention that every standard weight as well as those given in [4, (4.4)–(4.6)] are regular. It is clear that $\omega \in \mathscr{R}$ if and only if for each $s \in [0,1)$ there exists a constant $C = C(s,\omega) > 1$ such that

$$C^{-1}\omega(t) \le \omega(r) \le C\omega(t), \quad 0 \le r \le t \le r + s(1-r) < 1, \tag{4}$$

and

$$\frac{1}{1-r}\int_r^1 \omega(s)\,ds \lesssim \omega(r), \quad 0 \le r < 1.$$

The definition of regular weights used here is slightly more general than that in [26], but the main properties are essentially the same by Lemma 2.1 below and [26, Chapter 1].

A radial continuous weight ω is called rapidly increasing, denoted by $\omega \in \mathscr{I}$, if

$$\lim_{r\to 1^-} \frac{1}{\omega(r)(1-r)}\int_r^1 \omega(s)\,ds = \infty.$$

It follows from [26, Lemma 1.1] that $\mathscr{I} \subset \widehat{\mathscr{D}}$. Typical examples of rapidly increasing weights are

$$v_\alpha(r) = \left((1-r)\left(\log\frac{e}{1-r}\right)^\alpha\right)^{-1}, \quad 1 < \alpha < \infty.$$

Despite their name, rapidly increasing weights may admit a strong oscillatory behavior. Indeed, the weight

$$\omega(r) = \left|\sin\left(\log\frac{1}{1-r}\right)\right| v_\alpha(r) + 1, \quad 1 < \alpha < \infty,$$

belongs to $\mathscr{I}$ but it does not satisfy (4) [26, p. 7].

A radial continuous weight ω is called rapidly decreasing if

$$\lim_{r\to 1^-} \frac{1}{\omega(r)(1-r)}\int_r^1 \omega(s)\,ds = 0.$$

The exponential type weights

$$\omega_{\gamma,\alpha}(r) = (1-r)^\gamma \exp\left(\frac{-c}{(1-r)^\alpha}\right), \quad \gamma \ge 0, \quad \alpha, c > 0,$$

are rapidly decreasing. For further information on these classes see [26, Chapter 1] and the references therein.

The following characterizations of the class $\widehat{\mathscr{D}}$ will be frequently used from here on.

Lemma 2.1. *Let ω be a radial weight. Then the following conditions are equivalent:*

(i) $\omega \in \widehat{\mathcal{D}}$;

(ii) There exist $C = C(\omega) > 0$ *and* $\beta_0 = \beta_0(\omega) > 0$ *such that*

$$\widehat{\omega}(r) \leq C\left(\frac{1-r}{1-t}\right)^{\beta} \widehat{\omega}(t), \quad 0 \leq r \leq t < 1,$$

for all $\beta \geq \beta_0$;

(iii) There exist $C = C(\omega) > 0$ *and* $\gamma_0 = \gamma_0(\omega) > 0$ *such that*

$$\int_0^t \left(\frac{1-t}{1-s}\right)^{\gamma} \omega(s)\, ds \leq C\widehat{\omega}(t), \quad 0 \leq t < 1,$$

for all $\gamma \geq \gamma_0$;

(iv) There exists $C = C(\omega) > 0$ *such that*

$$\int_0^t s^{\frac{1}{1-t}} \omega(s)\, ds \leq C\widehat{\omega}(t), \quad 0 \leq t < 1.$$

(v) There exists $C = C(\omega) > 0$ *such that*

$$\widehat{\omega}(r) \leq Cr^{-\frac{1}{1-t}} \widehat{\omega}(t), \quad 0 \leq r \leq t < 1.$$

(vi) The asymptotic equality

$$\omega_x = \int_0^1 s^x \omega(s)\, ds \asymp \widehat{\omega}\left(1 - \frac{1}{x}\right),$$

is valid for any $x \geq 1$.

Proof. We are going to prove (i)⇔(ii)⇔(iii)⇒(iv)⇒(v)⇒(i) and (iv)⇔(vi).

Let $\omega \in \widehat{\mathcal{D}}$. If $0 \leq r \leq t < 1$ and $r_n = 1 - 2^{-n}$ for all $n \in \mathbb{N} \cup \{0\}$, then there exist k and m such that $r_k \leq r < r_{k+1}$ and $r_m \leq t < r_{m+1}$. Hence

$$\begin{aligned}
\widehat{\omega}(r) &\leq \widehat{\omega}(r_k) \leq C\widehat{\omega}(r_{k+1}) \leq \cdots \leq C^{m-k+1}\widehat{\omega}(r_{m+1}) \leq C^{m-k+1}\widehat{\omega}(t) \\
&= C^2 2^{(m-k-1)\log_2 C}\widehat{\omega}(t) \leq C^2\left(\frac{1-r}{1-t}\right)^{\log_2 C} \widehat{\omega}(t), \quad 0 \leq r \leq t < 1.
\end{aligned}$$

On the other hand, it is clear that (ii) implies that $\omega \in \widehat{\mathcal{D}}$. So, we have proved (i)⇔(ii).

If (ii) is satisfied and $\gamma > \beta$, then, for $0 \leq t < 1$,

$$\begin{aligned}
\int_0^t \left(\frac{1-t}{1-s}\right)^{\gamma} \omega(s)\, ds &\leq C^{\frac{\gamma}{\beta}} \int_0^t \left(\frac{\widehat{\omega}(t)}{\widehat{\omega}(s)}\right)^{\frac{\gamma}{\beta}} \omega(s)\, ds \\
&= C^{\frac{\gamma}{\beta}} \widehat{\omega}(t)^{\frac{\gamma}{\beta}} \int_0^t \frac{\omega(s)}{\left(\widehat{\omega}(s)\right)^{\frac{\gamma}{\beta}}}\, ds \\
&\leq \frac{\beta}{\gamma-\beta} C^{\frac{\gamma}{\beta}} \widehat{\omega}(t).
\end{aligned}$$

Conversely, if (iii) is satisfied, then an integration by parts yields

$$
\begin{aligned}
C\widehat{\omega}(t) &\geq \int_0^t \left(\frac{1-t}{1-s}\right)^{\gamma} \omega(s)\,ds \\
&= -\widehat{\omega}(t) + (1-t)^{\gamma}\widehat{\omega}(0) + \gamma(1-t)^{\gamma} \int_0^t \frac{\widehat{\omega}(s)}{(1-s)^{\gamma+1}}\,ds \\
&\geq -\widehat{\omega}(t) + (1-t)^{\gamma}\widehat{\omega}(0) + \gamma(1-t)^{\gamma}\widehat{\omega}(r) \int_0^r \frac{ds}{(1-s)^{\gamma+1}} \\
&= -\widehat{\omega}(t) + (1-t)^{\gamma}(\widehat{\omega}(0) - \widehat{\omega}(r)) + \left(\frac{1-t}{1-r}\right)^{\gamma} \widehat{\omega}(r) \\
&\geq \left(\frac{1-t}{1-r}\right)^{\gamma} \widehat{\omega}(r) - \widehat{\omega}(t), \quad 0 \leq r \leq t < 1,
\end{aligned}
$$

and therefore (ii)⇔(iii).

Although the proof of [26, Lemma 1.3] shows that (iii) implies (iv), we include a proof for the sake of completeness. A simple calculation shows that for all $s \in (0,1)$ and $x > 1$,

$$
s^{x-1}(1-s)^{\gamma} \leq \left(\frac{x-1}{x-1+\gamma}\right)^{x-1} \left(\frac{\gamma}{x-1+\gamma}\right)^{\gamma} \leq \left(\frac{\gamma}{x-1+\gamma}\right)^{\gamma}.
$$

Therefore (iii), with $t = 1 - \frac{1}{x}$, yields

$$
\begin{aligned}
\int_0^{1-\frac{1}{x}} s^x \omega(s)\,ds &\leq \left(\frac{\gamma x}{x-1+\gamma}\right)^{\gamma} \int_0^{1-\frac{1}{x}} \frac{\omega(s)}{x^{\gamma}(1-s)^{\gamma}} s\,ds \\
&\lesssim \int_{1-\frac{1}{x}}^1 \omega(s)\,ds, \quad x > 1,
\end{aligned}
$$

which is equivalent to (iv).

On the other hand, if (iv) is satisfied and $0 \leq r \leq t < 1$, then an integration by parts yields

$$
\begin{aligned}
C\widehat{\omega}(t) &\geq \int_0^t s^{\frac{1}{1-t}} \omega(s)\,ds = -\widehat{\omega}(t) t^{\frac{1}{1-t}} + \frac{1}{1-t} \int_0^t \widehat{\omega}(s) s^{\frac{t}{1-t}}\,ds \\
&\geq -\widehat{\omega}(t) t^{\frac{1}{1-t}} + \frac{1}{1-t} \int_0^r \widehat{\omega}(s) s^{\frac{t}{1-t}}\,ds \\
&\geq -\widehat{\omega}(t) t^{\frac{1}{1-t}} + \frac{\widehat{\omega}(r)}{1-t} \int_0^r s^{\frac{t}{1-t}}\,ds = -\widehat{\omega}(t) t^{\frac{1}{1-t}} + r^{\frac{t}{1-t}} \widehat{\omega}(r),
\end{aligned}
$$

and thus

$$
r^{\frac{1}{1-t}} \widehat{\omega}(r) \leq \left(C + t^{\frac{1}{1-t}}\right) \widehat{\omega}(t), \quad 0 \leq r \leq t < 1.
$$

This implies (v), and by choosing $t = \frac{1+r}{2}$ in (v), we deduce $\omega \in \widehat{\mathcal{D}}$. Finally, it is clear that (iv) is equivalent to (vi). □

3. Equivalent norms

In this section we shall present several equivalent norms on weighted Bergman spaces. In particular we shall give a detailed proof of a decomposition norm theorem for A^p_ω when $\omega \in \widehat{\mathscr{D}}$ and $1 < p < \infty$.

It is well-known that a choice of an appropriate norm is often a key step when solving a problem on a space of analytic functions. For instance, in the study of the integration operators

$$T_g(f)(z) = \int_0^z f(\zeta)\,g'(\zeta)\,d\zeta, \quad z \in \mathbb{D}, \quad g \in \mathscr{H}(\mathbb{D}),$$

one wants to get rid of the integral symbol, so one looks for norms in terms of the first derivative. It is worth mentioning that the operator T_g began to be extensively studied after the appearance of the works by Aleman, Cima and Siskakis [1, 4]. A description of its resolvent set on Hardy and standard Bergman spaces is strongly connected with the classical theory of the Muckenhoupt weights and the Bekollé-Bonami weights [2, 3].

3.1. *Norms in terms of the derivative*

Following Siskakis [30], the distortion function of a radial weight ω is defined by

$$\psi_\omega(z) = \frac{1}{\omega(|z|)} \int_{|z|}^1 \omega(s)\,ds, \quad z \in \mathbb{D}.$$

For a large class of radial weights, which includes any differentiable decreasing weight and all the standard ones, the most appropriate way to obtain a useful norm involving the first derivative is to establish a kind of Littlewood-Paley type formula [24, Theorem 1.1].

Theorem 3.1. *Suppose that ω is a radial differentiable weight, and there is $L > 0$ such that*

$$\sup_{0<r<1} \frac{\omega'(r)}{\omega(r)^2} \int_r^1 \omega(x)\,dx \le L.$$

Then, for each $p \in (0, \infty)$,

$$\int_{\mathbb{D}} |f(z)|^p \omega(z)\,dA(z) \asymp |f(0)|^p + \int_{\mathbb{D}} |f'(z)|^p\, \psi^p_\omega(z)\,\omega(z)\,dA(z), \quad f \in \mathscr{H}(\mathbb{D}).$$

If $\omega \in \mathscr{I}$ and $p \neq 2$, a result analogous to Theorem 3.1 cannot be obtained in general [26, Proposition 4.2].

Proposition 3.1. *Let $p \neq 2$. Then there exists $\omega \in \mathscr{I}$ such that, for any function $\varphi : [0,1) \to (0,\infty)$, the relation*

$$\|f\|_{A^p_\omega}^p \asymp \int_{\mathbb{D}} |f'(z)|^p \varphi(|z|)^p \omega(z)\,dA(z) + |f(0)|^p$$

can not be valid for all $f \in \mathcal{H}(\mathbb{D})$.

As for a Littlewood-Paley formula for A^p_ω, the following result was proved in [2, Theorem 3.1].

Theorem 3.2. *Suppose that ω is a weight such that $\frac{\omega(z)}{(1-|z|)^\eta}$ satisfies the Bekollé-Bonami condition $B_{p_0}(\eta)$ for some $p_0 > 0$ and some $\eta > -1$. Then, for each $p \in (0,\infty)$,*

$$\int_{\mathbb{D}} |f(z)|^p \omega(z)\,dA(z) \asymp |f(0)|^p + \int_{\mathbb{D}} |f'(z)|^p (1-|z|)^p \omega(z)\,dA(z), \tag{5}$$

for all $f \in \mathcal{H}(\mathbb{D})$.

We remark that whenever $\omega \in C^1(\mathbb{D})$ and $(1-|z|)|\nabla\omega(z)| \lesssim \omega(z)$, $z \in \mathbb{D}$, then (5) is equivalent to a Bekollé-Bonami condition [2, Theorem 3.1].

Now, we consider the non-tangential approach regions

$$\Gamma(\zeta) = \left\{ z \in \mathbb{D} : |\theta - \arg z| < \frac{1}{2}\left(1 - \tfrac{|z|}{r}\right) \right\}, \quad \zeta = re^{i\theta} \in \mathbb{D} \setminus \{0\},$$

and the related tents $T(z) = \{\zeta \in \mathbb{D} : z \in \Gamma(\zeta)\}$.

Whenever ω is a radial weight, A^p_ω can be equipped with other norms which are inherited from the classical Fefferman-Stein estimate [14] and the Hardy-Stein-Spencer identity [16] for the H^p-norm. Here

$$\omega^\star(z) = \int_{|z|}^1 \omega(s) \log \tfrac{s}{|z|}\, s\,ds, \quad z \in \mathbb{D} \setminus \{0\}.$$

Theorem 3.3. *Let $0 < p < \infty$, $n \in \mathbb{N}$ and $f \in \mathcal{H}(\mathbb{D})$, and let ω be a radial weight. Then*

$$\|f\|_{A^p_\omega}^p = p^2 \int_{\mathbb{D}} |f(z)|^{p-2} |f'(z)|^2 \omega^\star(z)\,dA(z) + \omega(\mathbb{D})|f(0)|^p,$$

and

$$\|f\|_{A^p_\omega}^p \asymp \int_{\mathbb{D}} \left(\int_{\Gamma(u)} |f^{(n)}(z)|^2 \left(1 - \left|\tfrac{z}{u}\right|\right)^{2n-2} dA(z) \right)^{\frac{p}{2}} \omega(u)\,dA(u) + \sum_{j=0}^{n-1} |f^{(j)}(0)|^p,$$

where the constants of comparison depend only on p, n and ω. In particular,

$$\|f\|_{A^2_\omega}^2 = 4\|f'\|_{A^2_{\omega^\star}}^2 + \omega(\mathbb{D})|f(0)|^2.$$

Next, we present an equivalent norm for weighted Bergman spaces which has been very recently used to describe the q-Carleson measures for A^p_ω when $\omega \in \widehat{\mathscr{D}}$ [27].

The *non-tangential maximal function* of $f \in \mathscr{H}(\mathbb{D})$ in the (punctured) unit disc is defined by

$$N(f)(u) = \sup_{z \in \Gamma(u)} |f(z)|, \quad u \in \mathbb{D} \setminus \{0\}.$$

Lemma 3.1 (Lemma 4.4 in [26]). *Let $0 < p < \infty$ and let ω be a radial weight. Then there exists a constant $C > 0$ such that*

$$\|f\|^p_{A^p_\omega} \le \|N(f)\|^p_{L^p_\omega} \le C\|f\|^p_{A^p_\omega}, \quad \textit{for all } f \in \mathscr{H}(\mathbb{D}).$$

Proof. It follows from [16, Theorem 3.1 on p. 57] that there exists a constant $C > 0$ such that the classical non-tangential maximal function

$$f^\star(\zeta) = \sup_{z \in \Gamma(\zeta)} |f(z)|, \quad \zeta \in \mathbb{T},$$

satisfies

$$\|f^\star\|^p_{L^p(\mathbb{T})} \le C\|f\|^p_{H^p}$$

for all $0 < p < \infty$ and $f \in \mathscr{H}(\mathbb{D})$. Therefore

$$\begin{aligned}\|f\|^p_{A^p_\omega} \le \|N(f)\|^p_{L^p_\omega} &= \int_{\mathbb{D}} (N(f)(u))^p \omega(u)\, dA(u)\\ &= \int_0^1 \omega(r) r \int_{\mathbb{T}} ((f_r)^\star(\zeta))^p\, |d\zeta|\, dr\\ &\le C \int_0^1 \omega(r) r \int_{\mathbb{T}} f(r\zeta)^p\, |d\zeta|\, dr = C\|f\|^p_{A^p_\omega},\end{aligned}$$

and the assertion is proved. □

3.2. *Decomposition norm theorems*

The main purpose of this section is to extend [25, Theorem 4] to the case $\omega \in \widehat{\mathscr{D}}$. Decomposition norm theorems have been obtained previously in [20–22] for several type of mixed norm spaces. For $0 < p \le \infty$, $0 < q < \infty$, and a radial weight ω, the mixed norm space $H(p,q,\omega)$ consists of those $g \in \mathscr{H}(\mathbb{D})$ such that

$$\|g\|^q_{H(p,q,\omega)} = \int_0^1 M^q_p(r,g)\omega(r)\, dr < \infty.$$

If in addition $-\infty < \beta < \infty$, we will denote $g \in H(p,\infty,\widehat{\omega}^\beta)$, whenever

$$\|g\|_{H(p,\infty,\widehat{\omega}^\beta)} = \sup_{0<r<1} M_p(r,g)\widehat{\omega}(r)^\beta < \infty.$$

It is clear that $H(p,p,\omega) = A^p_\omega$. The mixed norm spaces play an essential role in the closely related question of studying the coefficient multipliers and the generalized Hilbert operator

$$\mathcal{H}_g(f)(z) = \int_0^1 f(t)g'(tz)\,dt, \quad g \in \mathcal{H}(\mathbb{D}),$$

on Hardy and weighted Bergman spaces [5, 15, 25].

In order to give the precise statement of the main result of this section, we need to introduce some more notation. To do this, let ω be a radial weight such that $\int_0^1 \omega(r)\,dr = 1$. For each $\alpha > 0$ and $n \in \mathbb{N} \cup \{0\}$, let $r_n = r_n(\omega,\alpha) \in [0,1)$ be defined by

$$\widehat{\omega}(r_n) = \int_{r_n}^1 \omega(r)\,dr = \frac{1}{2^{n\alpha}}. \tag{6}$$

Clearly, $\{r_n\}_{n=0}^\infty$ is an increasing sequence of distinct points on $[0,1)$ such that $r_0 = 0$ and $r_n \to 1^-$, as $n \to \infty$. For $x \in [0,\infty)$, let $E(x)$ denote the integer such that $E(x) \le x < E(x)+1$, and set $M_n = E\left(\frac{1}{1-r_n}\right)$ for short. Write

$$I(0) = I_{\omega,\alpha}(0) = \left\{k \in \mathbb{N} \cup \{0\} : k < M_1\right\}$$

and

$$I(n) = I_{\omega,\alpha}(n) = \left\{k \in \mathbb{N} : M_n \le k < M_{n+1}\right\}$$

for all $n \in \mathbb{N}$. If $f(z) = \sum_{n=0}^\infty a_n z^n$ is analytic in $\mathbb{D}$, define the polynomials $\Delta_n^{\omega,\alpha} f$ by

$$\Delta_n^{\omega,\alpha} f(z) = \sum_{k \in I_{\omega,\alpha}(n)} a_k z^k, \quad n \in \mathbb{N} \cup \{0\}.$$

Theorem 3.4. *Let $1 < p < \infty$, $0 < \alpha < \infty$ and $\omega \in \widehat{\mathcal{D}}$ such that $\int_0^1 \omega(r)\,dr = 1$, and let $f \in \mathcal{H}(\mathbb{D})$.*

(i) If $0 < q < \infty$, then $f \in H(p,q,\omega)$ if and only if

$$\sum_{n=0}^\infty 2^{-n\alpha} \|\Delta_n^{\omega,\alpha} f\|_{H^p}^q < \infty.$$

Moreover,

$$\|f\|_{H(p,q,\omega)} \asymp \left(\sum_{n=0}^\infty 2^{-n\alpha} \|\Delta_n^{\omega,\alpha} f\|_{H^p}^q\right)^{1/q}.$$

(ii) *If* $0<\beta<\infty$, *then* $f\in H(p,\infty,\widehat{\omega}^{\beta})$ *if and only if*

$$\sup_n 2^{-n\alpha\beta}\|\Delta_n^{\omega,\alpha}f\|_{H^p}<\infty.$$

Moreover,

$$\|f\|_{H(p,\infty,\widehat{\omega}^{\beta})}\asymp\sup_n 2^{-n\alpha\beta}\|\Delta_n^{\omega,\alpha}f\|_{H^p}.$$

The proof of Theorem 3.4 follows that of [26, Theorem 4], and it only distinguishes from it because of the technicalities of extending the class $\mathcal{R}\cup\mathcal{I}$ to $\widehat{\mathcal{D}}$. Some previous results are needed. Recall that a function h is called essentially decreasing if there exists a positive constant $C=C(h)$ such that $h(x)\le Ch(y)$ whenever $y\le x$. Essentially increasing functions are defined in an analogous manner.

Lemma 3.2. *Let* $\omega\in\widehat{\mathcal{D}}$ *such that* $\int_0^1\omega(r)\,dr=1$. *For each* $\alpha>0$ *and* $n\in\mathbb{N}\cup\{0\}$, *let* $r_n=r_n(\omega,\alpha)\in[0,1)$ *be defined by* (6). *Then the following assertions hold:*

(i) *For each* $\gamma>0$, *there exists* $C=C(\alpha,\gamma,\omega)>0$ *such that*

$$\eta_\gamma(r)=\sum_{n=0}^{\infty}2^{n\gamma}r^{M_n}\le C\widehat{\omega}(r)^{-\frac{\gamma}{\alpha}},\quad 0\le r<1. \tag{7}$$

(ii) *For each* $0<\beta<1$, *there exists* $C=C(\alpha,\beta,\omega)>0$ *such that*

$$2^{-n\alpha\beta}\int_0^1\frac{r^{M_n}\omega(r)}{\widehat{\omega}(r)^{\beta}}\,dr\le C\int_0^1 r^{M_n}\omega(r)\,dr. \tag{8}$$

Proof. (i). We will begin with proving (7) for $r=r_N$, where $N\in\mathbb{N}$. To do this, note first that

$$\sum_{n=0}^{N}2^{n\gamma}r_N^{M_n}\le\frac{2^{\gamma}}{2^{\gamma}-1}\widehat{\omega}(r_N)^{-\frac{\gamma}{\alpha}}, \tag{9}$$

by (6). To deal with the remainder of the sum, we apply Lemma 2.1(ii) and (6) to find $\beta=\beta(\omega)>0$ and $C=C(\beta,\omega)>0$ such that

$$\frac{1-r_n}{1-r_{n+j}}\ge C\left(\frac{\widehat{\omega}(r_n)}{\widehat{\omega}(r_{n+j})}\right)^{1/\beta}=C2^{\frac{j\alpha}{\beta}},\quad n,j\in\mathbb{N}\cup\{0\}.$$

This, the inequality $\log\frac{1}{x}\ge 1-x$, $0<x\le 1$, and (6) give

$$\begin{aligned}\sum_{n=N+1}^{\infty}2^{n\gamma}r_N^{M_n}&\le 2^{N\gamma}\sum_{j=1}^{\infty}2^{j\gamma}e^{-r_{N+j}\frac{1-r_N}{1-r_{N+j}}}\le 2^{N\gamma}\sum_{j=1}^{\infty}2^{j\gamma}e^{-r_2C2^{\frac{j\alpha}{\beta}}}\\&=C(\beta,\alpha,\gamma,\omega)\,\widehat{\omega}(r_N)^{-\frac{\gamma}{\alpha}}.\end{aligned}$$

Since $\beta = \beta(\omega)$, this together with (9) gives (7) for $r = r_N$, where $N \in \mathbb{N}$. Now, using standard arguments, it implies (7) for any $r \in (0,1)$.

(ii). Let us write $\widetilde{\omega}(r) = \frac{\omega(r)}{\widehat{\omega}(r)^\beta}$. Clearly,

$$2^{-n\alpha\beta}\int_0^{r_n} r^{M_n}\widetilde{\omega}(r)\,dr \le \frac{2^{-n\alpha\beta}}{\widehat{\omega}(r_n)^\beta}\int_0^{r_n} r^{M_n}\omega(r)\,dr \le \int_0^1 r^{M_n}\omega(r)\,dr. \tag{10}$$

Moreover, [26, Lemma 1.4 (iii)] yields

$$\begin{aligned} 2^{-n\alpha\beta}\int_{r_n}^1 r^{M_n}\widetilde{\omega}(r)\,dr &\le 2^{-n\alpha\beta}\widetilde{\omega}(r_n)\psi_{\widetilde{\omega}}(r_n) = \frac{2^{-n\alpha\beta}}{1-\beta}\widetilde{\omega}(r_n)\psi_\omega(r_n) \\ &= \frac{1}{1-\beta}\int_{r_n}^1 \omega(r)\,dr \le C(\beta,\alpha,\omega)\int_{r_n}^1 r^{M_n}\omega(r)\,dr. \end{aligned} \tag{11}$$

By combining (10) and (11) we obtain (ii). □

We now present a result on power series with positive coefficients. This result will play a crucial role in the proof of Theorem 3.4.

Proposition 3.2. *Let $0 < p, \alpha < \infty$ and $\omega \in \widehat{\mathcal{D}}$ such that $\int_0^1 \omega(r)\,dr = 1$. Let $f(r) = \sum_{k=0}^\infty a_k r^k$, where $a_k \ge 0$ for all $k \in \mathbb{N}\cup\{0\}$, and denote $t_n = \sum_{k\in I_{\omega,\alpha}(n)} a_k$. Then there exists a constant $C = C(p,\alpha,\omega) > 0$ such that*

$$\frac{1}{C}\sum_{n=0}^\infty 2^{-n\alpha}t_n^p \le \int_0^1 f(r)^p\omega(r)\,dr \le C\sum_{n=0}^\infty 2^{-n\alpha}t_n^p. \tag{12}$$

Proof. We will use ideas from the proof of [19, Theorem 6]. The definition (6) yields

$$\begin{aligned} \int_0^1 f(r)^p\omega(r)\,dr &\ge \sum_{n=0}^\infty \int_{r_{n+1}}^{r_{n+2}}\left(\sum_{k=0}^\infty t_k r^{M_{k+1}}\right)^p\omega(r)\,dr \\ &\ge \sum_{n=0}^\infty\left(\sum_{k=0}^n t_k r_{n+1}^{M_{k+1}}\right)^p\int_{r_{n+1}}^{r_{n+2}}\omega(r)\,dr \\ &\ge \left(1-\frac{1}{2^\alpha}\right)\sum_{n=0}^\infty t_n^p r_{n+1}^{pM_{n+1}}2^{(-n-1)\alpha} \ge C\sum_{n=0}^\infty t_n^p 2^{-n\alpha}, \end{aligned}$$

where $C = C(p,\alpha,\omega) > 0$ is a constant. This gives the first inequality in (12).

To prove the second inequality in (12), let first $p > 1$ and take $0 < \gamma < \frac{\alpha}{p-1}$. Then Hölder's inequality gives

$$f(r)^p \le \left(\sum_{n=0}^\infty t_n r^{M_n}\right)^p \le \eta_\gamma(r)^{p-1}\sum_{n=0}^\infty 2^{-n\gamma(p-1)}t_n^p r^{M_n}.$$

Therefore, by (7) and (8) in Lemma 3.2 and Lemma 2.1(vi) there exist constants $C_1 = C_1(\alpha,\gamma,p,\omega) > 0$, $C_2 = C_2(\alpha,\gamma,p,\omega) > 0$ and $C_3 = C_3(\alpha,\gamma,p,\omega) > 0$ such that

$$\begin{aligned}\int_0^1 f(r)^p\omega(r)\,dr &\le \sum_{n=0}^{\infty} 2^{-n\gamma(p-1)} t_n^p \int_0^1 r^{M_n}\eta_\gamma(r)^{p-1}\omega(r)\,dr\\ &\le C_1\sum_{n=0}^{\infty} 2^{-n\gamma(p-1)} t_n^p \int_0^1 \frac{r^{M_n}\omega(r)}{\widehat{\omega}(r)^{\frac{\gamma(p-1)}{\alpha}}}\,dr\\ &\le C_2\sum_{n=0}^{\infty} t_n^p \int_0^1 r^{M_n}\omega(r)\,dr\\ &\le C_3\sum_{n=0}^{\infty} t_n^p\,\widehat{\omega}(r_n)\,dr = C_3\sum_{n=0}^{\infty} t_n^p 2^{-n\alpha}.\end{aligned}$$

Since $\gamma=\gamma(\alpha,p)$, this gives the assertion for $1<p<\infty$.

If $0<p\le 1$, then

$$f(r)^p \le \Big(\sum_{n=0}^{\infty} t_n r^{M_n}\Big)^p \le \sum_{n=0}^{\infty} t_n^p r^{M_n p},$$

so using Lemma 2.1(vi) and (ii), there exists a constant $C_1 = C_1(\alpha,\gamma,p,\omega)>0$ such that

$$\begin{aligned}\int_0^1 f(r)^p\omega(r)\,dr &\le \sum_{n=0}^{\infty} t_n^p \int_0^1 r^{pM_n}\omega(r)\,dr\\ &\le C_1\sum_{n=0}^{\infty} t_n^p\,\widehat{\omega}(r_n) = C_1\sum_{n=0}^{\infty} t_n^p 2^{-n\alpha}.\end{aligned}$$

This finishes the proof. □

Next, for $g(z)=\sum_{k=0}^{\infty} b_k z^k \in \mathcal{H}(\mathbb{D})$ and $n_1,n_2\in\mathbb{N}\cup\{0\}$, we set

$$S_{n_1,n_2}g(z) = \sum_{k=n_1}^{n_2-1} b_k z^k, \quad n_1<n_2.$$

The chain of inequalities

$$r^{n_2}\|S_{n_1,n_2}g\|_{H^p} \le M_p(r,S_{n_1,n_2}g) \le r^{n_1}\|S_{n_1,n_2}g\|_{H^p}, \quad 0<r<1, \tag{13}$$

follows from [20, Lemma 3.1].

Lemma 3.3. *Let $0<p\le\infty$ and $n_1,n_2\in\mathbb{N}$ with $n_1<n_2$. If $g(z)=\sum_{k=0}^{\infty} c_k z^k \in \mathcal{H}(\mathbb{D})$, then*

$$\|S_{n_1,n_2}g\|_{H^p} \asymp M_p\Big(1-\frac{1}{n_2}, S_{n_1,n_2}g\Big).$$

Proof of Theorem 3.4. (i). The M. Riesz projection theorem, and (13), yield

$$\begin{aligned}\|f\|_{H(p,q,\omega)} &\gtrsim \sum_{n=0}^{\infty} \|\Delta_n^{\omega,\alpha} f\|_{H^p}^q \int_{r_{n+1}}^{r_{n+2}} r^{qM_{n+1}} \omega(r)\,dr \\ &\asymp \sum_{n=0}^{\infty} \|\Delta_n^{\omega,\alpha} f\|_{H^p}^q \int_{r_{n+1}}^{r_{n+2}} \omega(r)\,dr \asymp \sum_{n=0}^{\infty} 2^{-n\alpha} \|\Delta_n^{\omega,\alpha} f\|_{H^p}^q.\end{aligned}$$

On the other hand, Minkowski's inequality and (13) give

$$M_p(r,f) \le \sum_{n=0}^{\infty} M_p(r,\Delta_n^{\omega,\alpha} f) \le \sum_{n=0}^{\infty} r^{M_n} \|\Delta_n^{\omega,\alpha} f\|_{H^p}, \tag{14}$$

and hence Proposition 3.2 yields

$$\|f\|_{H(p,q,\omega)} \le \int_0^1 \left(\sum_{n=0}^{\infty} r^{M_n} \|\Delta_n^{\omega,\alpha} f\|_{H^p}\right)^q \omega(r)\,dr \asymp \sum_{n=0}^{\infty} 2^{-n\alpha} \|\Delta_n^{\omega,\alpha} f\|_{H^p}^q.$$

(ii). Using again the M. Riesz projection theorem and (13) we deduce

$$\sup_{0<r<1} M_p(r,f)\,\widehat{\omega}(r)^\beta \gtrsim r_{n+1}^{M_{n+1}} \|\Delta_n^{\omega,\alpha} f\|_{H^p} 2^{-n\alpha\beta}, \quad n \in \mathbb{N}\cup\{0\},$$

and hence

$$\|f\|_{H(p,\infty,\widehat{\omega}^\beta)} \gtrsim \sup_n 2^{-n\alpha\beta} \|\Delta_n^{\omega,\alpha} f\|_{H^p}.$$

Conversely, assume that $M = \sup_n 2^{-n\alpha\beta} \|\Delta_n^{\omega,\alpha} f\|_{H^p} < \infty$. Then (14) and Lemma 3.2(i) yield

$$M_p(r,f) \le \sum_{n=0}^{\infty} r^{M_n} \|\Delta_n^{\omega,\alpha} f\|_{H^p} \le M \sum_{n=0}^{\infty} 2^{n\alpha\beta} r^{M_n} \lesssim M\widehat{\omega}(r)^{-\beta}.$$

This finishes the proof. □

It is worth mentioning that Theorem 3.4 does not remain valid for $0 < p \le 1$. But the part that is true in this case is contained in the next result.

Proposition 3.3. *Let $0 < p \le 1$, $0 < \alpha < \infty$ and $\omega \in \widehat{\mathcal{D}}$ such that $\int_0^1 \omega(r)\,dr = 1$.*

(i) If $0 < q < \infty$, then

$$\|f\|_{H(p,q,\omega)} \lesssim \left(\sum_{n=0}^{\infty} 2^{-n\alpha} \|\Delta_n^{\omega,\alpha} f\|_{H^p}^q\right)^{1/q}, \quad f \in \mathcal{H}(\mathbb{D}).$$

(ii) If $0 < \beta < \infty$, then

$$\|f\|_{H(p,\infty,\widehat{\omega}^\beta)} \lesssim \sup_n 2^{-n\alpha\beta} \|\Delta_n^{\omega,\alpha} f\|_{H^p}, \quad f \in \mathcal{H}(\mathbb{D}).$$

Proposition 3.3 follows from the inequality

$$M_p^p(r,f) \le \sum_{n=0}^{\infty} M_p^p(r,\Delta_n^{\omega} f) \le \sum_{n=0}^{\infty} r^{pM_n} \|\Delta_n^{\omega} f\|_{H^p}^p,$$

(13) and Proposition 3.2. See also [28, Lemma 8].

4. Bergman projection

4.1. *One weight inequality*

The boundedness of projections on L^p-spaces is an intriguing topic which has attracted a lot of attention in recent years [9–11, 17, 28, 32, 33]. In fact, as far as we know, to characterize those radial weights for which $P_\omega : L^p_\omega \to L^p_\omega$ is bounded, is still an open problem [10, p. 116].

For the class of standard weights, the Bergman projection P_α (as well as P^+_α) is bounded on L^p_α if and only if $1 < p < \infty$ [33, Theorem 4.24]. As for $p = \infty$, P_α is bounded and onto from L^∞ to $\mathcal{B}$. Here $\mathcal{B}$ [33, Chapter 5] denotes the Bloch space that consists of $f \in \mathcal{H}(\mathbb{D})$ such that

$$\|f\|_{\mathcal{B}} = \sup_{z\in\mathbb{D}} |f'(z)|(1-|z|^2) + |f(0)| < \infty.$$

These results have been recently extended to the class of regular weights [28].

Theorem 4.1. *Let* $1 < p < \infty$.

(i) If $\omega \in \mathcal{R}$, *then* $P^+_\omega : L^p_\omega \to L^p_\omega$ *is bounded. In particular,* $P_\omega : L^p_\omega \to A^p_\omega$ *is bounded.*
(ii) If $\omega \in \mathcal{R}$, *then* $P_\omega : L^\infty(\mathbb{D}) \to \mathcal{B}$ *is bounded.*

In the original source [28], Theorem 4.1 (i) is obtained as a consequence of Theorem 4.4 below. Here, we shall offer a simple proof of this result. Both arguments use strongly precise L^p-estimates of the Bergman reproducing kernels [28].

Theorem 4.2. *Let* $0 < p < \infty$, $\omega \in \widehat{\mathcal{D}}$ *and* $N \in \mathbb{N} \cup \{0\}$. *Then the following assertions hold:*

(i) $M^p_p\left(r, \left(B^\omega_a\right)^{(N)}\right) \asymp \int_0^{|a|r} \frac{dt}{\widehat{\omega}(t)^p(1-t)^{p(N+1)}}, \quad r, |a| \to 1^-.$

(ii) If $v \in \widehat{\mathcal{D}}$, *then*

$$\left\| \left(B^\omega_a\right)^{(N)} \right\|^p_{A^p_v} \asymp \int_0^{|a|} \frac{\widehat{v}(t)}{\widehat{\omega}(t)^p(1-t)^{p(N+1)}}\, dt, \quad |a| \to 1^-.$$

We would like to mention that Theorem 3.1 and [25, Theorem 4] play important roles in the proof of this result. Besides, we use strongly Lemma 2.1, in particular the description of the class $\widehat{\mathcal{D}}$ in terms of the moments of the weights

$$\int_0^1 s^x \omega(s)\, ds \asymp \widehat{\omega}\left(1 - \frac{1}{x}\right), \quad x \in [1, \infty).$$

Now, we offer a simple proof of the one weight inequality for regular weights.

Proof of Theorem 4.1 (i). Let $1<p<\infty$ and $\omega\in\mathcal{R}$. Let $h=\widehat{\omega}^{-\frac{1}{pp'}}$, where $\frac{1}{p}+\frac{1}{p'}=1$. Since $p>1$, [26, Lemma 1.4(iii)] shows that $h^{p'}\omega$ is a weight with $\psi_{h^{p'}\omega}=\frac{p}{p-1}\psi_\omega$, and thus $h^{p'}\omega\in\mathcal{R}$. Since $\omega\in\widehat{\mathcal{D}}$, by Lemma 2.1(ii) there exists $\beta=\beta(\omega)$ such that $\widehat{\omega}(s)(1-s)^{-\beta}$ is essentially increasing on $[0,1)$. On the other hand, since $\omega\in\mathcal{R}$ there is $\alpha=\alpha(\omega)>0$ with $\alpha\le\beta$ such that $\widehat{\omega}(s)(1-s)^{-\alpha}$ is essentially decreasing, see [26, (ii) p. 10]. By using this and $h^{p'}\omega\in\mathcal{R}$ we deduce

$$\int_0^r \frac{\widehat{h^{p'}\omega}(s)}{\widehat{\omega}(s)(1-s)}\,ds \asymp \int_0^r \frac{ds}{\widehat{\omega}(s)^{\frac{1}{p}}(1-s)} \asymp \frac{1}{\widehat{\omega}(r)^{\frac{1}{p}}} = h^{p'}(r), \quad r\ge\frac{1}{2}. \tag{15}$$

By symmetry, a similar reasoning applies when p' is replaced by p, and hence we may use Theorem 4.2(ii) and (15) to deduce

$$\int_{\mathbb{D}} |B^\omega(z,\zeta)|h^{p'}(\zeta)\omega(\zeta)\,dA(\zeta)\asymp h^{p'}(z), \quad z\in\mathbb{D},$$

and

$$\int_{\mathbb{D}} |B^\omega(z,\zeta)|h^{p}(z)\omega(z)\,dA(z)\asymp h^{p}(\zeta), \quad \zeta\in\mathbb{D}.$$

It follows from Schur's test [33, Theorem 3.6] that $P_\omega^+: L^p_\omega\to L^p_\omega$ is bounded. □

The situation is different for $\omega\in\mathcal{I}$ because then P_ω^+ is not bounded on L^p_ω [28]. This result points out that many finer function-theoretic properties of A^p_α do not carry over to A^p_ω induced by $\omega\in\mathcal{I}$.

Concerning rapidly decreasing weights, Dostanic [10] proved that the Bergman projection is bounded on L^p_v only for $p=2$ in the case of Bergman spaces with the exponential type weights $w(r)=(1-r^2)^A\exp\left(\frac{-B}{(1-r^2)^\alpha}\right)$, $A\in\mathbb{R}$, $B,\alpha>0$. The next result proves that it is a general phenomenon which holds for rapidly decreasing and smooth weights [9, 32].

Proposition 4.1. *Assume that $\omega(r)=e^{-2\varphi(r)}$ is a radial weight such that φ is a positive C^∞-function, φ' is positive on $[0,1)$, $\lim_{r\to1^-}\varphi(r)=\lim_{r\to1^-}\varphi'(r)=+\infty$ and*

$$\lim_{r\to1^-}\frac{\varphi^{(n)}(r)}{\left(\varphi'(r)\right)^n}=0, \quad \textit{for any } n\in\mathbb{N}\setminus\{1\}. \tag{16}$$

Then, the Bergman projection is bounded from L^p_ω to L^p_ω only for $p=2$.

Consequently, if ω a rapidly decreasing weight, it is natural to look for a substitute for the boundedness of the Bergman projection P_ω. Inspired by the Fock space setting, the following result has been proved for a canonical example [9].

Theorem 4.3. *Let* $\omega(r) = \exp\left(-\frac{\alpha}{1-r}\right)$, $\alpha > 0$, *and* $1 \le p < \infty$. *Then, the Bergman projection* P_ω *is bounded from* $L^p_{\omega^{p/2}}$ *to* $A^p_{\omega^{p/2}}$.

The approach to prove this result relies on an instance of Schur's test and accurate estimates for the integral means of order one of the corresponding Bergman reproducing kernel [9, Proposition 5].

Proposition 4.2. *Let* $\omega(r) = \exp\left(-\frac{\alpha}{1-r}\right)$, $\alpha > 0$, *and let* $K(z) = \sum_{n=0}^{\infty} \frac{z^n}{2\omega_{2n+1}}$. *Then,*

$$M_1(r,K) \asymp \frac{\exp\left(\frac{\alpha}{1-\sqrt{r}}\right)}{(1-r)^{\frac{3}{2}}}, \quad r \to 1^-.$$

These estimates are obtained by using two key tools; the sharp asymptotic estimates obtained in [18] for the moments of the weight in terms of the Legendre-Fenchel transform, and an upper estimate of $M_1(r,K)$ by the l^1-norm of the H^1-norms of the Hadamard product of K_r with certain smooth polynomials. We refer to [11, 31] for other results concerning the particular case $\omega = v$ in (1).

4.2. *Two weight inequality*

As it has been commented before, the weights v satisfying (1) when ω is an standard weight and $1 < p < \infty$, were characterized by Bekollé and Bonami [6, 8]. Recently [29], it has been proved the following quantitative version of this result

$$\|P^+_\alpha(f)\|_{L^p_{v(1-|z|^2)^\alpha}} \le C(p,\alpha) B_{p,\alpha}\left(v\right) \|f\|_{L^p_{v(1-|z|^2)^\alpha}},$$

where $B_{p,\alpha}(v)$ was defined in (2). With regard to the case $p = 1$, we define the weighted maximal function

$$M_\alpha(\omega)(z) = \sup_{z \in D(a,r)} \frac{\omega\left(D(a,r) \cap \mathbb{D}\right)}{A_\alpha\left(D(a,r) \cap \mathbb{D}\right)}, \quad z \in \mathbb{D},$$

where the supremum runs on all the Euclidean discs $D(a,r)$ which contain the point z. It is known [6, 7] that the weak $(1,1)$ inequality

$$\omega\left(\{z \in \mathbb{D} : |P_\alpha(f)(z)| > \lambda\}\right) \le C_{\alpha,\omega} \frac{\|f\|_{L^1_\omega}}{\lambda}$$

holds (and its analogue replacing P_α by P^+_α) if and only if the weighted maximal function satisfies

$$M_\alpha(\omega)(z) \le C \frac{\omega(z)}{(1-|z|)^\alpha}, \quad z \in \mathbb{D}.$$

As far as we know, apart from Bekollé-Bonami's results [6, 8] on the standard Bergman projection P_α, very little is known about (1) when $\omega \neq v$. We note that [9, Theorem 1] may be seen as a positive example for (1) in the context of rapidly increasing weights. A recent result [28] describes those regular weights ω and v for which (1) holds for $1 < p < \infty$.

Theorem 4.4. *Let $1 < p < \infty$ and $\omega, v \in \mathcal{R}$. Then the following conditions are equivalent:*

(a) $P^+_\omega : L^p_v \to L^p_v$ *is bounded;*
(b) $P_\omega : L^p_v \to L^p_v$ *is bounded;*
(c) $\displaystyle\sup_{0<r<1} \frac{\widehat{v}(r)^{\frac{1}{p}}\left(\int_r^1 \left(\frac{\omega(s)}{v(s)}\right)^{p'} v(s)\,ds\right)^{\frac{1}{p'}}}{\widehat{\omega}(r)} < \infty;$
(d) $\displaystyle\sup_{0<r<1} \frac{\omega(r)^p(1-r)^{p-1}}{v(r)} \int_0^r \frac{v(s)}{\omega(s)^p(1-s)^p}\,ds < \infty;$
(e) $\displaystyle\sup_{0<r<1} \left(\int_0^r \frac{v(s)}{\omega(s)^p(1-s)^p}\,ds\right)^{\frac{1}{p}} \left(\int_r^1 \left(\frac{\omega(s)}{v(s)}\right)^{p'} v(s)\,ds\right)^{\frac{1}{p'}} < \infty;$
(f) $\displaystyle\sup_{0<r<1} \frac{\widehat{v}(r)^{\frac{1}{p}} \int_r^1 \omega(s)\big((1-s)v(s)\big)^{-1/p}\,ds}{\widehat{\omega}(r)} < \infty;$
(g) $\displaystyle\sup_{0<r<1} \frac{\omega(r)(1-r)^{\frac{1}{p'}}}{v(r)^{1/p}} \int_0^r \frac{v(s)^{\frac{1}{p}}}{\omega(s)(1-s)^{1+\frac{1}{p'}}}\,ds < \infty.$

It is worth noticing that condition (f) above makes sense also for $p = 1$, and it turns out to be the condition that describes those regular weights such that P_ω is bounded on L^1_v [28].

Theorem 4.5. *Let $\omega, v \in \mathbb{R}$. Then the following conditions are equivalent:*

(a) $P_\omega : L^1_v \to L^1_v$ *is bounded;*
(b) $P^+_\omega : L^1_v \to L^1_v$ *is bounded;*
(c) $\displaystyle\sup_{0<r<1} \frac{\omega(r)}{v(r)} \int_0^r \frac{\widehat{v}(s)}{\widehat{\omega}(s)(1-s)}\,ds < \infty;$
(d) $\displaystyle\sup_{0<r<1} \frac{\widehat{v}(r)}{\widehat{\omega}(r)} \int_r^1 \frac{\omega(s)}{v(s)(1-s)}\,ds < \infty.$

References

[1] A. Aleman and J. A. Cima, An integral operator on H^p and Hardy's inequality, *J. Anal. Math.* **85**, pp. 157–176 (2001).
[2] A. Aleman and O. Constantin, Spectra of integration operators on weighted Bergman spaces, *J. Anal. Math.* **109**, pp. 199–231 (2009).

[3] A. Aleman and J. Á. Peláez, Spectra of integration operators and weighted square functions, *Indiana Univ. Math. J.* **61**, pp. 775–793 (2012).

[4] A. Aleman and A. G. Siskakis, Integration operators on Bergman spaces, *Indiana Univ. Math. J.* **46**, pp. 337–356 (1997).

[5] M. Arsenovic, M. Jevtić and D. Vukotić, Taylor coefficients of analytic functions and coefficients multipliers. Preprint.

[6] D. Bekollé, Inégalité à poids pour le projecteur de Bergman dans la boule unité de $\mathbf{C}^n$, *Studia Math.* **71**, pp. 305–323 (1981/82).

[7] D. Békollé, Projections sur des espaces de fonctions holomorphes dans des domaines plans, *Canad. J. Math.* **38**, pp. 127–157 (1986).

[8] D. Bekollé and A. Bonami, Inégalités à poids pour le noyau de Bergman, *C. R. Acad. Sci. Paris Sér. A-B* **286**, pp. A775–A778 (1978).

[9] O. Constantin and J. Á. Peláez, Boundedness of the Bergman projection on L^p-spaces with exponential weights, *Bull. Sci. Math.* **139**, pp. 245–268 (2015).

[10] M. R. Dostanić, Unboundedness of the Bergman projections on L^p spaces with exponential weights, *Proc. Edinb. Math. Soc. (2)* **47**, pp. 111–117 (2004).

[11] M. Dostanić, Boundedness of the Bergman projections on L^p spaces with radial weights, *Publ. Inst. Math. (Beograd) (N.S.)* **86(100)**, pp. 5–20 (2009).

[12] P. Duren and A. Schuster, *Bergman spaces*, Mathematical Surveys and Monographs, Vol. 100, pp. x+318 (American Mathematical Society, Providence, RI, 2004).

[13] P. L. Duren, *Theory of H^p spaces*, Pure and Applied Mathematics, Vol. 38, pp. xii+258 (Academic Press, New York-London, 1970).

[14] C. Fefferman and E. M. Stein, H^p spaces of several variables, *Acta Math.* **129**, pp. 137–193 (1972).

[15] P. Galanopoulos, D. Girela, J. Á. Peláez and A. G. Siskakis, Generalized Hilbert operators, *Ann. Acad. Sci. Fenn. Math.* **39**, pp. 231–258 (2014).

[16] J. B. Garnett, *Bounded analytic functions*, Pure and Applied Mathematics, Vol. 96, pp. xvi+467 (Academic Press, Inc. [Harcourt Brace Jovanovich, Publishers], New York-London, 1981).

[17] H. Hedenmalm, B. Korenblum and K. Zhu, *Theory of Bergman spaces*, Graduate Texts in Mathematics, Vol. 199, pp. x+286 (Springer-Verlag, New York, 2000).

[18] T. L. Kriete, Laplace transform asymptotics, Bergman kernels and composition operators, in *Reproducing kernel spaces and applications*, Oper. Theory Adv. Appl., Vol. 143, pp. 255–272 (Birkhäuser, Basel, 2003).

[19] M. Mateljević and M. Pavlović, L^p-behavior of power series with positive coefficients and Hardy spaces, *Proc. Amer. Math. Soc.* **87**, pp. 309–316 (1983).

[20] M. Mateljević and M. Pavlović, L^p-behaviour of the integral means of analytic functions, *Studia Math.* **77**, pp. 219–237 (1984).

[21] M. Pavlović, Mixed norm spaces of analytic and harmonic functions. I, *Publ. Inst. Math. (Beograd) (N.S.)* **40(54)**, pp. 117–141 (1986).

[22] M. Pavlović, Mixed norm spaces of analytic and harmonic functions. II, *Publ. Inst. Math. (Beograd) (N.S.)* **41(55)**, pp. 97–110 (1987).

[23] M. Pavlović, *Introduction to function spaces on the disk*, Posebna Izdanja [Special Editions], Vol. 20, pp. vi+184 (Matematički Institut SANU, Belgrade, 2004).

[24] M. Pavlović and J. Á. Peláez, An equivalence for weighted integrals of an analytic function and its derivative, *Math. Nachr.* **281**, pp. 1612–1623 (2008).

[25] J. Á. Peláez and J. Rättyä, Generalized Hilbert operators on weighted Bergman spaces, *Adv. Math.* **240**, pp. 227–267 (2013).

[26] J. Á. Peláez and J. Rättyä, Weighted Bergman spaces induced by rapidly increasing weights, *Mem. Amer. Math. Soc.* **227**, pp. vi+124 (2014).

[27] J. Á. Peláez and J. Rättyä, Embedding theorems for Bergman spaces via harmonic analysis, *Math. Ann.* **362**, pp. 205–239 (2015).

[28] J. Á. Peláez and J. Rättyä, Two weight inequality for Bergman projection, *J. Math. Pures Appl. (9)* **105**, pp. 102–130 (2016).

[29] S. Pott and M. C. Reguera, Sharp Békollé estimates for the Bergman projection, *J. Funct. Anal.* **265**, pp. 3233–3244 (2013).

[30] A. G. Siskakis, Weighted integrals of analytic functions, *Acta Sci. Math. (Szeged)* **66**, pp. 651–664 (2000).

[31] Y. E. Zeytuncu, L^p regularity of some weighted Bergman projections on the unit disc, *Turkish J. Math.* **36**, pp. 386–394 (2012).

[32] Y. E. Zeytuncu, L^p regularity of weighted Bergman projections, *Trans. Amer. Math. Soc.* **365**, pp. 2959–2976 (2013).

[33] K. Zhu, *Operator theory in function spaces*, Mathematical Surveys and Monographs, Vol. 138, second edn., pp. xvi+348 (American Mathematical Society, Providence, RI, 2007).

PART B

Talks

Meanings of "algebra" and "analysis" between two Encyclopedias: From the Enlightenment to the Great War

L. Español

Departamento de Matemáticas y Computación,
Universidad de La Rioja, Logroño, La Rioja, 26004, Spain
E-mail: luis.espanol@unirioja.es

The meanings given to the terms "algebra" and "analysis", separately or jointly in expressions like "algebraic analysis", have changed over time, even simultaneously being used with significant differentiating shades.

I will focus on the period between the publication, from 1751 on, of the *Encyclopédie* edited by Diderot and D'Alembert, and the publication from 1899 on, of the *Encyklopädie der Mathematischen Wissenschaften mit Einschluss ihrer Anwendungen*, driven by F. Klein. The latter was translated into French with slight additions, under the direction of J. Molk, from 1904 until 1915, when the process was interrupted by the Great War (1914-1918), which also marks the end of the period covered in this paper.

Keywords: History of mathematics, algebra, analysis, algebraic analysis, 18th, 19th and 20th centuries.

Introduction

At the beginning of the 18th century, the appearance of the word "analysis" in a text on mathematics referred to the meaning of the term in the methodological duality analysis - synthesis (to loosen up - to put together) that goes back to classical Greek philosophy. Analysis means to break down a whole into components, or to proceed from effects to causes, and it is associated with the invention or the discovery. In the opposite sense, synthesis means to combine components in order to form a whole, or to proceed from causes to effects, and it is associated with the explanation or the justification (see Ritchey [22]).

The paradigm of the synthetic method in mathematics is Euclid's *Elements*, and its alternative analytical arrived with Descartes's *Géométrie* (1637). The analytic method of Descartes consists in the use of algebra. When Michel Rolle (1652-1719) called "analytic geometry" (1709)[a] to Descartes's geometry, he was

[a]This name was not widespread until Biot's *Essai de géométrie analytique*, 1802. Before, in 1788, Lagrange's *Mécanique analytique* had been published, whose title indicates the methodological difference with the work of Newton, strongly supported by Euclid's synthetic geometry.

pointing out the opposite way to Euclid.

Although Descartes already referred to the methodological distinction between analysis and synthesis (see the article by G. Israel [11]), we illustrate that distinction with a quote of a letter from Leibniz to H. Corning:

> Synthesis ... is the process in which we begin from principles and [proceed to] build up theorems and problems, ... while analysis is the process in which we begin with a given conclusion or proposed problem and seek the principles. (Ref. [17, p. 187])

In that period, algebra was the science of equations. The Arabic term "Al-jabr wal-muqãbala" means "reduction and confrontation" of the terms of an equation in order to solve it.

Algebraic methods led mathematicians of the Italian Renaissance to solve equations of degree no more than four in a general way, i.e. expressing roots depending on the coefficients. To understand what happens beyond this point was not possible until the 19th century. Alternatively, the numerical resolution of particular equations was developed, at first being based on the processes of bounding, separation and approximation of roots, associated respectively to Descartes, Rolle and Newton, among others. Both the numerical and universal resolutions were methods of analysis, not synthetic ones.

Years after Descartes used algebra to study curves, the Marquis de L'Hôpital (1661-1704) published *L'Analyse des Infiniment Petits pour l'intelligence des Lignes Courbes* (under the dictates of J. Bernoulli), in 1696. In this work the "analysis of the infinitesimals" is the algebraic calculus with these quantities that looked like numbers, but were not actually. The usual algebra only studies curves given by polynomials, and so it will continue, but the algebra of the infinitesimals also will reach curves given by other functions. Infinitesimals had a mathematical sense, but also a certain philosophical or metaphysical meaning. The following statement can be read in L'Hôpital's book: "A quantity that increases or decreases only by a quantity that can be considered infinitely small remains the same".

In Section 1, we will see that these schemes are kept during the 18th century, notably in the texts written by d'Alembert in the French *Encyclopédie*. The invention of analytic geometry and infinitesimal calculus make these new branches of mathematics main roads to innovation in this old science, while the synthetic geometry is having a minor role. The way to understand algebra also suffered changes throughout the eighteenth century, as we shall see in Section 1. The analysis went from being a term referring to a method that is confronted with the synthesis to a term which, after its identification as algebra, is turning into the algebra of the infinitesimals (L'Hôpital), which is different to the usual universal algebra (Descartes), the first one concerning general functions and the second one being limited to polynomials. Hence the algebra was for a time the analysis with polynomials. In this transit the "algebraic analysis" was conceived. Its nature and evolution will be explained in Section 2. In the early years of the 20th century, two ways of understanding the algebraic analysis co-exist. One of them lives on the undergraduate level of university education, and can be seen in text books. Another lives on a higher level, and was collected in the German *Encyklopädie der matematischen Wissenchaften.* Section 2 will end with a mention of the reception in Spain of the European algebraic analysis.

1. Algebra and analysis in the 18th century

The aim of this section is to offer a personal comment of the next quote by M. Kline:

> It will be helpful, in appreciating the work and arguments of the eighteenth-century thinkers, to kep in mind that they did not distinguish between algebra and analysis. Because they failed to recognize the problems introduced by the use of infinite series, they naively regarded the calculus as an extension of algebra. (Ref. [14, p. 401])

For glossing this assertion, we will look at the work of D'Alembert in the French *Encyclopédie ou Dictionnaire raisonné des sciences, des arts et des métiers.* Earlier, it should be pointed out that the algebraic calculus with infinitesimal "quantities" had dubious logical consistency, and this fact originated discussions on the "methaphysique of the infinitesimals", which were not exempt from ideological intentions. The first and notorious case was that of the Bishop George Berkeley (1685-1753) with *The analyst: or a discourse addressed to an infidel mathematician* (1734), in which the author showed the inconsistency of the calculus with infinitesimals, although he recognizes its efficiency in applications, what occurred, said Berkeley, by compensation of errors. The Bishop wanted to discredit the Newtonian rationalists. His criticism is very present in the prologue of Colin Maclaurin (1698-1746) to his *Treatise of fluxions* (1742), and forced the expository style to be maintained, as Newton did, on the synthetic-geometric line, leaving the analytic-algebraic style in a relegated position.

Shortly before the appearance of the *Encyclopédie,* two influential works referred to analysis as a methodology were published. Maria Gaetana Agnesi (1718-1799) wrote *Instituzioni analítiche ad use della gioventú Italiana*

(1748-49), and Gabriel Cramer (1704-1752), *Introduction à l'Analyse des Lignes Courbes Algébriques* (1750). The analysis is a method of work in both titles. Agnesi said that when it works with finite quantities, then the analysis is the "algebra of Cartesio". Cramer, on his part, uses analysis when studying curves given by polynomials.

1.1. *The French* Encyclopédie

Jean le Rhond D'Alembert (1717-1783) was an Enlightenment's man: mathematician, physicist, philosopher, theorist of music, etc. He was working on *Encyclopédie*'s mathematical contents during the period 1751-1759, regarding analysis and algebra as synonymous terms. He said that Cartesian geometry is "the application of the algebra or the analysis to the geometry", and that "analysis is properly the method of solving mathematical problems in reducing them to equations".

The classification of knowledge that is embodied in the *Encyclopédie*, inspired by Roger Bacon (1561-1626) and Chambers's *Cyclopaedia*, includes mathematics between the natural sciences, which are part of the knowledge that comes from the understanding through the reason and the philosophy. The natural sciences are divided into metaphysics of bodies or general physics, (that is: the extension, the infinitesimality, the movement, the vacuum, etc,), mathematics and particular physics. Mathematics is pure, mixed (mechanics, astronomy, optics, etc.[b]) and also physics-mathematics.

According to the *Encyclopédie* (see Novy [20]), the object of pure mathematics is the study of abstract quantity. The two parts of pure mathematics are the traditional arithmetic and geometry, corresponding to the two kinds of the abstract quantity: countable and continuous. Arithmetic can be numerical by numbers or algebra (universal arithmetics) by letters, the parts of the latter being:

$$\text{Algebra} \begin{cases} \text{Elementary} \\ \text{Infinitesimal} \begin{cases} \text{Differential} \\ \text{Integral} \end{cases} \end{cases}$$

Hence, "analysis" does not appear in this classification. All the problems in algebra must be solved by analysis, involving an algebraic calculus with letters which represent numbers or infinitesimals. On the other hand, the geometry

[b] Including the "analysis of random".

is elementary or transcendent. The latter is dedicated to the theory of curves, including the analytic study of them, like in Descartes or L'Hôpital works.

Infinitesimals appear in the *Encyclopédie* as quantities to be studied by the algebra, and also as a concept to be clarified by the metaphysics. This double character would remain until the 19th century, when its mathematical nature was rigorously determined by means of the concept of limit. This concept was already used in an incipient way by d'Alembert, who said that "the theory of limits is the basis of the true metaphysics of the infinitesimal calculus". It suffices to recall the work of Lazare Carnot (1753-1823) *Reflexions sur la métaphysique du calcul infinitésimal* (1797) to confirm the validity of these considerations throughout the 18th century.

The sense of "analysis" in the cited work of L'Hôpital appears at the end of the 18th century in the work of Gaspar Monge (1746-1818) *Sheets of analysis applied to geometry* (1798), devoted to the geometry of surfaces of first and second degree.[c] A sample of the identification between algebra and analysis is offered by Monge in the prologue of *Géométrie descriptive* (1798):

> It is not irrelevant that we compare here the descriptive geometry to the algebra; these two sciences have the most intimate relations. There is no construction of descriptive geometry that could not be translated into analysis; and when issues do not include more than three unknowns, each analytical operation may be regarded as the writing of a show in geometry.
> It would be desirable than those two sciences were cultivated together: geometry would contribute giving its characteristic evidence to the most complicated analytical operations, and, in turn, analysis would contribute giving its own generality to the geometry.[d]

1.2. *Two images of the algebra*

From its Arabic origin, and for many centuries, algebra denotes the manipulations to be performed with the parts of an equation to solve it. But over time this calculus with numbers or letters (specious or universal arithmetic) was extended to consider the same operations but with new entities, such as imaginaries, infinitesimals, etc. This vision of algebra as a "calculus of quantities"

[c] *Feuilles d'analyse appliquée à la Géométrie. Application de l'analyse à la géométrie des surfaces du premier et du deuxième degrée.*

[d] Translated from French by the author, like other quotes that appear later with the original in French or Spanish.

was present in authors such as Newton, D'Alembert, Euler and Lagrange. In the second half of the 18th century the algebra becomes more than solving equations, it emerges as a calculus of quantities that can be applied to the equations or other problems of mathematics and physics.[e] Leonhard Euler (1707-1783) worked with imaginary quantities, and described algebra as "the theory of the calculations with quantities of different kinds" in *Introduction to algebra* (1770). We will use determinants to illustrate the emergence of this modern image of the algebra.

Prior to receiving this name, determinants arose as algebraic expressions that appear in the resolution of the linear systems of small size and an equal number of equations than unknowns, carried out by traditional methods. In *Intoduction à l'Analyse des Lignes Courbes Algébriques* (1750), Cramer solves linear systems with unknowns "z, y, x, v, etc., & the same number of equations":

$$\begin{cases} A^1 = Z^1 z + Y^1 y + X^1 x + V^1 v + etc. \\ A^2 = Z^2 z + Y^2 y + X^2 x + V^2 v + etc. \\ A^3 = Z^3 z + Y^3 y + X^3 x + V^3 v + etc. \\ A^4 = Z^4 z + Y^4 y + X^4 x + V^4 v + etc. \\ \qquad\qquad\qquad etc. \end{cases}$$

In the case of three unknowns z, y, x, eliminating x, y (traditional algebra) it results:

$$z = \frac{A^1Y^2X^3 - A^1Y^3X^2 + A^2Y^1X^3 - A^2Y^3X^1 + A^3Y^1X^2 - A^3Y^2X^1}{Z^1Y^2X^3 - Z^1Y^3X^2 + Z^2Y^1X^3 - Z^2Y^3X^1 + Z^3Y^1X^2 - Z^3Y^2X^1},$$

then a similar expression $y = [\dots], x = [\dots]$ can be given for the eliminated unknowns. After a regularity in the formation of the expression of solutions was observed, a direct description of this regularity is sought through an algorithm with the coefficients of the system as input. In this quest, problems of combinatorics and symbolic language must be solved. Cramer proposed the solution to this problem for linear systems with any number of unknowns.[f]

The description of the solutions of the linear systems by quotients of determinants was perfected by Bézout (1764) and Laplace (1772). Bézout needed determinants for the theory of elimination, and Laplace for the integration of linear systems of differential equations.[g] Also in 1772, Alexandre-Théophile Van-

[e]This double aspect was analyzed by L. Novy [19].

[f]A few years earlier, the same task had been carried out by Colin Maclaurin (1698-1742) in his lectures, which were collected in *Treatise on algebra* (1748).

[g]Étienne Bézout (1730-1783), *Recherches sur le degré des équations résultants de l'évanouissement des inconnues.* Pierre-Simon de Laplace (1749-1827), *Recherches sur le calcul intégral et sur le sistème du monde.*

dermonde (1735-1796), in *Mémoire sur l'élimination*, established for the first time[h], a theory of determinants which had the resolution of linear systems as its first application.

We are going from Paris to Berlin, where a year later Joseph-Louis Lagrange (1736-1813) used those expressions which we now call determinants to solve a problem of geometry motivated by another of mechanics. He did it in two memoirs (1773):

> –*Nouvelle solution du problème du mouvement de rotation d'un corps de figure quelconque qui n'est animé par aucune force accélératrice.*
> –*Solutions analytiques de quelques problèmes sur les pyramides triangulaires.*

In *Nouvelle solution*, Lagrange calculates the volume of a triangular pyramid (an elementary rigid body) as a function of the lengths of its edges. In *Solutions analytiques* he goes back into this subject with much more detail, to pose an "algebraic method" which operates as well: from the analysis of the problem he extracted some autonomous calculations which then are applied to the initial problem, or to other problems. Lagrange's algebraic calculus is a symphony of determinants of order three, a very calculus with quantities, where quantities are determinants. He considered a pyramid with the vertex in the origin of coordinates, and a triangular base with vertices $M(x,y,z)$, $M'(x',y',z')$, $M''(x'',y'',z'')$. But he immediately forgets their geometrical meaning as coordinates and considers "any nine quantities" $x,y,z,x',y',z',x'',y'',z''$, which verify the equality

$$\begin{aligned}
&(xy'z''+yz'x''+zx'y''-xz'y''-yx'z''-zy'x'')^2\\
&=(x^2+y^2+z^2)(x'^2+y'^2+z'^2)(x''^2+y''^2+z''^2)\\
&\quad+2(xx'+yy'+zz')(xx''+yy''+zz'')(x'x''+y'y''+z'z'')\\
&\quad-(x^2+y^2+z^2)(x'x''+y'y''+z'z'')^2\\
&\quad-(x'^2+y'^2+z'^2)(xx''+yy''+zz'')^2\\
&\quad-(x''^2+y''^2+z''^2)(xx'+yy'+zz')^2,
\end{aligned}$$

and denotes $\Delta = xy'z''+yz'x''+zx'y''-xz'y''-yx'z''-zy'x''$.

In order to explain the always long Lagrange's expressions, we shall denote $[x]$ the above nine quantities set, and $\Delta = \Delta[x]$. Then he defines a new set $[\xi]$

[h]With a public character, because there was the theory developed by Leibniz, then unpublished. See E. Knobloch [15]. See the Spanish translation of Leibniz's manuscripts on determinants in Leibniz [18].

of nine quantities: $\xi,\eta,\zeta,\xi',\eta',\zeta',\xi'',\eta'',\zeta''$, given by $\xi = y'z'' - z'y''$, etc., and obtains the "very notable equation" $\Delta[\xi] = \Delta^2$. Lagrange also consider a set $[a]$ of six quantities: a, a', a'', b, b', b'', given by $a = x^2 + y^2 + z^2$, etc. With these six quantities, Lagrange form a new quantity

$$aa'a'' + 2bb'b'' - ab^2 - a'b'^2 - a''b''^2,$$

that we denote $\delta[a]$[i]. Then Lagrange proves that $\delta[a] = \Delta^2$.

In this way, determinants are quantities with its own algebra, and can be applied to solve linear systems, to calculate volumes, etc. The determinants began as a tool to solve linear systems, and became an object of study in itself, giving rise to a theory with its own applications. Similar processes, "from tool to object", have occurred repeatedly in the history of mathematics.

In the introduction to *Solutions analytiques*, Lagrange write that he can calculate the volume of the pyramid depending on their edges

> with the help of some quite remarkable reductions and transformations that I expose at the beginning of this memoir, and that also may be of the greatest use in many other cases. Regardless of the direct usefulness that these solutions may have on several occasions, they will serve mainly to show that algebraic method can be used with much ease and success in issues that seem to be left to the geometry itself.

Lagrange's "algebraic method" proceeds by "reductions and transformations" (algebra in its original meaning, universal algebra), carried out not with numbers but other kind of quantities, determinants in this case.

1.3. *Euler and Lagrange on analysis*

The analysis, with either finite or infinitesimal quantities, had been addressed to solve problems on curves or surfaces, but Euler considered infinitesimal algebra/analysis in itself, regardless of the geometry, giving rise to a theory of functions that would later have geometric or other applications.

While D'Alembert incorporated the mathematics to the *Encyclopédie*, Euler wrote the works in which he laid out his vision of the "infinitesimal algebra":

–*Introductio in analysis infinitorum* (1748)
–*Institutiones calculi differentialis / integralis* (1755/1768-74)

[i]Lagrange doesn't realize that $\delta = \Delta$ if the set $[a]$ is written as a nine quantities set in the form $a, b'', b', b'', a', b, b', b, a''$. The matrix notation would clarify this fact in the following century.

Euler's *Introductio* gathers the necessary knowledge to address the differential and integral calculus. It contains numbers, algebra, polynomials, and the functions given by analytical expressions, including expansion in power series, considered in a formal way despite the difficulties arising in numerical calculations.

Lagrange took the algebrization of functions to the extreme, avoiding the use of infinitesimals, that still were subject to metaphysical discussions. He considered functions directly given by their power series expansions in his famous works:

–*Théorie des fonctions analytiques* (1797)
–*Leçons sur le calcul des fonctions* (1801)

These first attempts to present a markedly algebraic theory of functions did not have a widespread effect, as they were modified by the requirements of rigor established in the nineteenth century. Its main features are summarized in this quote from Fraser [9]:

> The algebraic viewpoint of Euler and Lagrange is global... The idea behind the proof is always algebraic. It is invariably understood that the theorem in question is generally correct, true everywhere except possibly at isolated exceptional values. The failure of the theorem at such values is not considered significant. The primary fact, the meaning of the theorem, derives always from the underlying algebra. (Ref. [9, p. 328])

2. The algebraic analysis

Euler's *Introductio* started what later was called *algebraic analysis*, which is the algebraic treatment of functions proposed by Euler prior to the study of the differential and integral calculus.

Cauchy significantly changed the algebraic approaches of Euler and Lagrange, but previously these authors have had a strong influence on the educational reform held in Germany by von Humboldt. Around 1800, a trend among mathematicians, philosophers, and educators arose to overcome the linkage of mathematics to the intuition of space and time (applications to geometry and natural sciences). The objects of mathematics are purely mental constructions produced by a faculty of man's reason. From elementary arithmetic up to the binomial theorem, the whole field comprised a remarkable unity.

Algebraic analysis became the core of school teaching because it was seen as an elementary model of pure mathematics. This was the program implemented by the called Combinatorial School in Germany, where Carl F. Hindenburg (1739-1808) and Martin Ohm (1792-1872) played an important role (see Jahnke [12, 13]).

In his lectures at the École Polytechnique, Cauchy put aside the theory of functions of Lagrange and returned to Euler's approach, but substituting infinitesimals by limits. Hence the convergence and the continuous functions became the ground of his course with two levels:

$$\textit{Cours d'analyse} \begin{cases} \textit{Analyse algébrique}, 1821 \\ \textit{Calcul différentiel}, 1829 \end{cases}$$

Once settled this last approach, discussions on the "metaphysic of calculus" finished among mathematicians, but continued alive among philosophers of science.[j]

Since Euler, the study of functions was independent of geometry, and hence the analysis, referred to this subject, lost its connection with the synthesis, and grew up as the name of a field of mathematics. From Cauchy, the term "analysis" was identified with the content of his course, and by extension, with all the tremendous growth that the function theory, which was already synonymous of mathematical analysis, had during the nineteenth century, also incorporating the complex variable. The analytic-synthetic antithesis continued throughout the nineteenth century in the field of geometry, with the relevant novelty of projective geometry.

The algebraic analysis made by the German Combinatorial School joined Cauchy's point of view after the book by O. Schlomilch *Handbuch der algebraischen analysis* (1845), which had numerous issues during the second half of the nineteenth century. Then, when the arithmetization of the analysis was formulated, the step by step extensions of number domains, according to the "principle of permanence" introduced by H. Hankel in 1867 [10], became the core of the whole arithmetic-algebraic syllabus.

In the next pages we will see the persistence of algebraic analysis on the advanced level of mathematics and on the first level of university mathematical teaching.

---

[j]However, later in the nineteenth century, they became philosophical discussions about the theory of the infinite by Cantor.

2.1. *The German* Encyklopädie

The state of mathematics circa 1900 was presented by D. Struik with these words:

> When the twentieth century opened, mathematics was in flourishing condition, although creative mathematics was still in the main confined to one section of the world, was in most cases an academic profession, and was restricted, with few exceptions, to white males of European stock. The leading countries remained France and Germany. In France the center was Paris; in Germany, less centralized, it was Göttingen, with other universities such as Berlin running pretty close. (Ref. [23, p. 189])

Since 1885, Felix Klein (1849-1925) exerted from the University of Göttingen an influential activity to boost mathematics on the national and international levels, and at the spheres of research and education. By the end of the century, International Congress of Mathematicians were initiated[k], and, at the same time, appeared the *Encyklopädie der Mathematischen Wissenschaften mit Einschluss ihrer Anwendungen*[l] (*EMW*) (1898), which was promoted by Klein and published under the auspices of several German academies of sciences.[m]

The *EMW* was organized into tomes, volumes, and articles, with the participation of numerous authors. Its objectives included making available the great advances of the nineteenth century with a unified language, to cohere together mathematics itself and its applications, and to fight against esoteric superspecialization. It was a German project with international participation of English, Italian, and French mathematicians. But the Berlin school and its main subjects (algebra and number theory) were marginalized.

It had a French translation from 1904 on, entitled *Encyclopédie des sciences mathématiques pures et appliquées* (*ESM*). The original articles were translated until the French edition was interrupted in 1915 by the Great War. Each item had a French translator who expanded the article and was added as an author. Thus the references to French mathematicians increased, and also became a greater emphasis on historical comments. From now on we shall cite the French edition.

In this great collection of articles on nineteenth century mathematics, limits appear but not infinitesimals, and the alternative analysis/synthesis remains

[k]Zürich 1897, Paris 1900, Hamburg 1904, Rome 1908, Cambridge 1912, etc.

[l]*Encyclopedia of Mathematical Sciences with the inclusion of their applications.*

[m]Academies of Göttingen, Leipzig, München, Vienna sponsored, but not that of Berlin.

only in geometry, with the article by G. Fano and S. Carrus [8] entitled *Parallel exposure of the development of synthetic geometry and analytic geometry during the 19th century.*

In the *EMW/ESM*, mathematics are pure or applied (mechanics, physics, geodesy and astronomy). Parts of pure mathematics are no longer the traditional couple "arithmetic and geometry", but:

Pure mathematics { I. Arithmetic and algebra; II. Analysis; III. Geometry }

The "Tome II. Analysis" is divided into six volumes: 1– Functions of real variables. 2– Functions of complex variables. 3– Ordinary differential equations. 4– Partial differential equations. 5– Series expansions. 6– Calculus of variations. Complements.

We are interested in this particular content:

Vol. 2: Functions of complex variables { II-7. Algebraic analysis; II-8. Analytic functions }

In the article "Algebraic analysis" signed by A. Pringsheim, G. Faber, and J. Molk, the following definition is given:

> Under the name of algebraic analysis we can understand, nowadays, the study of algorithms unlimited of real or complex numbers, and the special methods for representing by using series, infinite products and continued fractions, the elementary functions. (Ref. [21, p. 4])

The authors state that the appropriate framework for this study is that of functions of complex variables, where functions of real variables should only be regarded as particular cases. They also note that separating this first chapter of analytic functions has advantages "at least from the point of view of teaching".

Let us note in passing that the following article on analytic functions was written in the original version by W. F. Osgood. Then P. Boutroux and J. Chazy were in charge of the French translation and additions, but their work was published only partially, until 1915, remaining unfinished. After the last page one can read: "The end of the article has not been published because of the war".

Pierre Léon Boutroux (1880-1922)[n] presented at the ICM Rome 1908 a communication entitled "On the relationship between algebra and mathematical analysis" [4], in which he noted the difficulties in distinguishing between algebra and analysis. The author began with this question: "What exactly is the

[n]French mathematician working also in history and philosophy of mathematics.

mathematical analysis, and how it differs from algebra?" (Ref. [4, p. 380]). After making a digression based on historical and contemporary aspects, Boutroux opts for an appreciation of the analysis as a science for the "understanding [...] of mathematical laws or correspondences", and of the algebra as a "science of the calculus", as an "instrument", because, he concludes: "like the physician makes physics with mathematics, the analyst makes analysis with algebra". (Ref. [4, p. 383]).

Now we return to the algebraic analysis. The reduction of its meaning by Pringsheim, Faber and Molk, to be confined as the elemental complex analysis, aimed to expand elementary functions by power series, is well understood from the point of view of advanced mathematics. Nevertheless the algebraic analysis had also an existence in elementary mathematics whose evolution we must find.

Pringsheim, Faber and Molk note that "in several recently published treatises it appears the tendency to significantly be limited the study to the elementary functions of real variable" (Ref. [21, p. 4]). They give as example the textbooks of E. Cesàro, *Corso di analisi algebrica* (1894), and of H. Burkhardt, *Algebraische analysis* (1903). These books are heirs of the algebraic analysis that started with Euler and continued with the German Combinatorial School, at the level of teaching mathematics in high school (gymnasium) and universities. This most basic algebraic analysis is also present in the German-French *Encyclopedia*, but its name is hidden: elementary algebraic analysis covers much of the Tome I – "Arithmetic and Algebra" of the *Encyclopedia*. In fact, this tome has four volumes:

1– Arithmetic. 2– Algebra. 3– Number theory. 4– Probability calculus. Error theory. Various applications.

Algebraic analysis is concerned with 1, 2, and the first article of 3, entitled "Elementary propositions of number theory".

To be complete, we include the articles of volumes 1 and 2 consecutively listed as divisions of Tome I:

Vol. 1. Arithmetic: 1– Fundamental principles of arithmetics. 2– Combinatorial analysis and determinant theory. 3– Irrational numbers and notion of limit. 4– Unlimited algorithms. 5– Complex numbers. 6– Unlimited algorithms of complex numbers. 7– Set theory. 8– On discontinuous finite groups.

Vol. 2. Algebra: 9– Rational functions. 10– General properties of fields and algebraic varieties. 11– Theory of forms and invariants.[o]

It must be noted that circa 1900 "fields" are not abstract fields but fields of numbers or rational functions.

Burkhardt's book focuses on the construction of number systems by the genetic method and unlimited algorithms with real and complex numbers, but the work of Cesàro is completed with other issues such as determinants and linear systems, the classical algebra of numerical resolution of algebraic equations, and also the algebraic resolution previous to Galois theory. Moreover, Cesàro, as well as other Italian authors of similar books in this period, laments having to include in the algebraic analysis the differential calculus, a subject that is not its own but is required by the official syllabus of the universities. In these books, algebra appears as the analysis restricted to polynomials, considered a particular kind of functions.

Perhaps the most complete book of this kind was that of de Italian A. Capelli, *Instituzioni di analisi algebrica* (1909), which had a first edition in 1902, and a joint work with Garbieri prior to it.[p] Capelli understands algebraic analysis in a broad sense, alongside traditional themes and differential calculus, also including algebraic functions and the theory of invariants. This tradition of textbooks on algebraic analysis ended in the years of the Great War, Capelli's book was his last great exponent.

During the war, in 1916, the Italian Beppo Levi published a textbook [16] that broke with the genetic tradition of the algebraic analysis by incorporating abstract algebra ("teorie formali", said Levi, referring to module theory and commutative algebra), something very new then. In the circumstances of his time, the book went unnoticed (see Álvarez *et al* [1]).

2.2. *Algebraic analysis in Spain*

Around 1900, the university teaching of mathematical analysis in Spain followed the traditional European scheme: first algebraic analysis and then infinitesimal calculus. The elementary algebraic analysis was taught during the first two university courses, in subjects called "Mathematical Analysis 1st and 2nd". In the third year the subject "Elements of infinitesimal calculus" arrived. Along with other books by French authors, that of the German Richard Baltzer (1818-1887), which had an Spanish translation [3], was very influential

---

[o]The article 11 is not always included in books of algebraic analysis.

[p]In his book above cited, Cesàro gave the Capelli-Garbieri as reference to the most basic questions of algebraic analysis he was not considering.

for Spanish authors of "treatises of mathematical analysis". When the young Julio Rey Pastor (1888-1962) won the chair in Madrid, a major renovation was operated, because he was trained in Germany (Berlin and Göttingen)[q], and also because Capelli's orientation was chosen by him to guide their lectures. However, Rey Pastor gave a remarkable personal style to his work.[r]

Rey Pastor wrote a first draft of their lessons during the academic years 1914-15 and 1915-16, from which three books were printed: *Elements of algebraic analysis* (*EAA*, 1917), *Real functions, Lectures on algebra* (1924)[s]. The last two books include topics which are characteristic of the algebraic analysis: in *Real functions*, expansions in power series of elementary functions (even with complex variable in Pringhsheim's way) are found, meanwhile *Lectures on algebra* covers classical resolution of algebraic equations, interpolation and elimination.

Taken together, these three books cover much of the content of the extensive *Instituzioni* of Capelli, but Rey Pastor reserved the name "algebraic analysis" only for his book addressed to the first university year. Therefore, he introduced a definition of the subject restricted in some way:

> The algebraic analysis is the study of the operations with numbers, and various algorithms that result from combining these, even algebraic and unlimited algorithms.

From a functional point of view, for Rey Pastor, the algebraic analysis is the mathematical analysis with only polynomials (and its quotients, the rational functions), and not with the more general algebraic functions as Capelli pointed out. Rey Pastor's algebraic algorithm is the divisibility of polynomials in one or several variables.

In the classic spirit of algebraic analysis, Rey Pastor organized his book following the genetic introduction of different kinds of numbers: natural, rational, real and complex ones, giving the definition of each from the previous one. Thus, the properties of the operations of addition and multiplication, and its existing inverses, stressing the principle of permanence of formal laws of the arithmetic. In addition, each type of numbers carries certain issues of its own, as shown below:

---

[q] This fact was very new in Spain in that time.

[r] See in Español [6] the biography of the young Rey Pastor. For his contribution to the algebraic analysis in Spain, see Español *et al.* [7] and Àlvarez *et al.* [2].

[s] In Spanish: *Elementos de análisis algebraico, Funciones reales, Lecciones de álgebra.* They had numerous editions.

Part I. The natural number.
Finite sets, divisibility, combinatory, permutation groups
Part II. The rational number.
Polynomial and divisibility, determinants and linear systems
Part III. The real number.
Limits, series, continuous fractions
Part IV. The complex number.
Limits, series, and the final theorem of arithmetic[t]

There is no place to the integer number in *EAA*. Indeed, negative numbers is introduced jointly with rational numbers. When Rey Pastor studied polynomials and determinants as topics owns of rational numbers, he notes that these theories remain also valid for real and complex numbers, because the field properties satisfied by the rational numbers are preserved when the system of numbers is extended. There are few specific topics of complex numbers in *EAA*, because it is a textbook for the first undergraduate year, namely (i) the expansion of elementary functions in power series (elementary analytic functions), and (ii) the fundamental theorem of algebra and solving equations (classical algebra). They were taught in the second undergraduate year, and were incorporated, respectively, to *Real functions* —despite its title—, and *Lectures on algebra.*

Rey Pastor was influenced by Burkhardt and above mentioned Italian authors, mainly Capelli. He also knew of the abstract work of Levy, but he did not follow it. On the one hand, he learned about Levy's book when he had passed two years developing its project on algebraic analysis with the above scheme and, on the other hand, he did not support the emerging abstract algebra, as it is shown in Ref. [5]. Thus, *EAA* was the last of this type of textbooks, but it had a lasting presence in Spain and Latin America.

Acknowledgements

I would like to acknowledge to the organizers of the VI CIDAMA 2014 for including a talk on history of mathematics in the program, and for inviting me to give it.

[t]It is the name given by Rey Pastor to the result saying that complex numbers are the unique extension of real numbers with commutative product.

References

[1] Y. Álvarez Polo and L. Español González, Álgebra en el libro de Análisis Matemático de Beppo Levi (1916), in *Actas del XI Congreso SEHCYT, 8-10 sep. 2011, Gipuzkoa*, ed. J. M. Urkía (Donostia, RSBAP, 2012).

[2] Y. Álvarez Polo and L. Español González, Algoritmos algebraicos lineales en el primer libro de texto (1917) de Julio Rey Pastor, *LLULL* **35**, pp. 13–36 (2012), https://documat.unirioja.es/descarga/articulo/3943896.pdf.

[3] R. Baltzer, *Elementos de Matemáticas, 5 vols.* (F. Góngora y Cía, Madrid, 1879–81), Translated by E. Jiménez and M. Melero from the first German Edition (Leipzig, 1860).

[4] P. Boutroux, Sur la relation de l'algébre à l'analyse mathématique, in *Atti del 4 Congresso Internazionale dei Matematici, Rome 1908*, (Nendeln/Leichtenstein, Kraus Reprint, 1967). Sexione IV: Questioni Filosofiche, Storiche, Didattiche.

[5] L. Español González, Julio Rey Pastor ante los cambios en el álgebra de su tiempo, in *Matemática y Región: La Rioja. Sobre matemáticos riojanos y matemática en La Rioja*,

[6] L. Español González, Julio Rey Pastor. Primeros años españoles: hasta 1920, *Gac. R. Soc. Mat. Esp.* **9**, pp. 546–585 (2006), http://gaceta.rsme.es/abrir.php?id=572.

[7] L. Español González, M. Á. Martínez García, Y. Álvarez Polo and C. Vela, Julio Rey Pastor y el análisis algebraico: de los apuntes de 1914-16 a tres libros de texto (1917-1925), *Zubía* **28**, pp. 139–166 (2010), https://documat.unirioja.es/descarga/articulo/3611411.pdf.

[8] G. Fano and S. Carrus, III.3 Exposé paralèlle du développement de la géométrie synthétique et de la géométrie analytique pendant le xixe siècle, in *Encyclopédie des sciences mathématiques pures et appliquées*, ed. J. Molk, (1), pp. 185–259 (Jacques Gabay, Reprinted from the 1904-16 Gauthier-Villars Edition, 1991). http://gallica.bnf.fr/ark:/12148/bpt6k29100t.

[9] C. G. Fraser, The calculus as algebraic analysis: some observations on mathematical analysis in the 18th century, *Arch. Hist. Exact Sci.* **39**, pp. 317–335 (1989).

[10] H. Hankel, *Theorie der complexen Zahlensysteme: insbesondere der gemeinen imaginären Zahlen und der Hamilton'schen Quaternionen, nebst ihrer geometrischen Darstellung*, Vorlsungen über die complexen Zahlen und Functionen, pp. 196 (Leopold Voss, Leipzig, 1867), https://books.google.es/books?id=MkttAAAAMAAJ.

[11] G. Israel, The analytical method in Descartes' *géométrie*, in *Analysis and synthesis in mathematics*, Boston Stud. Philos. Sci., Vol. 196, pp. 3–34 (Kluwer Acad. Publ., Dordrecht, 1997).

[12] H. N. Jahnke, Algebraic analysis in Germany, 1780–1840: some mathematical and philosophical issues, *Historia Math.* **20**, pp. 265–284 (1993).

[13] H. N. Jahnke, Algebraic analysis in the 18th century, in *A history of analysis*, Hist. Math., Vol. 24, pp. 105–136 (Amer. Math. Soc., Providence, RI, 2003).

[14] M. Kline, *Mathematical thought from ancient to modern times*, pp. xvii+1238 (Oxford University Press, New York, 1972).

[15] E. Knobloch, First European theory of determinants, in *Gottfried Wilhelm Leibniz: The Work of the Great Universal Scholar as Philosopher, Mathematician, Physicist,*

Engineer; [anlässlich Der Leibniz-Ausstellung 2000], eds. K. Popp and E. Stein, pp. 56–64 (Schlütersche, Hannover, 2000).

[16] B. Levi, *Introduzione alla analisi matematica*, pp. 482 (A. Hermann & Fils, 1916).

[17] L. E. Loemker, *G.W. Leibniz: Philosophical Papers and Letters. A Selection*, Synthese Historical Library, Vol. 2, 2 edn., pp. xii+736 (Springer Netherlands, 1989).

[18] M. S. de Mora Charles, *G.W. Leibniz: obras filosóficas y científicas*, Escritos Matemáticos, Vol. 7A, pp. 496 (Comares, 2015).

[19] L. Nový, *Origins of modern algebra*, pp. viii+252 (Noordhoff International Publishing, Leyden; Academia [Publishing House of the Czechoslovak Academy of Sciences], Prague, 1973), Translated from the Czech by Jaroslav Tauer.

[20] L. Nový, Las matemáticas en la enciclopedia de Diderot y d'Alembert, *Llull, Revista de la SEHCYT* **16**, pp. 265–284 (1993).

[21] A. Pringsheim and G. Faber, II.7 Analyse algébrique, in *Encyclopédie des sciences mathématiques pures et appliquées*, ed. J. Molk, (2), pp. 1–93 (Jacques Gabay, Reprinted from the 1911 Gauthier-Villars Edition, 1992). http://gallica.bnf.fr/ark:/12148/bpt6k202581m.

[22] T. Ritchey, Analysis and synthesis: On scientific method – based on a study by Bernhard Riemann, *Syst. Res.* **8**, pp. 21–41 (1991), 1996 revised version downloadable from the Swedish Morphological Society (http://www.swemorph.com/pdf /anaeng-r.pdf.

[23] D. J. Struik, *A concise history of mathematics*, fourth edn., pp. xiv+228 (Dover Publications, Inc., New York, 1987).

A weak 2-weight problem for the Poisson-Hermite semigroup

G. Garrigós

Departamento de Matemáticas, Universidad de Murcia,
30100 Murcia, Spain
E-mail: gustavo.garrigos@um.es
http://webs.um.es/gustavo.garrigos

This survey is a slightly extended version of the lecture given by the author at the *VI International Course of Mathematical Analysis in Andalucía* (CIDAMA), in September 2014. Most results form part of the paper [3], written jointly with S. Hartzstein, T. Signes, J.L. Torrea and B. Viviani.

Keywords: Hermite semigroup, Poisson integral, weighted inequalities, fractional laplacian.

1. Introduction

Consider the following integral identity

$$e^{-t\sqrt{L}} = \frac{t}{\sqrt{4\pi}} \int_0^\infty e^{-\frac{t^2}{4s}} e^{-sL} \frac{ds}{s^{3/2}}, \quad t > 0 \tag{1}$$

valid for all real numbers $L > 0$. If we allow L be the infinitesimal generator of a "heat" semigroup $\{e^{-sL}\}_{s>0}$ in $L^2(\mathbb{R}^d)$, then (1) defines, using the terminology in Stein's book [9, Chapter II.2], a *subordinated Poisson semigroup*. Moreover, for suitably "good" functions $f : \mathbb{R}^d \to \mathbb{C}$, the formal *Poisson integral* $u(t,\cdot) = e^{-t\sqrt{L}} f$ solves the partial differential equation

$$u_{tt} = Lu, \quad (t,x) \in (0,\infty) \times \mathbb{R}^d, \quad \text{with } u(0) = f.$$

A relevant question is then to find, for each operator L, the most general class of functions f for which the Poisson integrals $u(t,x) = e^{-t\sqrt{L}} f(x)$ satisfy

(i) $u(t,x)$ is well-defined and belongs to $C^\infty((0,\infty) \times \mathbb{R}^d)$,
(ii) $u(t,x)$ satisfies the pde $u_{tt} = Lu$ in $(0,\infty) \times \mathbb{R}^d$,
(iii) there exists $\lim_{t\to 0^+} u(t,x) = f(x)$, for a.e. $x \in \mathbb{R}^d$.

In the classical setting, corresponding to the Laplace operator $L = -\Delta$ in $\mathbb{R}^d$, the largest class of admissible initial data f is the weighted space

$$L^1(\varphi) = \Big\{ f : \int_{\mathbb{R}^d} |f(x)| \varphi(x)\, dx < \infty \Big\}, \tag{2}$$

with $\varphi(x) = (1+|x|)^{-(d+1)}$, and the assertions (i)-(iii) can easily be proved from the explicit form of the Poisson kernel.

For general operators L, however, the kernel will not be so explicit, and investigating such results requires very precise estimates of the subordinated integrals in (1), as well as of the associated maximal operators.

In this work we take up this question for a collection of Hermite operators in $\mathbb{R}^d$

$$L = -\Delta + |x|^2 + m, \quad \text{with } m \geq -d. \tag{3}$$

We shall also consider a slightly more general family of partial differential equations:

$$u_{tt} + \tfrac{1-2\nu}{t}\, u_t = Lu, \quad (t,x) \in (0,\infty) \times \mathbb{R}^d, \quad \text{with } \nu > 0. \tag{4}$$

The parameters m and ν allow us to include various interesting cases, which can all be covered with essentially the same estimates. In particular, $m = 0$ corresponds to the usual Hermite operator, while $m = -d$ leads to an operator L which can be transformed[a] into the Ornstein-Uhlenbeck operator $-\Delta + 2x \cdot \nabla$. Likewise, the parameter $\nu = 1/2$ in (4) gives the usual Poisson equation, while for general ν it leads to a pde appearing in the theory of *fractional laplacians*[b].

Our goal in this work is to give the most general conditions on a function $f : \mathbb{R}^d \to \mathbb{C}$ so that a meaningful solution to (4) is given by the *Poisson-like integral*

$$P_t f(x) := \tfrac{t^{2\nu}}{4^\nu \Gamma(\nu)} \int_0^\infty e^{-\frac{t^2}{4u}} \left[e^{-uL} f\right](x) \, \frac{du}{u^{1+\nu}}, \quad t > 0. \tag{5}$$

This subordinated integral is slightly more general than (1), and we justify its expression in §2 below. In our results, which we are about to state, the following function will play a crucial role

$$\varphi(y) = \begin{cases} \dfrac{e^{-|y|^2/2}}{(1+|y|)^{\frac{d+m}{2}} [\ln(e+|y|)]^{1+\nu}}, & \text{if } m > -d, \\[2ex] \dfrac{e^{-|y|^2/2}}{[\ln(e+|y|)]^{\nu}}, & \text{if } m = -d. \end{cases} \tag{6}$$

Theorem 1.1. *For every $f \in L^1(\varphi)$ the function $u(t,x) = P_t f(x)$ in (5) is defined by an absolutely convergent integral such that*

(i) $u(t,x) \in C^\infty((0,\infty) \times \mathbb{R}^d)$,

---

[a] Note that $e^{|x|^2/2} L[e^{-|x|^2/2} u] = -\Delta u + 2x \cdot \nabla u + (m+d)u$.

[b] The fractional operator L^ν can be recovered from (4) and $u(0,x) = f(x)$ by the formula $L^\nu f(x) = c_\nu \lim_{t \to 0} t^{1-2\nu} u_t(t,x)$, at least for suitably good f; see [10, Thm 1.1].

(ii) *$u(t,x)$ satisfies the pde* (4),

(iii) *For a.e. $x \in \mathbb{R}^d$, it holds* $\lim_{t\to 0^+} u(t,x) = f(x)$.

Conversely, if a function $f \geq 0$ is such that the integral in (5) *is finite for some* $(t,x) \in (0,\infty)\times\mathbb{R}^d$, *then f must necessarily belong to* $L^1(\varphi)$.

In particular, the function $f(y) = e^{\frac{|y|^2}{2}}/[(1+|y|)^d \ln(e+|y|)]$ has nicely convergent Poisson integrals, for all $m \geq -d$ and $\nu > 0$. This is in contrast with the classical case $L = -\Delta$ for which only a mild sublinear growth is allowed; see (2). It also illustrates that $L^1(\varphi)$ is strictly larger than the "gaussian" space $L^1(\mathbb{R}^d, e^{-\frac{|y|^2}{2}}dy)$, which was the natural domain for Poisson integrals considered by Muckenhoupt in [7] (in the special case $\nu = 1/2$, $m = -d$ and $d = 1$).

Our second goal is to investigate the following local maximal operators

$$P_a^* f(x) := \sup_{0<t<a} \left|P_t f(x)\right|, \quad \text{with } a > 0 \text{ fixed.} \tag{7}$$

These operators arise naturally in the a.e.-pointwise convergence of $P_t f(x) \to f(x)$ as $t \to 0$. In fact, the natural strategy to prove such convergence for all f in a Banach space $\mathbb{X}$, is to establish first the result in a dense class, and next prove the boundedness of P_a^* from $\mathbb{X}$ into $L^{p,\infty}(v)$ (or even better into $L^p(v)$) for some weight $v > 0$. It turns out that we can prove Theorem 1.1 without appeal to such maximal operators, but it still makes sense to consider the following

Problem 1.1. ***A weak 2-weight problem for the operator** P_a^*.* *Given $a > 0$ and $1 < p < \infty$, characterize the weights $w(x) > 0$ for which there exists some other weight $v(x) > 0$ such that*

$$P_a^* : L^p(w) \to L^p(v) \quad \textit{boundedly.} \tag{8}$$

We named the problem "weak" in contrast with the "strong" (and more difficult) question of characterizing all pairs of weights (w, v) for which (8) holds. Such weak 2-weight problems, for various classical operators, were considered in the early 80s by Rubio de Francia [8] and Carleson and Jones [1], who found explicit answers for the Hardy-Littlewood maximal operator and the Hilbert transform.

Our second main result in [3] gives an answer to Problem 1.1.

Theorem 1.2. *Let $1 < p < \infty$ and $a > 0$ be fixed. Then, for a weight $w(x) > 0$ the condition*

$$\left\| w^{-\frac{1}{p}} \varphi \right\|_{L^{p'}(\mathbb{R}^d)} < \infty \tag{9}$$

is equivalent to the existence of some other weight $v(x) > 0$ such that (8) *holds.*

Condition (9) is easily seen to be equivalent to $L^p(w) \subset L^1(\varphi)$. So, the necessity of (9) in Theorem 1.2 is a consequence of the last sentence in Theorem 1.1. Concerning the sufficiency of (9), we first point out that, from Theorem 1.1 (iii) and abstract results due to Nikishin, there always exists a weight $u(x) > 0$ such that

$$P_a^* : L^p(w) \to L^{p,\infty}(u) \quad \text{boundedly.} \tag{10}$$

The main contribution of Theorem 1.2 is to show that the weak-space $L^{p,\infty}(u)$ in (10) can be replaced by the strong space $L^p(v)$ (with perhaps another weight v). This is the main difficulty in the 2-weight Problem 1.1 described above, and requires additional estimates to those needed in Theorem 1.1.

A last question regards the explicit form of the weight $v(x)$, whose existence, under the condition (9), is asserted in Theorem 1.2. In [3] we used a non-constructive procedure which nevertheless provided a size estimate. Namely, for every $\sigma < 1$ a weight $v = v_\sigma$ can be chosen such that

$$\left\| v^{-\frac{\sigma}{p}} \varphi \right\|_{L^{p'}(\mathbb{R}^d)} < \infty. \tag{11}$$

Notice that this is "almost" the same integrability condition that $w(x)$ satisfies. Here we state a new result, which provides an explicit expression for $v(x)$, and recovers in particular the property (11). We shall use the following *local Hardy-Littlewood maximal operator*

$$\mathcal{M}^{\text{loc}} f(x) = \sup_{r>0} \frac{1}{|B_r|} \int_{B_r(x)} |f(y)|\, \chi_{\{|y| \le 3\max(|x|,1)\}}\, dy. \tag{12}$$

Theorem 1.3. *Let $1 < p < \infty$ be fixed, and let $w(x) > 0$ be a weight satisfying* (9). *Then a family of weights $v(x)$ such that* (8) *holds for all $a > 0$ is given explicitly by*

$$v(x) = \left[\mathcal{M}^{\text{loc}}\left(w^{-\frac{p'}{p}} e^{-\frac{p'|y|^2}{2}} \right)(x) \right]^{-\frac{\alpha}{p'/p}} e^{-\frac{p|x|^2}{2}} (1+|x|)^{-N}, \tag{13}$$

provided $\alpha > 1$ and $N > N_0$, for some $N_0 = N_0(\alpha, p, d, m, v)$.

We finally remark that, via the elementary identity

$$e^{|x|^2/2} L[e^{-|\cdot|^2/2} u] = -\Delta u + 2x \cdot \nabla u + (m+d)u =: \mathcal{O},$$

all the results in this paper admit corresponding statements with L replaced by the Ornstein-Uhlenbek type operator $\mathcal{O}$. These essentially amount to replace the exponentials $e^{-|y|^2/2}$ (as in (6) or (13)) by the gaussians $e^{-|y|^2}$. We leave the simple verification to the interested reader.

The proof of Theorems 1.1 and 1.2 was given in [3], but we outline the main steps below. Namely, in §2 we justify why the integral formula in (5) gives a solution to the pde (4). In §3 we state the optimal kernel estimates which are behind these theorems, and outline the proof of Theorem 1.1, slightly modified with respect to [3]. The new results appear in §4, where we solve a weak 2-weight problem for $\mathcal{M}^{\rm loc}$, and present the proof of Theorem 1.3 (which in turn implies Theorem 1.2).

2. The subordinated integral

For $\nu \in \mathbb{R}$, consider the following real-valued function

$$F_\nu(z) := \int_0^\infty e^{-u-\frac{z^2}{4u}}\, u^{\nu-1}\, du, \quad z>0, \tag{14}$$

where the integral is absolutely convergent (actually for all $\nu \in \mathbb{C}$ and $\mathrm{Re}(z^2) > 0$). This integral is well-known in the theory of special functions, as it appears in the definition of the so-called modified Bessel function of the third kind $K_\nu(z)$. Namely, they are related by

$$F_\nu(z) = 2\,(z/2)^\nu\, K_\nu(z); \tag{15}$$

see e.g. [12, p. 183] or [4, p. 119]. In particular, F_ν satisfies the ordinary differential equation

$$F_\nu''(z) + \frac{1-2\nu}{z} F_\nu'(z) \;=\; F_\nu(z). \tag{16}$$

We give next the elementary proof of (16), which does not depend on the properties of K_ν. Integrating by parts in (14) we can write

$$F_\nu(z) = \int_0^\infty e^{-u}\Big(e^{-\frac{z^2}{4u}}\, u^{\nu-1}\Big)'\, du = \int_0^\infty e^{-u}\big(\tfrac{z^2}{4u^2} + \tfrac{\nu-1}{u}\big)\, e^{-\frac{z^2}{4u}}\, u^{\nu-1}\, du.$$

Taking derivatives inside the integral in (14) we also have

$$F_\nu'(z) = \int_0^\infty e^{-u-\frac{z^2}{4u}}\,\big(-\tfrac{z}{2u}\big)\, u^{\nu-1}\, du$$

and

$$F_\nu''(z) = \int_0^\infty e^{-u-\frac{z^2}{4u}}\,\big(\tfrac{z^2}{4u^2} - \tfrac{1}{2u}\big)\, u^{\nu-1}\, du.$$

From these identities (16) follows easily. Moreover we have the following

Lemma 2.1. *Let ν and L be positive real numbers. Then, the function $u(t) = \frac{1}{\Gamma(\nu)} F_\nu(t\sqrt{L})$, $t>0$, satisfies the differential equation*

$$u''(t) + \frac{1-2\nu}{t}\, u'(t) \;=\; Lu(t), \quad \textit{with} \ \lim_{t\to 0^+} u(t) = 1. \tag{17}$$

Moreover, the function $u(t)$ can also be written as

$$u(t) = \frac{(t/2)^{2\nu}}{\Gamma(\nu)} \int_0^\infty e^{-\frac{t^2}{4v}-Lv} \frac{dv}{v^{1+\nu}}, \quad t>0. \tag{18}$$

Proof. If $\nu > 0$, from (14) and dominated convergence it follows that

$$\lim_{z\to 0^+} F_\nu(z) = \int_0^\infty e^{-u} u^{\nu-1}\,du = \Gamma(\nu).$$

It is then straightforward to derive (17) from this observation and (16). To obtain the integral expression in (18), first set $z^2 = t^2 L$ in (14), and then change variables $v = \frac{t^2}{4u}$. □

When L is a positive self-adjoint differential operator which generates a semigroup $\{e^{-sL}\}_{s>0}$ in $L^2(\mathbb{R}^d)$, we may then consider the function

$$u(t,x) = \tfrac{(t/2)^{2\nu}}{\Gamma(\nu)} \int_0^\infty e^{-\frac{t^2}{4v}} \left[e^{-vL} f\right](x) \frac{dv}{v^{1+\nu}}, \quad t>0.$$

In view of (17), this is a natural candidate to solve the pde

$$u_{tt} + \tfrac{1-2\nu}{t} u_t = Lu, \quad (t,x) \in (0,\infty)\times\mathbb{R}^d, \quad \text{with } u(0,\cdot) = f$$

(and coincides with the definition we used in (5) for the Poisson integral $P_t f(x)$ associated with L). Theorem 1.1 will give a rigorous proof of this formal statement, at least for the Hermite operators in (3). We refer to [10] for more on this kind of arguments for general operators L.

3. Estimates on the Poisson kernels

Suppose L is the infinitesimal generator of a semigroup of operators in $L^2(\mathbb{R}^d)$, say $\{h_t = e^{-tL}\}_{t>0}$, and that these are given by the integrals

$$h_t f(x) = \int_{\mathbb{R}^d} h_t(x,y) f(y)\,dy, \tag{19}$$

for suitable positive kernels $h_t(x,y)$. Then, the family of subordinated operators $\{P_t\}_{t>0}$ defined in (5) can be written as

$$P_t f(x) = \int_{\mathbb{R}^d} p_t(x,y) f(y)\,dy,$$

with the corresponding kernels given by the integrals

$$p_t(x,y) = \tfrac{(t/2)^{2\nu}}{\Gamma(\nu)} \int_0^\infty e^{-\frac{t^2}{4v}} h_v(x,y) \frac{dv}{v^{1+\nu}}. \tag{20}$$

If one is interested in *optimal* estimates for such kernels $p_t(x,y)$, two things become necessary: first, a precise *a priori* knowledge of $h_v(x,y)$, and next a careful analysis of the integrals (20).

Such tasks are difficult to carry in full generality, so in this work we have considered the special case of the Hermite operators $L = -\Delta + |x|^2 + m$, for which we can start with an *explicit* expression of the associated heat kernels

$$h_v(x,y) = e^{-vm}\,[2\pi\sinh(2v)]^{-\frac{d}{2}}\,e^{-\frac{|x-y|^2}{2\tanh(2v)}-\tanh(v)\,x\cdot y}, \quad v>0; \tag{21}$$

see e.g. [11, (4.3.14)][c]. To make this expression more manageable it is common to use the new variable $s = \tanh(v)$ (or equivalently $v = \frac{1}{2}\log(\frac{1+s}{1-s})$), which after elementary computations allows us to write

$$h_v(x,y) = \frac{(1-s)^{\frac{m+d}{2}}}{(1+s)^{\frac{m-d}{2}}}\;\frac{e^{-\frac{1}{4}(\frac{|x-y|^2}{s}+s|x+y|^2)}}{(4\pi s)^{\frac{d}{2}}}.$$

Inserting this into the integral (20) (with $dv = \frac{ds}{1-s^2}$) one obtains the expression

$$p_t(x,y) = \frac{(\frac{t}{2})^{2\nu}}{(4\pi)^{\frac{d}{2}}\Gamma(\nu)}\int_0^1 \frac{e^{-\frac{t^2}{2\ln\frac{1+s}{1-s}}}\,(1-s)^{\frac{m+d}{2}-1}\,e^{-\frac{1}{4}(\frac{|x-y|^2}{s}+s|x+y|^2)}}{s^{\frac{d}{2}}\,(1+s)^{\frac{m-d}{2}+1}\,\left(\frac{1}{2}\ln\frac{1+s}{1-s}\right)^{1+\nu}}\,ds. \tag{22}$$

This will be our starting formula for $p_t(x,y)$, from which we shall derive the necessary estimates needed for Theorems 1.1, 1.2 and 1.3. These are summarized in the next two lemmas.

The first one gives, for fixed t and x, the *optimal* decay of $y \mapsto p_t(x,y)$ in terms of the function $\varphi(y)$ defined in (6). We shall sketch its proof in §3.2 below.

Lemma 3.1. *Given $t>0$ and $x \in \mathbb{R}^d$, there exist $c_1(t,x)>0$ and $c_2(t,x)>0$ such that*

$$c_1(t,x)\,\varphi(y) \le p_t(x,y) \le c_2(t,x)\,\varphi(y), \quad \forall\, y\in\mathbb{R}^d. \tag{23}$$

The second lemma is a refinement of the upper bound in (23) with two main advantages: it is uniform in the variable t, and it restricts to the "local part" the singularities of the kernel $p_t(x,y)$. The proof of this more precise lemma is sketched in §3.3.

Lemma 3.2. *Given $x \in \mathbb{R}^d$, the following estimate holds for all $t>0$ and $y \in \mathbb{R}^d$*

$$p_t(x,y) \le \frac{C_1(x)\,t^{2\nu}\,e^{-\frac{|y|^2}{2}}}{(t+|x-y|)^{d+2\nu}}\,\chi_{\{|y|\le 3\max\{|x|,1\}\}} + C_2(x)\,t^{2\nu}\,\varphi(y), \tag{24}$$

for some positive functions $C_1(x) \lesssim (1+|x|)^{2\nu+d-1}e^{\frac{|x|^2}{2}}$ and $C_2(x) \lesssim 1/\varphi(x)$.

---

[c] Note that $e^{-vL} = e^{-v(-\Delta+|x|^2)}e^{-vm}$, and the case $m=0$ corresponds to the usual Mehler kernel.

Observe that, as a consequence of (24), we obtain the following bound for the maximal operators P_a^* in (7)

$$P_a^* f(x) \lesssim C_1(x)\, \mathcal{M}^{\mathrm{loc}}\big(f e^{-\frac{|y|^2}{2}}\big)(x) + C_2(x)\, a^{2\nu} \int_{\mathbb{R}^d} |f(y)|\, \varphi(y)\, dy, \tag{25}$$

where $\mathcal{M}^{\mathrm{loc}}$ denotes the local Hardy-Littlewood maximal operator defined in (12).

3.1. *Proof of Theorem 1.1*

Assuming the lemmas, we can sketch the proof of Theorem 1.1. First of all, it is a direct consequence of Lemma 3.1 that $P_t|f|(x) < \infty$ for some (or all) $t > 0$ and $x \in \mathbb{R}^d$ if and only if $f \in L^1(\varphi)$. This justifies that $f \in L^1(\varphi)$ is the right setting for this problem. Observe also that taking derivatives with respect to t in $p_t(x, y)$ slightly improves the decay of the kernel, and from here it is not difficult to deduce that (i) and (ii) must hold; see the details in [3, Proposition 4.4].

We shall be a bit more precise about the proof of (iii), that is the pointwise convergence

$$\lim_{t \to 0^+} P_t f(x) = f(x), \quad \text{a.e. } x \in \mathbb{R}^d \tag{26}$$

for all $f \in L^1(\varphi)$. We first claim that such convergence holds in the dense set $\mathcal{D} = \mathrm{span}\{h_{\mathbf{k}}\}_{\mathbf{k} \in \mathbb{N}^d}$, where $h_{\mathbf{k}}(x)$ denote the d-dimensional Hermite functions (as in [11, p.5]). These are eigenfunctions of $L = -\Delta + |x|^2 + m$ with

$$L h_{\mathbf{k}} = (2|\mathbf{k}| + d + m)\, h_{\mathbf{k}}, \quad \text{if } |\mathbf{k}| = k_1 + \ldots + k_d \geq 0;$$

see [11, (1.1.28)]. Recall that the operators $h_t = e^{-tL}$ from the Hermite semigroup can be represented in two ways: as in (19) with the Mehler kernel (21), or equivalently as

$$h_t f = \sum_{\mathbf{k} \in \mathbb{N}^d} e^{-(2|\mathbf{k}| + d + m)t} \langle f, h_{\mathbf{k}} \rangle\, h_{\mathbf{k}},$$

at least for $f \in \mathcal{D}$; see [11, (4.1.1)]. From this last formula and the results in §2 one also deduces that

$$P_t f = \frac{1}{\Gamma(\nu)} \sum_{\mathbf{k} \in \mathbb{N}^d} F_\nu\big(t\sqrt{2|\mathbf{k}| + d + m}\big) \langle f, h_{\mathbf{k}} \rangle\, h_{\mathbf{k}}, \quad f \in \mathcal{D}. \tag{27}$$

This clearly implies (26) when $f \in \mathcal{D}$.

To extend this convergence to all $f \in L^1(\varphi)$ we shall argue as in [3, Proposition 4.5]. Namely, it suffices to show (26) for a.e. $|x| \leq R$, for every fixed $R \geq 1$. We split $f \in L^1(\varphi)$ by

$$f = f\chi_{\{|y| \leq 3R\}} + f\chi_{\{|y| > 3R\}} = f_0 + f_1.$$

Using Lemma 3.2 we see that, for every $|x| \le R$

$$\begin{aligned} \big|P_t f_1(x)\big| &\le \int_{|y|>3R} p_t(x,y)\,|f(y)|\,dy \\ &\le C_R\, t^{2\nu} \int_{\mathbb{R}^d} |f(y)|\varphi(y)\,dy \to 0, \quad \text{as } t \to 0^+. \end{aligned} \tag{28}$$

On the other hand, Lemma 3.2 (or rather its consequence in (25)) also implies that

$$\sup_{0<t\le 1} |P_t f_0(x)| \le C_R\Big(M f_0(x) + \int |f_0|\varphi\Big), \quad |x| \le R,$$

where M denotes the usual Hardy-Littlewood maximal operator. Since the right-hand side is a bounded operator from $L^1(B_{3R}(0)) \to L^{1,\infty}(B_R(0))$, a classical procedure[d] then gives, from the validity of (26) in the dense class $\mathscr{D}$, the existence of $\lim_{t\to 0^+} P_t f_0(x) = f(x)$ for a.e $|x| \le R$. This completes the proof of Theorem 1.1. □

Remark 3.1. *Notice that the series representation of* $P_t f$ *in* (27) *allows us to reformulate* (26) *as a result on pointwise convergence of Hermite expansions by a "summability method" (based on the function* F_ν *and the parameter* m*). This is in the same spirit as the Poisson summability for Hermite expansions considered by Muckenhoupt in [7] (in the special case* $\nu = 1/2$, $m = -d$ *and* $d = 1$*). Notice, however, that the integral representation of* $P_t f(x)$ *in* (5) *is much more versatile, as it exists for functions in* $f \in L^1(\varphi)$ *which may have* $\langle f, h_{\mathbf{k}}\rangle = \infty$ *for all* $\mathbf{k}$.

3.2. *Proof of Lemma 3.1*

For the sake of originality, we shall use a slightly different approach than the one given in [3, Lemma 4.1]. In the Mehler kernel $h_v(x,y)$ we shall consider the "more natural" variable $r = e^{-2v}$ (or equivalently $v = \frac{1}{2}\ln\frac{1}{r}$), which leads to the formula[e]

$$h_v(x,y) = \frac{r^{\frac{m+d}{2}}\, e^{-\frac{|x-ry|^2}{1-r^2}}}{[\pi(1-r^2)]^{d/2}}\; e^{\frac{|x|^2-|y|^2}{2}}.$$

Inserting this into the integral defining $p_t(x,y)$ (with $dv = dr/(2r)$) one obtains the expression

$$p_t(x,y) = \frac{t^{2\nu}}{2^\nu \pi^{\frac{d}{2}}\Gamma(\nu)} \int_0^1 \frac{e^{-\frac{t^2}{2\ln\frac{1}{r}}}\, r^{\frac{m+d}{2}}\, e^{-\frac{|x-ry|^2}{1-r^2}}}{(1-r^2)^{\frac{d}{2}}\left(\ln\frac{1}{r}\right)^{1+\nu}}\,\frac{dr}{r}\; e^{\frac{|x|^2-|y|^2}{2}}. \tag{29}$$

---

[d]See e.g. [2, Theorem 2.2].

[e]This change of variables is common in the Ornstein-Uhlenbeck setting; see e.g. [7, (3.3)] or [6].

Starting from this formula, we now argue as in [3, Lemma 4.1]. We may assume that $|y| \geq 3\max\{1,|x|\}$ (since in the region $|y| \leq 3\max\{1,|x|\}$ one can bound $p_t(x,y)/\varphi(y)$ above and below by positive functions of t,x).

The main difficulty is to determine the values of r which carry the main contribution of the integral in (29). The leading term will be the exponential in the numerator, and as we shall see, it becomes largest when $r \approx |x|/|y|$.

Consider first the region $1/2 < r < 1$. Since $|y| \geq 3|x|$, we have

$$|ry - x| \geq |y|/2 - |x| \geq |y|/6 \quad \Longrightarrow \quad e^{-\frac{|x-ry|^2}{1-r^2}} \leq e^{-\frac{|y|^2}{36}},$$

so the leading exponential becomes quite small in this part. Since we also have

$$\ln\frac{1}{r} \approx 1-r, \quad r \in [1/2,1], \tag{30}$$

we can estimate the corresponding integral in (29) by

$$\begin{aligned} \int_{1/2}^{1} \cdots &\lesssim e^{\frac{|x|^2-|y|^2}{2}} e^{-\frac{|y|^2}{36}} t^{2\nu} \int_{1/2}^{1} \frac{e^{-\frac{ct^2}{1-r}}}{(1-r)^{\frac{d}{2}+\nu+1}}\,dr \qquad [u = \tfrac{t^2}{1-r}] \\ &\leq e^{\frac{|x|^2-|y|^2}{2}} e^{-\frac{|y|^2}{36}} t^{-d} \int_0^\infty e^{-cu} u^{\frac{d}{2}+\nu-1}\,du \\ &\lesssim e^{\frac{|x|^2}{2}} t^{-d} \varphi(y). \end{aligned} \tag{31}$$

Suppose now that $0 < r < \frac{1}{2}$. As we shall see, the main contribution occurs here, precisely when $r \approx \frac{|x|}{|y|}$. We first consider the range $2\frac{|x|}{|y|} < r < 1/2$, where we can estimate

$$|x - ry| \geq r|y| - |x| \geq r|y|/2 \quad \Longrightarrow \quad e^{-\frac{|x-ry|^2}{1-r^2}} \leq e^{-\frac{r^2|y|^2}{4}}.$$

Thus

$$\begin{aligned} \int_{2\frac{|x|}{|y|}}^{1/2} \cdots \quad &\lesssim e^{\frac{|x|^2-|y|^2}{2}} t^{2\nu} \int_{2\frac{|x|}{|y|}}^{1/2} \frac{r^{\frac{m+d}{2}} e^{-\frac{r^2|y|^2}{4}}}{\left(\ln\frac{1}{r}\right)^{1+\nu}} \frac{dr}{r} \\ {\scriptstyle [u=r|y|]} \quad &\leq t^{2\nu} e^{\frac{|x|^2-|y|^2}{2}} |y|^{-\frac{m+d}{2}} \int_0^{\frac{|y|}{2}} \frac{u^{\frac{m+d}{2}} e^{-u^2/4}}{\left(\ln\frac{|y|}{u}\right)^{1+\nu}} \frac{du}{u}. \end{aligned}$$

Note that when $m + d > 0$ the last integral[f] has its major contribution at $u \approx 1$, so it can be estimated by $1/(\ln|y|)^{1+\nu}$. Thus, overall one obtains

$$\int_{2\frac{|x|}{|y|}}^{1/2} \cdots \lesssim t^{2\nu} e^{\frac{|x|^2}{2}} \varphi(y). \tag{32}$$

---

[f]When $m = -d$, the integral still converges and is controlled by $1/(\ln|y|)^{\nu}$.

Finally, in the range $0 < r < 2\frac{|x|}{|y|}$ we disregard the exponential $e^{-\frac{|x-ry|^2}{1-r^2}}$ in (29) to obtain

$$\int_0^{2\frac{|x|}{|y|}} \dots \lesssim t^{2\nu} e^{\frac{|x|^2-|y|^2}{2}} \int_0^{2\frac{|x|}{|y|}} \frac{r^{\frac{m+d}{2}}}{\left(\ln\frac{1}{r}\right)^{1+\nu}} \frac{dr}{r}.$$

This last integral can be estimated by $(|x|/|y|)^{\frac{d+m}{2}}/[\ln(|y|/|x|)]^{1+\nu}$ when[g] $m + d > 0$. Now, using elementary bounds on logarithms (see Lemma A.1 in the Appendix) one concludes that

$$\int_0^{2\frac{|x|}{|y|}} \dots \lesssim t^{2\nu} e^{\frac{|x|^2}{2}} |x|^{\frac{d+m}{2}} [\ln(|x|+e)]^{1+\nu} \varphi(y). \tag{33}$$

Combining (31), (32) and (33) one obtains the upper bound in (23).

To establish the lower bound, it suffices to integrate in the range $0 < r \le 2\frac{|x|}{|y|}$. Note that $|y| \ge 3|x|$ also implies $r \le 2/3$, so we obtain

$$|x - ry| \le |x| + r|y| \le 3|x| \quad \Longrightarrow \quad e^{-\frac{|x-ry|^2}{1-r^2}} \ge e^{-18|x|^2}$$

(using $1 - r^2 \ge 1/2$). The first exponential in (29) can be handled simply by

$$\exp\left(-\frac{t^2}{2\ln\frac{1}{r}}\right) \ge \exp\left(-\frac{t^2}{2\ln(3/2)}\right), \quad r \in [0, \tfrac{2}{3}],$$

so all together we conclude that

$$\begin{aligned} p_t(x,y) &\gtrsim t^{2\nu} e^{-ct^2} e^{-18|x|^2-\frac{|y|^2}{2}} \int_0^{2\frac{|x|}{|y|}} \frac{r^{\frac{m+d}{2}}}{\left(\ln\frac{1}{r}\right)^{1+\nu}} \frac{dr}{r} \\ &\approx t^{2\nu} e^{-ct^2} e^{-18|x|^2-\frac{|y|^2}{2}} \frac{(|x|/|y|)^{\frac{m+d}{2}}}{[\ln(|y|/|x|)]^{1+\nu}} \gtrsim c_1(t,x)\varphi(y), \end{aligned}$$

for some positive function $c_1(t,x)$. □

3.3. *Proof of Lemma 3.2*

We split the integral defining $p_t(x,y)$, as

$$p_t(x,y) = \int_0^{\frac{1}{2}} \dots + \int_{\frac{1}{2}}^1 \dots \le I_0 + I_1.$$

The singularity of kernel lies in the first piece I_0, and in order to find a good estimate it will be crucial to use the formula in (22)[h]. Suppose we are in the

[g] When $m = -d$, the integral is bounded by $1/[\ln(|y|/|x|)]^{\nu}$.

[h] The formula for $p_t(x,y)$ in (29) does not make so explicit the term $|x-y|$ in the leading exponential.

local region $|y| \leq 3\max\{|x|,1\}$. Then, using (30), we can estimate I_0 by

$$\begin{aligned} I_0 &\lesssim t^{2\nu} \int_0^{\frac{1}{2}} \frac{e^{-\frac{ct^2+|x-y|^2}{4s}}\, e^{-\frac{s|x+y|^2}{4}}}{s^{\frac{d}{2}+1+\nu}}\, ds \\ &\approx \frac{t^{2\nu}}{(ct^2+|x-y|^2)^{\frac{d}{2}+\nu}} \int_{\frac{ct^2+|x-y|^2}{2}}^{\infty} e^{-u}\, e^{-\frac{(ct^2+|x-y|^2)|x+y|^2}{16u}}\, u^{\frac{d}{2}+\nu-1}\, du, \end{aligned} \tag{34}$$

where we have changed variables $u = (ct^2+|x-y|^2)/(4s)$. In the last integral we can disregard t in the exponential, and overall estimate it crudely by

$$J := \int_0^{\infty} e^{-u}\, e^{-\frac{(|x-y||x+y|)^2}{16u}}\, u^{\frac{d}{2}+\nu-1}\, du = F_{\frac{d}{2}+\nu}\Big(\frac{|x-y||x+y|}{2}\Big),$$

where $F_\sigma(z)$ was defined in (14). As we noticed in (15) we can write

$$F_\sigma(z) = 2^{1-\sigma} z^\sigma K_\sigma(z) \lesssim (1+z)^{\sigma-\frac{1}{2}} e^{-z}, \quad z > 0,$$

by the standard asymptotics of K_σ; see e.g. [4, p. 136]. Thus, we obtain

$$J \lesssim (1+|x-y||x+y|)^{\nu+\frac{d-1}{2}}\, e^{-\frac{|x-y||x+y|}{2}}.$$

Now, $|x-y||x+y| \geq |\langle x+y, x-y\rangle| \geq -|x|^2+|y|^2$, so in the region $|y| \leq 3\max\{|x|,1\}$ we have

$$J \lesssim e^{\frac{|x|^2}{2}}\, e^{-\frac{|y|^2}{2}}\, (1+|x+y|\,|x-y|)^{\nu+\frac{d-1}{2}} \lesssim (1+|x|)^{2\nu+d-1}\, e^{\frac{|x|^2}{2}}\, e^{-\frac{|y|^2}{2}}.$$

Inserting this into (34) we obtain the bound for the local part asserted in the statement of the lemma.

The estimate of I_0 when $|y| \geq 3\max\{|x|,1\}$ is much better, of the order $I_0 \lesssim t^{2\nu} e^{-(\frac{1}{2}+\gamma)|y|^2}$ for some $\gamma > 0$; see [3, Lemma 4.2] for details. Likewise, from the arguments we already used in Lemma 3.1 one obtains a bound for $I_1 \leq C_2(x) t^{2\nu} \varphi(y)$ with $C_2(x) \lesssim 1/\varphi(x)$. We again refer to [3] for details. This completes the proof of Lemma 3.2. □

4. Proof of Theorems 1.2 and 1.3

As we mentioned in the introduction, it suffices to give a proof of Theorem 1.3. That is, assuming $w \in D_p(\varphi)$, by which we mean

$$\|w\|_{D_p(\varphi)} := \big\| w^{-\frac{1}{p}} \varphi \big\|_{L^{p'}(\mathbb{R}^d)} < \infty,$$

we must show that the weights $v(x)$ defined in (13) are such that P_a^* maps $L^p(w) \to L^p(v)$ boundedly, for all $a > 0$. We shall use the bound for P_a^* in (25), namely

$$\begin{aligned} P_a^* f(x) &\lesssim C_1(x)\, \mathcal{M}^{\mathrm{loc}}\big(f e^{-\frac{|y|^2}{2}}\big)(x) + C_2(x)\, a^{2\nu} \int_{\mathbb{R}^d} |f(y)|\varphi(y)\, dy \\ &= I(x) + II(x), \end{aligned} \tag{35}$$

with $C_1(x)$ and $C_2(x)$ given explicitly in Lemma 3.2. We treat first the last term, which by Hölder's inequality is bounded by

$$II(x) \le C_2(x)\, a^{2\nu}\, \|f\|_{L^p(w)}\, \|w\|_{D_p(\varphi)}\,.$$

So using $C_2(x) = 1/\varphi(x)$, we will have

$$\|II\|_{L^p(v)} \le a^{2\nu}\, \|w\|_{D_p(\varphi)}\, \|f\|_{L^p(w)} \left[\int_{\mathbb{R}^d} \frac{v(x)}{\varphi(x)^p}\, dx\right]^{\frac{1}{p}} \tag{36}$$

with the last integral being a finite expression provided we choose

$$v(x) \le v_1(x) := \frac{e^{-\frac{p}{2}|x|^2}}{(1+|x|)^M} \tag{37}$$

for any $M > N_1 = d + p(d+m)/2$. We remark that the weights $v(x)$ in (13) have this property, at least if N is sufficiently large. This is a consequence of the following elementary lemma.

Lemma 4.1. *If $f \in L^1_{\text{loc}}(\mathbb{R}^d)$ satisfies $f(x) > 0$, a.e. $x \in \mathbb{R}^d$, then*

$$\mathcal{M}^{loc} f(x) \ge c_f\,(1+|x|)^{-d}, \quad \forall\, x \in \mathbb{R}^d, \tag{38}$$

with $c_f = \frac{1}{|B_1|}\int_{B_1(0)} f > 0$.

Proof. Choosing $r = 1+|x|$, one trivially has $B_1(0) \subset B_r(x) \cap \{|y| \le 3\max(|x|,1)\}$, so (38) is immediate from the definition of $\mathcal{M}^{\text{loc}} f(x)$ in (12). □

Now, using the lemma, one sees that the weights $v(x)$ defined in (13) satisfy

$$v(x) \lesssim c'_w\,(1+|x|)^{d\alpha(p-1)}\, e^{-\frac{p}{2}|x|^2}\,(1+|x|)^{-N},$$

hence choosing $N > N_1 + d\alpha(p-1)$ ensures that (37) holds.

We now turn to main part $I(x)$ in (35). The following proposition will be crucial. The result is new, and the proof is based on arguments due to Carleson and Jones (see [1]).

Proposition 4.1. *Let $1 < p < \infty$ and $w(x) > 0$ such that $w^{-\frac{1}{p-1}} \in L^1_{\text{loc}}(\mathbb{R}^d)$. Define*

$$V_\alpha(x) := \frac{\left[\mathcal{M}^{loc}(w^{-\frac{1}{p-1}})(x)\right]^{-(p-1)\alpha}}{(1+|x|)^{(p-1)d\alpha}}, \quad \text{for } \alpha > 1. \tag{39}$$

Then

$$\mathcal{M}^{loc} : L^p(w) \to L^p(V_\alpha) \quad \text{boundedly.}$$

Moreover, given $\sigma < 1$, if we choose $1 < \alpha < 1/\sigma$, then we also have $V_\alpha^{-\frac{\sigma}{p-1}} \in L^1_{\text{loc}}(\mathbb{R}^d)$.

Proof. Note that $\mathcal{M}^{\mathrm{loc}}(w^{-\frac{1}{p-1}})(x) < \infty$, a.e. $x \in \mathbb{R}^d$, by the assumption $w^{-\frac{1}{p-1}} \in L^1_{\mathrm{loc}}$. This, together with Lemma 4.1, imply that

$$0 < V_\alpha(x) \le c_w, \quad \text{a.e. } x \in \mathbb{R}^d.$$

Now call $E_n = \{x \in \mathbb{R}^d : \mathcal{M}^{\mathrm{loc}}(w^{-\frac{1}{p-1}})(x) < 2^n\}$, $n = 0, 1, 2, \ldots$, and define the operators

$$T_n g(x) := \chi_{E_n} \mathcal{M}^{\mathrm{loc}}(w^{-\frac{1}{p-1}} g)(x). \tag{40}$$

Note that $T_n : L^1(w^{-\frac{1}{p-1}}) \to L^{1,\infty}(\mathbb{R}^d)$, with a uniform bound in n; in fact

$$\left|\left\{T_n g(x) > \lambda\right\}\right| \le \left|\left\{M(w^{-\frac{1}{p-1}} g)(x) > \lambda\right\}\right| \le \frac{c_0}{\lambda} \int_{\mathbb{R}^d} w^{-\frac{1}{p-1}} |g|. \tag{41}$$

Similarly, $T_n : L^\infty(w^{-\frac{1}{p-1}}) \to L^\infty(\mathbb{R}^d)$ with $\|T_n\| \le 2^n$, since

$$\|T_n g\|_\infty = \sup_{x \in E_n} \left|\mathcal{M}^{\mathrm{loc}}(w^{-\frac{1}{p-1}} g)(x)\right| \le 2^n \|g\|_\infty. \tag{42}$$

Thus, by Marcinkiewicz interpolation theorem we obtain

$$\int_{E_n} |T_n(g)|^p \le c_0 2^{\frac{np}{p'}} \int_{\mathbb{R}^d} |g|^p w^{-\frac{1}{p-1}}. \tag{43}$$

Setting $g = f w^{\frac{1}{p-1}}$ in the above inequality, this is the same as

$$\int_{E_n} |\mathcal{M}^{\mathrm{loc}}(f)|^p \le c_0 2^{n(p-1)} \int_{\mathbb{R}^d} |f|^p w. \tag{44}$$

Now, writing $\mathbb{R}^d = E_0 \cup [\cup_{n \ge 1} E_n \setminus E_{n-1}]$, and using

$$V_\alpha(x) \le c_w \text{ in } E_0, \quad \text{and} \quad V_\alpha(x) \le 2^{-(n-1)(p-1)\alpha} \text{ if } x \notin E_{n-1},$$

we obtain

$$\begin{aligned}
\int_{\mathbb{R}^d} |\mathcal{M}^{\mathrm{loc}} f|^p V_\alpha &\le c_w \int_{E_0} |\mathcal{M}^{\mathrm{loc}} f|^p + \sum_{n=1}^{\infty} 2^{-(n-1)(p-1)\alpha} \int_{E_n} |\mathcal{M}^{\mathrm{loc}} f|^p \\
{\scriptstyle (\text{by } (44))} &\le c_w c_0 \int_{\mathbb{R}^d} |f|^p w + 2^{(p-1)\alpha} \sum_{n=1}^{\infty} 2^{-n(\alpha-1)(p-1)} \int_{\mathbb{R}^d} |f|^p w \\
&\lesssim \int_{\mathbb{R}^d} |f|^p w,
\end{aligned}$$

as we wished to show. Finally note that, when $\alpha\sigma < 1$, the classical Kolmogorov inequality gives

$$V_\alpha^{-\frac{\sigma}{p-1}}(x) = \left[\mathcal{M}^{\mathrm{loc}}(w^{-\frac{1}{p-1}})(x)\right]^{\alpha\sigma} (1 + |x|)^{d\alpha\sigma} \in L^1_{\mathrm{loc}}(\mathbb{R}^d). \qquad \square$$

Remark 4.1. *The condition* $w^{-\frac{1}{p-1}} \in L^1_{\rm loc}(\mathbb{R}^d)$ *actually characterizes the property that* $\mathcal{M}^{\rm loc}$ *maps* $L^p(w) \to L^p(V)$ *for some weight* $V(x) > 0$. *We sketch a proof of the converse in Proposition A.1 below.*

Below we shall need a refinement of Proposition 4.1, which we state now.

Proposition 4.2. *In the conditions of Proposition 4.1, if* $w(x)$ *additionally satisfies*

$$\int_{\mathbb{R}^d} w^{-\frac{1}{p-1}}(x)\, e^{-a|x|^2}\, dx < \infty, \quad \forall\, a > 0, \tag{45}$$

then, for every $\sigma < 1$ *and* $1 < \alpha < 1/\sigma$, *the weight* $V_\alpha(x)$ *defined in* (39) *also satisfies*

$$\int_{\mathbb{R}^d} V_\alpha^{-\frac{\sigma}{p-1}}(x)\, e^{-b|x|^2}\, dx < \infty, \quad \forall\, b > 0. \tag{46}$$

Proof. Note that (46) is true if we only integrate in $B_1(0)$. So, we shall consider $S_j = \{2^j \le |x| < 2^{j+1}\}$, $j = 0, 1, 2, \dots$ Call $s = \alpha\sigma < 1$ and take any $b > 0$. Then

$$\begin{aligned}
I &= \sum_{j=0}^{\infty} \int_{S_j} \left|\mathcal{M}^{\rm loc}(w^{-\frac{1}{p-1}})(x)\right|^s (1+|x|)^{ds}\, e^{-b|x|^2}\, dx \\
&\lesssim \sum_{j=0}^{\infty} 2^{jds}\, e^{-b4^j} \int_{S_j} \left|\mathcal{M}^{\rm loc}\big(w^{-\frac{1}{p-1}} \chi_{B_{3\cdot 2^{j+1}}(0)}\big)(x)\right|^s dx \\
\text{by Kolmogorov ineq} &\lesssim \sum_{j=0}^{\infty} 2^{jds}\, e^{-b4^j}\, |S_j|^{1-s} \left\| M\big(w^{-\frac{1}{p-1}} \chi_{B_{3\cdot 2^{j+1}}(0)}\big)(x)\right\|^s_{L^{1,\infty}(\mathbb{R}^d)} \\
&\lesssim \sum_{j=0}^{\infty} 2^{jd}\, e^{-b4^j} \left[\int_{B_{3\cdot 2^{j+1}}(0)} w^{-\frac{1}{p-1}}(x) e^{a|x|^2} e^{-a|x|^2}\, dx\right]^s \\
&\lesssim \sum_{j=0}^{\infty} 2^{jd}\, e^{-b4^j}\, e^{6^2 as4^j} \left[\int_{\mathbb{R}^d} w^{-\frac{1}{p-1}}(x) e^{-a|x|^2}\, dx\right]^s,
\end{aligned}$$

and this is finite if we choose $a < b/(36s)$. □

4.1. *Conclusion of the proof of Theorem 1.3*

Suppose now that $w \in D_p(\varphi)$, that is $\int w^{-\frac{1}{p-1}}(x)\varphi(x)^{p'}\, dx < \infty$, for φ as in (6). This implies that $W(x) = w(x) e^{\frac{p}{2}|x|^2}$ satisfies

$$\int_{\mathbb{R}^d} W^{-\frac{1}{p-1}}(x)\, e^{-a|x|^2}\, dx = \int_{\mathbb{R}^d} w^{-\frac{1}{p-1}}(x)\, e^{-\frac{p'}{2}|x|^2}\, e^{-a|x|^2}\, dx < \infty, \tag{47}$$

for all $a > 0$. Now, by Proposition 4.1, $\mathcal{M}^{\mathrm{loc}}$ maps $L^p(W) \to L^p(V_\alpha)$ boundedly, if we set

$$V_\alpha(x) = \frac{\left[\mathcal{M}^{\mathrm{loc}}(W^{-\frac{1}{p-1}})(x)\right]^{-(p-1)\alpha}}{(1+|x|)^{(p-1)d\alpha}}, \quad \text{with} \quad \alpha > 1.$$

In particular, if $f \in L^p(w)$ and we write $\tilde{f}(y) = f(y)e^{-\frac{|y|^2}{2}} \in L^p(W)$, we have

$$\|\mathcal{M}^{\mathrm{loc}}\tilde{f}\|_{L^p(V_\alpha)} \lesssim \|\tilde{f}\|_{L^p(W)} = \|f\|_{L^p(w)}.$$

So, recalling the value of $C_1(x)$ in Lemma 3.2, and setting

$$v(x) \le v_0(x) := \frac{V_\alpha(x)\, e^{-\frac{p}{2}|x|^2}}{(1+|x|)^L}, \tag{48}$$

with $L \ge L_1 = (2\nu + d - 1)p$, we see that the term $I(x)$ in (35) is controlled by

$$\begin{aligned} \|I(x)\|^p_{L^p(v)} &\le \int_{\mathbb{R}^d} \frac{C_1(x)^p\, e^{-\frac{p}{2}|x|^2}}{(1+|x|)^L} \left|\mathcal{M}^{\mathrm{loc}}\tilde{f}(x)\right|^p V_\alpha(x)\, dx \\ &\lesssim \|f\|^p_{L^p(w)}. \end{aligned} \tag{49}$$

So, combining (35), (36) and (49) we have shown that $\|P^*_a f\|_{L^p(v)} \lesssim \|f\|_{L^p(w)}$, provided

$$v(x) \le \min\{v_0(x), v_1(x)\},$$

with $v_0(x)$ and $v_1(x)$ defined in (48) and (37). But this inequality is clearly satisfied by the weights $v(x)$ defined in (13), for every $\alpha > 1$, provided that

$$N > (p-1)d\alpha + \max\{(2\nu + d - 1)p, d + p\tfrac{d+m}{2}\} =: N_0.$$

Finally notice that, for such weights in (13), since (47) holds, if $\sigma < 1$ and $1 < \alpha < 1/\sigma$, we can use Proposition 4.2 to obtain

$$\int_{\mathbb{R}^d} v(x)^{-\frac{\sigma}{p-1}} \varphi(x)^{p'}\, dx \le \int_{\mathbb{R}^d} V_\alpha(x)^{-\frac{\sigma}{p-1}} e^{-(1-\sigma)\frac{p'|x|^2}{2}} (1+|x|)^{\frac{\sigma N}{p-1}}\, dx < \infty.$$

This proves that we can choose the weight $v(x)$ satisfying the inequality (11), as asserted in the introduction. It also gives a different proof (constructive) of [3, Theorem 1.3]. With a similar computation one can also show that, given $\varepsilon > 0$, the weights v in (13) belong to $D_{p+\varepsilon}(\varphi)$, whenever $1 < \alpha < 1 + \frac{\varepsilon}{p-1}$. □

Acknowledgments

I wish to thank S. Hartzstein, T. Signes, J.L. Torrea and B. Viviani for allowing me to use the results in our joint paper [3]. This presentation is strongly influenced by many conversations with them. Special thanks also to F.J. Martín Reyes and all the organizers of the *VI CIDAMA* conference for their hospitality and nice research environment enjoyed during the workshop.

Appendix A.

The following elementary estimates were used in the proof of Lemma 3.1.

Lemma A.1. *Let x, y be positive real numbers such that $y \geq \lambda \max\{x, 1\}$, for some $\lambda > 1$. Then there exist $c_\lambda, d_\lambda > 0$ such that*

$$\ln \frac{y}{x} \geq c_\lambda \frac{\ln(y+e)}{\ln(x+e)} \tag{A.1}$$

and

$$\ln \frac{y}{x} \leq d_\lambda \ln(y+e) \ln(\frac{1}{x}+e). \tag{A.2}$$

Proof. We first consider (A.1). Since $\ln y / \ln(y+e)$ is bounded above and below when $y \in [\lambda, \infty)$, it suffices to prove the weaker estimate

$$\ln \frac{y}{x} \geq c'_\lambda \frac{\ln y}{\ln(x+e)}. \tag{A.3}$$

Consider first the case $x \leq \sqrt{\lambda}$. Then $y \geq \lambda$ implies that $\sqrt{y} \geq x$ and hence

$$\ln \frac{y}{x} \geq \ln \sqrt{y} = \tfrac{1}{2} \ln y \geq \tfrac{1}{2} \frac{\ln y}{\ln(x+e)},$$

since $\ln(x+e) > 1$. This proves (A.3) with $c'_\lambda = 1/2$. Consider now the case $x \geq \sqrt{\lambda}$, and write

$$\ln y = \ln \tfrac{y}{x} + \ln x. \tag{A.4}$$

Observe that, if $a \geq a_0 > 0$ and $b \geq b_0 > 0$, then

$$\frac{a+b}{ab} = \frac{1}{a} + \frac{1}{b} \leq \frac{1}{a_0} + \frac{1}{b_0}. \tag{A.5}$$

So, using this fact in (A.4) we see that

$$\ln y \leq \left(\tfrac{1}{\ln \lambda} + \tfrac{1}{\ln \sqrt{\lambda}}\right) \ln \frac{y}{x} \ln x \leq \tfrac{3}{\ln \lambda} \ln \frac{y}{x} \ln(x+e),$$

which implies (A.3) with $c'_\lambda = (\ln \lambda)/3$. The proof of (A.2) is similar. If $x \geq 1/\lambda$ then

$$\ln \tfrac{y}{x} \leq \ln(\lambda y) \leq 2 \ln y \leq 2 \ln(y+e) \ln(e + \tfrac{1}{x}).$$

If $x \geq 1/\lambda$ then

$$\ln \tfrac{y}{x} = \ln y + \ln \tfrac{1}{x} \leq \tfrac{2}{\ln \lambda} \ln y \ln(\tfrac{1}{x}),$$

using in the last step the inequality (A.5). □

We give a proof of the converse of Proposition 4.1, whose validity we mentioned in Remark 4.1. The arguments are similar to those in [1].

Proposition A.1. *Let $1 < p < \infty$, and suppose that $w(x) > 0$ is such that $\mathcal{M}^{\rm loc}$ maps $L^p(w) \to L^p(V)$ for some weight $V(x) > 0$. Then, necessarily, $w^{-\frac{1}{p-1}} \in L^1_{\rm loc}(\mathbb{R}^d)$.*

Proof. Given $\varepsilon > 0$, we set $w_\varepsilon(x) = w(x) + \varepsilon$. Notice that every $f = w_\varepsilon^{-\frac{1}{p-1}} \chi_{B_R(0)}$ belongs to $L^p(w)$ since

$$\int_{\mathbb{R}^d} |f|^p w \leq \int_{B_R(0)} w_\varepsilon^{-p'} w_\varepsilon \leq |B_R(0)| \varepsilon^{-(p'-1)} < \infty.$$

Call $S_0 = B_1(0)$ and $S_j = \{2^{j-1} \leq |x| < 2^j\}$, $j = 1, 2, \ldots$, and consider the $L^p(w)$-functions $f_j = w_\varepsilon^{-\frac{1}{p-1}} \chi_{S_j}$. Then, it is easy to verify from the definition of $\mathcal{M}^{\rm loc}$ that

$$\mathcal{M}^{\rm loc} f_j(x) \gtrsim 2^{-jd} \int_{S_j} w_\varepsilon^{-\frac{1}{p-1}}, \quad \text{if } x \in S_j,$$

for each $j = 0, 1, 2, \ldots$ Indeed, it suffices to average over a ball $B_r(x)$ of radius $r = 2^j + |x|$, and observe that $S_j \subset B_r(x) \cap \{|y| \leq 3\max(|x|, 1)\}$ when $x \in S_j$. Thus, the assumed boundedness of $\mathcal{M}^{\rm loc}$ gives

$$V(S_j)^{\frac{1}{p}} 2^{-jd} \int_{S_j} w_\varepsilon^{-\frac{1}{p-1}} \lesssim \| \mathcal{M}^{\rm loc} f_j \|_{L^p(V)}$$

$$\leq C \| f_j \|_{L^p(w)} \leq C \Big(\int_{S_j} w_\varepsilon^{-(p'-1)} \Big)^{\frac{1}{p}},$$

with a constant C independent of ε. Since both sides are finite and $p' - 1 = 1/(p-1)$, we conclude that

$$\Big[\int_{S_j} w_\varepsilon^{-\frac{1}{p-1}} \Big]^{1/p'} \lesssim C 2^{jd} / V(S_j)^{\frac{1}{p}},$$

so letting $\varepsilon \to 0$ we obtain that $w^{-\frac{1}{p-1}} \in L^1(S_j)$ for each $j = 0, 1, 2, \ldots$ □

References

[1] L. Carleson and P. Jones, *Weighted norm inequalities and a theorem of Koosis*, Tech. Rep. 2, Mittag-Leffler Inst. (1981).

[2] J. Duoandikoetxea, *Fourier analysis*, Graduate Studies in Mathematics, Vol. 29, pp. xviii+222 (American Mathematical Society, Providence, RI, 2001), Translated and revised from the 1995 Spanish original by David Cruz-Uribe.

[3] G. Garrigós, S. Hartzstein, T. Signes, J. Torrea and B. Viviani, Pointwise convergence to initial data of heat and Laplace equations, *Trans. Amer. Math. Soc.* To appear.

[4] N. N. Lebedev, *Special functions and their applications*, Revised English edition. Translated and edited by Richard A. Silverman, pp. xii+308 (Prentice-Hall, Inc., Englewood Cliffs, N.J., 1965).

[5] B.-H. Li, Explicit relation between the solutions of the heat and the Hermite heat equation, *Z. Angew. Math. Phys.* **58**, pp. 959–968 (2007).

[6] L. Liu and P. Sjögren, A characterization of the Gaussian Lipschitz space and sharp estimates for the Ornstein-Uhlenbeck Poisson kernel, *Rev. Matem. Iberoam.* To appear. arXiv:1401.4288

[7] B. Muckenhoupt, Poisson integrals for Hermite and Laguerre expansions, *Trans. Amer. Math. Soc.* **139**, pp. 231–242 (1969).

[8] J. L. Rubio de Francia, Weighted norm inequalities and vector valued inequalities, in *Harmonic analysis (Minneapolis, Minn., 1981)*, Lecture Notes in Math., Vol. 908, pp. 86–101 (Springer, Berlin-New York, 1982).

[9] E. M. Stein, *Topics in harmonic analysis related to the Littlewood-Paley theory*, Annals of Mathematics Studies, No. 63, pp. viii+146 (Princeton University Press, Princeton, N.J.; University of Tokyo Press, Tokyo, 1970).

[10] P. R. Stinga and J. L. Torrea, Extension problem and Harnack's inequality for some fractional operators, *Comm. Partial Differential Equations* **35**, pp. 2092–2122 (2010).

[11] S. Thangavelu, *Lectures on Hermite and Laguerre expansions*, Mathematical Notes, Vol. 42, pp. xviii+195 (Princeton University Press, Princeton, NJ, 1993), With a preface by Robert S. Strichartz.

[12] G. N. Watson, *A Treatise on the Theory of Bessel Functions*, pp. vi+804 (Cambridge University Press, Cambridge, England; The Macmillan Company, New York, 1944).

Frequently hypercyclic operators: Recent advances and open problems

Karl-G. Grosse-Erdmann

Institut Complexys, Département de Mathématique, Université de Mons,
20 Place du Parc, 7000 Mons, Belgium
E-mail: kg.grosse-erdmann@umons.ac.be

We report on recent advances on one of the central notions in linear dynamics, that of a frequently hypercyclic operator. We include a list of ten open problems.

Keywords: Frequently hypercyclic operator, linear chaos, ergodic theory.

1. Introduction

Frequent hypercyclicity is one of the most fascinating notions in linear dynamics: it is a natural strengthening of the key concept in linear dynamics, that of hypercyclicity, and it is very close in spirit (though not, as we will see, in a strict sense) to that of linear chaos. Although initiated only ten years ago, the study of frequently hypercyclic operators has seen very deep and important developments, many of them quite recent. In this survey we will discuss some of these advances, and we will highlight several open problems.

Let us first recall the two central concepts in linear dynamics. Throughout this paper, X will denote a separable F-space, that is, a topological vector space whose topology is induced by a complete translation-invariant metric. The reader will lose very little in assuming that X is a separable Banach space. Moreover, $T : X \to X$ will denote a (continuous and linear) operator on X.

Definition 1.1. (a) An operator T is called *hypercyclic* if there exists some $x \in X$ whose orbit

$$\operatorname{orb}(x, T) = \{x, Tx, T^2x, T^3x, \dots\}$$

is dense in X. In this case, x is called a *hypercyclic vector* for T. The set of hypercyclic vectors is denoted by $HC(T)$.

(b) An operator T is called *chaotic* if it is hypercyclic and if the set of periodic points for T is dense in X.

For an introduction to linear dynamics we refer to the recent textbooks [8] and [44].

The concept of a frequently hypercyclic operator was introduced in 2004 by F. Bayart and S. Grivaux [5, 6]. The idea is to measure the size of the set

$$N(x,U) = \{n \geq 0 : T^n x \in U\}$$

of powers of T that send a given vector x into a non-empty open set $U \subset X$. For a vector x with a dense orbit, $N(x,U)$ is non-empty for any such set U, and therefore automatically infinite. For a periodic point x, $N(x,U)$ may be as large as a set of the form $\{n_0 + kp : k \geq 0\}$ for *some* sets U, but at the price that the orbit misses completely some others. For frequent hypercyclicity, Bayart and Grivaux demand that an orbit meets *every* non-empty open set *often* – in the sense of positive lower density.

Definition 1.2. An operator T is called *frequently hypercyclic* if there exists some $x \in X$ such that, for any non-empty open set $U \subset X$,

$$\underline{\text{dens}}\{n \geq 0 : T^n x \in U\} > 0.$$

In this case, x is called a *frequently hypercyclic vector* for T. The set of frequently hypercyclic vectors is denoted by $FHC(T)$.

We recall that for a set $A \subset \mathbb{N}_0$,

$$\underline{\text{dens}}\, A = \liminf_{N\to\infty} \frac{1}{N+1} \text{card}\{0 \leq n \leq N : n \in A\}.$$

There are two approaches to frequent hypercyclicity that have been studied in tandem right from the start: a topological approach and a probabilistic, that is ergodic theoretic, approach. These will be discussed in Sections 2 and 3. The subsequent sections follow roughly a chronological order. In Section 4 we ask whether any infinite-dimensional separable Banach space admits a frequently hypercyclic operator, in Section 5 we consider the rate of growth of frequently hypercyclic entire functions, Section 6 is concerned with the so-called frequently hypercyclic subspaces, and in Section 7 we report on the relationship between frequent hypercyclicity and linear chaos. Throughout the text we will draw attention to some open problems; additional ones will be listed in the final Section 8.

2. Topological approach to frequent hypercyclicity

The first question that comes to mind is whether there exist orbits that satisfy the rather strong condition of frequent hypercyclicity. In fact, as we will discuss below, a Baire category approach as in hypercyclicity is not at our disposal. However, in some cases one may obtain frequently hypercyclic orbits by a (countable) constructive procedure. Such a construction is feasible if the operator satisfies the so-called Frequent Hypercyclicity Criterion, see [5, 6, 27, 28].

Theorem 2.1 (Frequent Hypercyclicity Criterion). *Let T be an operator on a separable F-space. Suppose that there is a dense subset X_0 of X and a map $S: X_0 \to X_0$ such that, for each $x \in X_0$,*

(i) $\sum_{n=0}^{\infty} T^n x$ *converges unconditionally,*
(ii) $\sum_{n=0}^{\infty} S^n x$ *converges unconditionally,*
(iii) $TSx = x$.

Then T is frequently hypercyclic and chaotic.

Any new notion in linear dynamics is first tested on the weighted backward shifts. Let $w = (w_n)$ be a bounded sequence of non-zero scalars. Then the (unilateral) weighted backward shift B_w is defined on $X = \ell^p$, $1 \le p < \infty$, or c_0 by

$$B_w(x_n) = (w_2 x_2, w_3 x_3, w_4 x_4, \ldots).$$

Now, if

$$\Big(\frac{1}{|w_2 \cdots w_n|}\Big)_n \in X, \tag{1}$$

then one easily sees that B_w satisfies the Frequent Hypercyclicity Criterion and is therefore frequently hypercyclic. One need only take for X_0 the set of finitely non-zero sequences and for S the forward shift $S(x_n) = (0, x_1/w_2, x_2/w_3, x_3/w_4, \ldots)$. Incidentally, condition (1) characterizes when B_w is chaotic (see [41]).

One naturally wonders which condition on the weights characterizes frequent hypercyclicity of B_w. For $X = c_0$, it was shown by Bayart and Grivaux [7] that there are frequently hypercyclic weighted shifts that are not chaotic and therefore do not satisfy (1). Bayart and Ruzsa [10] recently gave a characterization, which, however, is rather technical.

The counter-example of Bayart and Grivaux also shows that not every frequently hypercyclic operator can satisfy the Frequent Hypercyclicity Criterion (since that criterion implies chaos). Even more, Badea and Grivaux [3] have constructed an operator that is frequently hypercyclic and chaotic while failing to be mixing (the latter says that, for any non-empty open sets $U, V \subset X$, there is some $N \in \mathbb{N}$ such that $T^n(U) \cap V \neq \varnothing$ for all $n \ge N$). Since the conditions of the Frequent Hypercyclicity Criterion also imply that T is mixing (see [44, Theorem 3.4]) we see that not even every frequently hypercyclic and chaotic operator needs to satisfy the Frequent Hypercyclicity Criterion. Thus one is naturally led to the following.

Problem 2.1. (a) *Which (strong) dynamical behaviour does the Frequent Hypercyclicity Criterion characterize?*

(b) *Does every chaotic, mixing and frequently hypercyclic operator satisfy the Frequent Hypercyclicity Criterion?*

Let us return to the characterization of frequently hypercyclic weighted shifts. The results for c_0 suggest that a characterizing condition is necessarily also complicated for the spaces ℓ^p. Surprisingly, this is not the case. Bayart and Ruzsa [10] have been able to show that the sufficient condition (1) is also necessary. Their proof is based on an improvement of the well-known result of Erdős and Sarközy on difference sets.

Theorem 2.2 (Bayart–Ruzsa). *Let B_w be a weighted backward shift on $X = \ell^p$, $1 \le p < \infty$. Then the following assertions are equivalent:*

(a) *B_w is frequently hypercyclic,*
(b) *B_w satisfies the Frequent Hypercyclicity Criterion,*
(c) $\sum_{n=2}^{\infty} \frac{1}{|w_2 w_3 \cdots w_n|^p} < \infty$, $\qquad$ ($\iff B_w$ *is chaotic*).

Let us briefly mention that frequently hypercyclic operators are also found among other classes of operators. For example, on the space $H(\mathbb{C})$ of entire functions the differentiation operator $D : H(\mathbb{C}) \to H(\mathbb{C})$, $Df = f'$, and the translation operator $T : H(\mathbb{C}) \to H(\mathbb{C})$, $Tf(z) = f(z+1)$, are frequently hypercyclic [6]. More generally, any non-scalar operator T on $H(\mathbb{C})$ that commutes with D is frequently hypercyclic [25]. For the frequent hypercyclicity of functions of weighted backward shifts, as well as differentiation operators, (weighted) composition operators and adjoint multipliers on Banach (or Fréchet) spaces of analytic or harmonic functions we refer to [6, 7, 15, 18, 19, 22, 23, 46, 51, 53]. The proof of frequent hypercyclicity in these papers often relies on a probabilistic approach, to which we turn in the next section.

In the study of (ordinary) hypercyclicity, the Baire category theorem enters in an unspectacular way. By definition, the set of hypercyclic vectors for an operator is given by

$$HC(T) = \bigcap_{\varnothing \neq U \text{ open}} \bigcup_{n=1}^{\infty} T^{-n}(U).$$

Now, if T is hypercyclic then every vector along a dense orbit is also hypercyclic, so that $HC(T)$ is dense. Moreover, since the space X has to be separable, the above intersection can be reduced to a countable one. This shows that $HC(T)$ is a dense G_δ-set, hence residual, as soon as T is hypercyclic. The importance of this simple observation for the study of hypercyclic operators cannot be overestimated.

A direct consequence is that every vector in X is the sum of two hypercyclic vectors for T:

$$X = HC(T) + HC(T).$$

Indeed, for any $x \in X$, the two residual sets $x - HC(T)$ and $HC(T)$ need to intersect.

It is natural to wonder if Baire continues to have the same impact on frequent hypercyclicity. The answer is a resounding 'no'. It was already observed in [6] and [27] that for many frequently hypercyclic operators the set

$$FHC(T)$$

of frequently hypercyclic vectors for T is of first Baire category. In [27, 42] Bonilla and the author asked if the set $FHC(T)$ is always of first Baire category. The positive answer was recently given independently by Bayart–Ruzsa [10] (for Banach spaces) and by Moothathu [50] and Grivaux–Matheron [39].

Theorem 2.3 (Moothathu, Bayart–Ruzsa, Grivaux–Matheron). *For any operator T on a separable F-space the set $FHC(T)$ of frequently hypercyclic vectors is of first Baire category.*

We give here a slight modification of the proof in [50].

Proof. Let $\|\cdot\|$ denote an F-norm that defines the topology of X, see [27]. Note, in particular, that for an F-norm we have that $\|cx\| \le n\|x\|$ if $c \in \mathbb{K}$ and $n \in \mathbb{N}$ with $|c| \le n$.

(I) We first show that there exists a non-empty open set V such that the sets

$$V, 2V, 2^2 V, 2^3 V, \ldots$$

are pairwise disjoint. Indeed, for an arbitrary vector $x \in X \setminus \{0\}$ we consider the open ball $V = B(x, \frac{\|x\|}{4})$. It then suffices to show that, for all $n \ge 1$, $V \cap 2^n V = \varnothing$. Now, if $x + w = 2^n(x + \widetilde{w}) \in V \cap 2^n V$ then $(1 - 2^{-n})x = 2^{-n}w - \widetilde{w}$. Since $\frac{1}{1-2^{-n}} \le 2$ we deduce that

$$\|x\| = \|\tfrac{1}{1-2^{-n}}(1-2^{-n})x\| \le 2\|(1-2^{-n})x\| = 2\|2^{-n}w - \widetilde{w}\| < 2\cdot 2\frac{\|x\|}{4} = \|x\|,$$

which is impossible.

(II) We now show that $FHC(T)$ is of first Baire category. Let U be a non-empty open set whose closure $\overline{U}$ is contained in a set V as given by (I). Then

$$FHC(T) \subset \{x \in X : \underline{\text{dens}}\{k \ge 0 : T^k x \in U\} > 0\} \subset \bigcup_{m=1}^{\infty} \bigcup_{N=0}^{\infty} A_{m,N},$$

where

$$A_{m,N} = \{x \in X : \text{for all } n \geq N, \tfrac{1}{n+1}\text{card}\{0 \leq k \leq n : T^k x \in \overline{U}\} \geq \tfrac{1}{m}\}.$$

It is easy to see that each set $A_{m,N}$ is closed.

To finish the proof it suffices to show that each set $A_{m,N}$ has empty interior. Suppose that this is not the case for some $m \geq 1$, $N \geq 0$. We may suppose that the operator T is hypercyclic; let x be a corresponding hypercyclic vector. Since, for any $j \geq 0$, also $2^{-j}x$ is hypercyclic and since $A_{m,N}$ has non-empty interior we can find some $n_j \geq 0$ such that

$$T^{n_j}(2^{-j}x) \in A_{m,N},$$

which implies that there is some $N_0 \geq N$ such that, for any $0 \leq j \leq 2m$,

$$\tfrac{1}{N_0+1}\text{card}\{0 \leq k \leq N_0 : T^k x \in 2^j\overline{U}\} \geq \tfrac{1}{2m}.$$

This is clearly impossible since the sets $\overline{U}, 2\overline{U}, 2^2\overline{U}, 2^3\overline{U}, \ldots$ are pairwise disjoint. □

In the light of this result, and unlike for hypercyclic operators, we can no longer be sure that

$$X = FHC(T) + FHC(T).$$

Indeed, Bonilla and the author [27] found that under mild conditions on the frequently hypercyclic operator T,

$$X \neq FHC(T) + FHC(T).$$

This is true, for example, for the multiples λB ($|\lambda| > 1$) of the backward shift on the spaces ℓ^p, $1 \leq p < \infty$, or c_0, and for the differentiation operator D on the space $H(\mathbb{C})$ of entire functions. However, the translation operator $T : H(\mathbb{C}) \to H(\mathbb{C})$, $Tf(z) = f(z+1)$ satisfies

$$H(\mathbb{C}) = FHC(T) + FHC(T).$$

This is, in the final analysis, due to the flexibility of the Runge approximation theorem. It is not clear if the same can happen in a Banach space. The following seems to be still open, see [27, 42].

Problem 2.2. *Is there an operator T on a Banach space X for which $X = FHC(T) + FHC(T)$ holds?*

3. Probabilistic approach to frequent hypercyclicity

In this section we will only briefly touch upon what is in fact the deepest part of the theory of frequently hypercyclic operators: the use of probabilistic techniques. After pioneering work by Flytzanis [34], a rich edifice of research has been erected by F. Bayart, S. Grivaux, and E. Matheron, see [6–9, 37–39].

We will limit ourselves here to stating what seems to us to be the most pertinent results, without defining the relevant notions of ergodic theory and of measure theory on Fréchet spaces. A starting point for the interested reader might be the earlier survey of the author [42] and Chapters 5 and 6 of the book of Bayart and Matheron [8]. A recent survey article by Bayart [4] is also highly recommended.

The use of ergodic theoretic techniques in linear dynamics goes back to Flytzanis [34]. He noted that if an operator T on a separable Hilbert space admits an ergodic measure of full support then the operator is hypercyclic. He then showed that, under certain assumptions on the operator T, a sufficiently large supply of unimodular eigenvectors (that is, eigenvectors to eigenvalues of modulus 1) will ensure the existence of a T-ergodic measure of full support. This has set the tone for all the research that followed.

Now, Bayart and Grivaux [5, 6] noted that, in view of the Birkhoff ergodic theorem, an ergodic measure of full support for T even leads to the operator being frequently hypercyclic. In fact, this observation motivated the notion of frequent hypercyclicity.

With their work, Bayart, Grivaux and Matheron have shed considerable light on the link between the existence of ergodic measures of full support, the set of unimodular eigenvectors, and the frequent hypercyclicity of an operator. Gaussian measures play a prominent (but not exclusive) role in these investigations. In the sequel, let $\mathbb{T}$ denote the unit circle.

The following has evolved over the course of an intensive ten years of research, see [6–9, 37–39].

Theorem 3.1 (Bayart, Grivaux, Matheron). *Let T be an operator on a separable complex Fréchet space. Consider the following:*

(a) *the unimodular eigenvectors are* perfectly spanning, *that is, for any countable set $D \subset \mathbb{T}$ the linear span of $\bigcup_{\lambda \in \mathbb{T}\setminus D} \ker(T-\lambda)$ is dense in X;*
(b) *there exists an ergodic Gaussian measure of full support for T;*
(c) *T is frequently hypercyclic.*

Then (a)$\Longrightarrow$(b)$\Longrightarrow$(c). *If X is a Hilbert space or, more generally, a Banach space of cotype* 2 *then* (a)$\Longleftrightarrow$(b).

There are some Banach spaces where condition (b) does not imply (a), see [7].

For applications it is useful to note that condition (a) is implied by (and in fact equivalent to) the existence of perfect sets $\Lambda_j \subset \mathbb{T}$ $(j \in \mathbb{N})$ and continuous maps $E_j : \Lambda_j \to X$ such that $TE_j(\lambda) = \lambda E_j(\lambda)$ for all $\lambda \in \Lambda_j$ $(j \in \mathbb{N})$ and such that the span of $\bigcup_{j\in\mathbb{N}} E_j(\Lambda_j)$ is dense in X; see [9, 39].

In particular, for weighted backward shifts B_w on $X = \ell^p$, $1 \le p < \infty$, conditions (a), (b) and (c) are equivalent, and they hold if and only if

$$\sum_{n=2}^{\infty} \frac{1}{|w_2 w_3 \cdots w_n|^p} < \infty;$$

see [9].

It is also interesting to note that every operator on a separable complex Fréchet space that satisfies the Frequent Hypercyclicity Criterion admits an ergodic Gaussian measure of full support, which then implies its frequent hypercyclicity, see [9]. If one is not necessarily demanding a Gaussian measure, a simpler proof of this fact was obtained by Murillo and Peris [52], which even works for (real or complex) F-spaces.

The investigations that were started by Bayart and Grivaux [5, 6] and that culminated in Theorem 3.1 lead naturally to two questions: can frequent hypercyclicity only arise in the presence of an ergodic (Gaussian) measure of full support? Can it only arise under the existence of sufficiently many unimodular eigenvectors? Both questions have a negative answer: there is a frequently hypercyclic operator on c_0 that has no unimodular eigenvalues and that does not admit any invariant Gaussian measure of full support, see [7]. The latter was substantially improved by Grivaux and Matheron [39] who showed that there is even a frequently hypercyclic operator on $c_0(\mathbb{Z})$ that does not admit any ergodic measure of full support. In fact, both operators are weighted backward shifts.

The best positive result so far is the following, see [39].

Theorem 3.2 (Grivaux–Matheron). *Let T be a frequently hypercyclic operator on a reflexive Banach space. Then T admits a continuous invariant measure of full support.*

But several natural questions have so far remained open, see [7, 9, 39].

Problem 3.1. (a) *Does every frequently hypercyclic operator on an arbitrary Banach space admit an invariant measure of full support?*

(b) *Does every frequently hypercyclic operator on a Hilbert space admit an ergodic measure of full support?*

(c) *Does every frequently hypercyclic operator on a Hilbert space admit an ergodic Gaussian measure of full support?*

(d) *Does every frequently hypercyclic operator on a Hilbert space have unimodular eigenvalues?*

Questions (b), (c) *and* (d) *may also be asked, more generally, for arbitrary reflexive Banach spaces.*

4. Existence of frequently hypercyclic operators, and the spectrum

By an important result of Ansari [1] and Bernal [14], every infinite-dimensional separable Banach space admits a hypercyclic operator. Bonet, Martínez and Peris [24] (see also [8, Theorem 6.36]) have subsequently shown that this result breaks down for chaos: no hereditarily indecomposable complex Banach space supports a chaotic operator. Recall that a Banach space is called hereditarily indecomposable if none of its closed subspaces is decomposable as a direct sum of infinite-dimensional closed subspaces. Such spaces were first constructed by Gowers and Maurey [36].

Shkarin [57] showed that, as for existence, frequent hypercyclicity behaves like chaos. For this, he first studied the spectrum of frequently hypercyclic operators.

Theorem 4.1 (Shkarin). *Let T be a frequently hypercyclic operator on a complex Banach space. Then its spectrum $\sigma(T)$ has no isolated points.*

The proof uses the theory of entire functions in an ingenious way. It now follows from Shkarin's result and the Riesz theory that no operator of the form

$$T = \lambda I + K, \quad K \text{ compact}$$

can be frequently hypercyclic. But the celebrated Argyros–Haydon theorem [2] shows that there are separable complex Banach spaces on which *every* operator is of this form. Thus, Argyros–Haydon spaces do not admit frequently hypercyclic operators. More generally, Shkarin [57] proved the following analogue of the Bonet–Martínez–Peris result on chaotic operators.

Theorem 4.2 (Shkarin). *No hereditarily indecomposable complex Banach space supports a frequently hypercyclic operator.*

In a way, the following is the positive counterpart of the results of Bonet, Martínez, Peris and Shkarin, see [31].

Theorem 4.3 (de la Rosa–Frerick–Grivaux–Peris). *Every infinite-dimensional* *separable complex Banach space with an unconditional Schauder decomposition admits an operator that is frequently hypercyclic and chaotic.*

Still, the following problem is open, see [31].

Problem 4.1. *Which Banach spaces admit a frequently hypercyclic operator? Which Banach spaces admit a chaotic operator?*

We will return to this question in the light of a very recent significant advance in frequent hypercyclicity, see Section 7.

Independently of the question of existence one may be interested in identifying the sets $K \subset \mathbb{C}$ that can arise as spectra of frequently hypercyclic operators. Shkarin [57] has shown that a non-empty compact subset K of $\mathbb{C}$ is the spectrum of some hypercyclic operator on a complex Hilbert (or Banach) space if and only if each of its connected components meets the unit circle. By Theorem 4.1 one needs to add the absence of isolated points in the case of frequent hypercyclicity. Is that enough? This was wrongly claimed in [44, Theorem 9.43], and the authors of that book take full responsibility for the blunder. Beise [12] recently showed that the whole truth is more complicated.

Theorem 4.4 (Beise). *Let C be a closed and open component of the spectrum of an operator T on a complex Banach space. If $C \subset \{z \in \mathbb{C} : |z| \leq 1\}$ and $C \cap \{z \in \mathbb{C} : |z| = 1\} = \{e^{i\alpha_1}, \ldots, e^{i\alpha_n}\}$, where $\alpha_1, \ldots, \alpha_n$ are linearly independent over the field of rational numbers, then T is not frequently hypercyclic.*

For example, since frequent hypercyclicity is preserved under a rotation $T \to \lambda T$, $|\lambda| = 1$, (see [8]), no connected compact set $K \subset \{z \in \mathbb{C} : |z| \leq 1\}$ that meets the unit circle in a single point can be the spectrum of a frequently hypercyclic operator. This excludes, in particular, the set $K = [0,1]$. It is interesting to note that Beise's proof links the dynamics of an arbitrary operator with the dynamics of the translation operator on the space of entire functions.

Thus we are left with the following, see [57].

Problem 4.2. *Which compact subsets of $\mathbb{C}$ can be the spectrum of a frequently hypercyclic operator on a complex Banach space?*

5. Rate of growth of frequently hypercyclic entire functions

The differentiation operator

$$D : H(\mathbb{C}) \to H(\mathbb{C}), \quad Df = f'$$

is one of the classical hypercyclic operators. The author [40] and Shkarin [56] independently obtained a sharp result on the possible rates of growth of corresponding hypercyclic entire functions: for any function $\varphi : \mathbb{R}_+ \to \mathbb{R}_+$ with $\varphi(r) \to \infty$ as $r \to \infty$ there is some D-hypercyclic entire function f such that

$$|f(z)| = O\Big(\varphi(r)\frac{\exp(r)}{r^{1/2}}\Big) \quad \text{as } |z| = r \to \infty, \tag{2}$$

while there is no D-hypercyclic entire function f that satisfies

$$|f(z)| = O\Big(\frac{\exp(r)}{r^{1/2}}\Big) \quad \text{as } |z| = r \to \infty.$$

Now, D is even frequently hypercyclic [6]. So, how about the corresponding possible rates of growth? Blasco, Bonilla and the author [22] showed that, for any function $\psi : \mathbb{R}_+ \to \mathbb{R}_+$ with $\psi(r) \to 0$ as $r \to \infty$ there is no D-frequently hypercyclic entire function f that satisfies

$$|f(z)| = O\Big(\psi(r)\frac{\exp(r)}{r^{1/4}}\Big) \quad \text{as } |z| = r \to \infty.$$

In a positive direction they were only able to obtain a rate of growth as in (2) with $r^{1/2}$ replaced by 1. The question of the optimal rate, see [22, 26], was settled by Drasin and Saksman [33].

Theorem 5.1 (Drasin–Saksman). *There is a D-frequently hypercyclic entire function f such that*

$$|f(z)| = O\Big(\frac{\exp(r)}{r^{1/4}}\Big) \quad \text{as } |z| = r \to \infty.$$

They obtained such an entire function by an explicit construction and complex analytic tools without any functional analytic machinery. Interestingly, Nikula [54] found a probabilistic approach to a result that is only slightly weaker than Drasin and Saksman's.

Theorem 5.2 (Nikula). *Let $(X_n)_{n\geq 0}$ be a sequence of independent identically distributed complex random variables whose law has support $\mathbb{C}$ and such that, for some $a > 0$,*

$$E(e^{t|X_n|}) = O(e^{at^2}) \quad \text{as } t \to \infty.$$

Then the random power series

$$f(z) = \sum_{n=0}^{\infty} \frac{X_n}{n!} z^n$$

almost surely represents a D-frequently hypercyclic entire function that satisfies

$$|f(z)| = O\Big(\sqrt{\log r}\,\frac{\exp(r)}{r^{1/4}}\Big) \quad \text{as } |z| = r \to \infty.$$

Mouze and Munnier [51] develop the ideas of Nikula and combine them with the Birkhoff ergodic theorem to obtain random frequently hypercyclic vectors for various operators.

The growth conditions considered so far are radial. Beise and Müller [13] have initiated the study of growth conditions along rays emanating from the origin. In particular they obtain the following.

Theorem 5.3 (Beise–Müller). *Let $K \subset \mathbb{C}$ be a convex compact set whose intersection with the unit circle contains a continuum. Then there is an entire function of exponential type that is D-frequently hypercyclic and such that, for any $\varepsilon > 0$,*

$$|f(z)| = O\big(\exp(H_K(z) + \varepsilon|z|)\big)$$

for all $z \in \mathbb{C}$; here, $H_K(z) = \sup\{\mathrm{Re}(zu) : u \in K\}$ is the support function of K.

For example, a D-frequently hypercyclic function can be of exponential type 1 and tend to zero exponentially in the sector $\{z : |\arg(z)| \geq \alpha\}$, with $\alpha > \frac{\pi}{2}$ fixed.

We mention that growth rates in terms of L^p-averages and growth rates for entire functions that are frequently hypercyclic for the translation operator $Tf(z) = f(z+1)$ can be found in [11, 22, 23, 33]. In [13] the authors consider arbitrary operators that commute with D and obtain corresponding non-radial growth rates.

6. Frequently hypercyclic subspaces

We have seen in Section 2 that the set $FHC(T)$ of frequently hypercyclic vectors for an operator T is always of first Baire category. But can this set nonetheless be *large* in an algebraic sense? One of the fundamental results in linear dynamics, due to Herrero [45] and Bourdon [30], states that the set $HC(T)$ contains a dense linear subspace of X (except, of course, the zero vector) as soon as T is hypercyclic. Their proof leads immediately to the same result for $FHC(T)$ if T is frequently hypercyclic, see Bayart and Grivaux [6]. Indeed, if x is a frequently hypercyclic vector for T then span orb(x, T) is a dense linear subspace contained in $FHC(T) \cup \{0\}$.

The following, however, seems to be open.

Problem 6.1. *Can the set $FHC(T) \cup \{0\}$ be a linear subspace of X if T is frequently hypercyclic?*

For hypercyclicity, the corresponding problem has a positive answer: by a deep result of Read [55] there exists an operator T on ℓ^1 for which every non-zero vector is hypercyclic, so that $HC(T) \cup \{0\} = X$. But by Section 2 the set $FHC(T) \cup \{0\}$ is always a proper subset of X.

Bernal and Montes [16, 49] studied a different notion of largeness: for a given operator T, does the set $HC(T)$ contain an infinite-dimensional closed subspace (except 0)? They showed that for some hypercyclic operators the answer is positive, while for others it is negative.

Such a subspace is nowadays called a *hypercyclic subspace*, and the set $HC(T)$ is called *spaceable*. We refer to an excellent recent survey by Bernal, Pellegrino and Seoane [17] on spaceability, lineability and related topics.

González, León and Montes [35] obtained the following characterization: if an operator T on a separable complex Banach spaces satisfies the Hypercyclicity Criterion (see [44, Theorem 3.12]) then it possesses a hypercyclic subspace if and only if there exists an infinite-dimensional closed subspace M_0 of X and an increasing sequence (n_k) of positive integers such that

$$T^{n_k}x \to 0 \quad \text{for all } x \in M_0.$$

Now, Bonilla and the author [29] introduced the corresponding notion of a frequently hypercyclic subspace, that is, an infinite-dimensional closed subspace consisting (except 0) of frequently hypercyclic vectors.

Theorem 6.1 ([29]). *Let X be a separable F-space with a continuous norm and T an operator on X. If*

(i) (a) *T satisfies the Frequent Hypercyclicity Criterion, or*
(b) *X is a complex Banach space, and the unimodular eigenvectors for T are perfectly spanning (see Section 3),*

(ii) *there exists an infinite-dimensional closed subspace M_0 of X such that*

$$T^n x \to 0 \quad \textit{for all } x \in M_0,$$

then T possesses a frequently hypercyclic subspace, that is, $FHC(T)$ is spaceable.

The assumption of the existence of the subspace M_0 is rather strong. To give a concrete example we note that the operator T on $C_0(\mathbb{R}_+)$, $Tf(x) = \lambda f(x+a)$ ($\lambda > 1$, $a > 0$) is easily seen to satisfy the assumptions of the theorem and thus possess a frequently hypercyclic subspace.

Menet [47] recently noted that if T satisfies the Frequent Hypercyclicity Criterion then condition (ii) may be weakened to the following: there exists an infinite-dimensional closed subspace M_0 of X and a set $A \subset \mathbb{N}$ of positive lower density such that

$$T^n x \underset{n\in A}{\longrightarrow} 0 \quad \text{for all } x \in M_0.$$

This allowed him to show that certain weighted backward shifts B_w on the space ℓ^p, $1 \le p < \infty$, possess a frequently hypercyclic subspace. But we are still very far from an analogue of the characterization of González, León and Montes, even for weighted shifts. Thus we have the following, see [47].

Problem 6.2. (a) *Characterize the weighted backward shifts B_w on ℓ^p, $1 \le p < \infty$, that possess a frequently hypercyclic subspace.*

(b) *More generally, characterize the operators on a separable (complex) Banach space that possess a frequently hypercyclic subspace.*

Menet [47] also found a necessary condition for an operator to possess a frequently hypercyclic subspace. This enabled him to solve affirmatively a problem posed in [29].

Theorem 6.2 (Menet). *There exists a frequently hypercyclic operator that possesses a hypercyclic subspace but not a frequently hypercyclic subspace.*

Such an operator can be taken to be a weighted backward shift B_w on ℓ^p, $1 \le p < \infty$. Some related results can be found in Bès and Menet [20].

7. Frequent hypercyclicity versus linear chaos

Since the introduction of frequent hypercyclicity by Bayart and Grivaux in 2004 the relationship of this notion with linear chaos has intrigued the researchers. It became quickly clear that the two concepts are not equivalent. By an intricate construction, Bayart and Grivaux [7] were able to show that there is a weighted backward shift on c_0 that is frequently hypercyclic but not chaotic. In fact, the shift does not even possess a single non-trivial periodic point and, worse, not even a unimodular eigenvalue. These authors also found a non-chaotic frequently hypercyclic operator on a Hilbert space. But the following question of Bayart and Grivaux [7] remained open:

Is every chaotic operator frequently hypercyclic?

This problem has since been iterated in various articles, and it has come to be considered as one of the major questions in linear dynamics; see, for example, [8, Chapter 6] and [44, Chapter 9]. Menet [48] very recently obtained a counter-example.

Theorem 7.1 (Menet). *There exists a chaotic operator that is not frequently hypercyclic.*

More precisely, he constructed such an operator on the spaces ℓ^p, $1 \leq p < \infty$, and c_0 as a perturbation of (essentially) a weighted forward shift by an upper-triangular matrix with very few entries.

On the positive side, Menet [48] showed that every chaotic operator satisfies a weak form of frequent hypercyclicity, called reiterative hypercyclicity, where the lower density is replaced by upper Banach density, see [21].

In view of the non-existence of chaotic or of frequently hypercyclic operators in certain Banach spaces one may be interested in the following, see [48].

Problem 7.1. *Are there Banach spaces that support a chaotic operator but no frequently hypercyclic operator, or vice versa?*

8. Further problems

We end this paper by recalling two major open problems in the theory of frequent hypercyclicity; they have been posed by Bayart and Grivaux [6].

First, it is well known, and easily proven by a Baire category argument, that if T is an invertible hypercyclic operator then T^{-1} is also hypercyclic. But since Baire loses its power in frequent hypercyclicity the following needs to be attacked by different means.

Problem 8.1. *If T is an invertible frequently hypercyclic operator, is then also T^{-1} frequently hypercyclic?*

Bayart and Ruzsa [10] have shown that T^{-1} enjoys a weak form of frequent hypercyclicity, called $\mathcal{U}$-frequent hypercyclicity.

For a long time Herrero's question on whether the direct sum $T \oplus T$ of a hypercyclic operator T is also hypercyclic was a driving force behind much of the research in linear dynamics. As Bayart and Grivaux [6, p. 5085] say themselves, this question motivated their introduction of the notion of frequent hypercyclicity. Herrero's question was answered in the negative by De la Rosa and Read [32]. The corresponding problem for frequent hypercyclicity, however, remains open.

Problem 8.2. *If T is a frequently hypercyclic operator, is then also $T \oplus T$ frequently hypercyclic?*

The intermediate problem if $T \oplus T$ is hypercyclic whenever T is frequently hypercyclic (see [6]) was answered positively by Peris and the author [43].

Acknowledgments

The paper contains an extended version of my talk at the VI International Course of Mathematical Analysis in Andalucía in September 2014. I wish to thank the organizers, in particular Professor Francisco Javier Martín Reyes and Professor Fernando León Saavedra, for the kind invitation to this stimulating conference.

References

[1] S. I. Ansari, Existence of hypercyclic operators on topological vector spaces, *J. Funct. Anal.* **148**, pp. 384–390 (1997).

[2] S. A. Argyros and R. G. Haydon, A hereditarily indecomposable $\mathscr{L}_\infty$-space that solves the scalar-plus-compact problem, *Acta Math.* **206**, pp. 1–54 (2011).

[3] C. Badea and S. Grivaux, Unimodular eigenvalues, uniformly distributed sequences and linear dynamics, *Adv. Math.* **211**, pp. 766–793 (2007).

[4] F. Bayart, Probabilistic methods in linear dynamics, in *Topics in functional and harmonic analysis*, Theta Ser. Adv. Math., Vol. 14, pp. 1–26 (Theta, Bucharest, 2013).

[5] F. Bayart and S. Grivaux, Hypercyclicité: le rôle du spectre ponctuel unimodulaire, *C. R. Math. Acad. Sci. Paris* **338**, pp. 703–708 (2004).

[6] F. Bayart and S. Grivaux, Frequently hypercyclic operators, *Trans. Amer. Math. Soc.* **358**, pp. 5083–5117 (electronic) (2006).

[7] F. Bayart and S. Grivaux, Invariant Gaussian measures for operators on Banach spaces and linear dynamics, *Proc. Lond. Math. Soc. (3)* **94**, pp. 181–210 (2007).

[8] F. Bayart and É. Matheron, *Dynamics of linear operators*, Cambridge Tracts in Mathematics, Vol. 179, pp. xiv+337 (Cambridge University Press, Cambridge, 2009).

[9] F. Bayart and É. Matheron, Mixing operators and small sets of the circle, *J. Reine Angew. Math.* To appear.

[10] F. Bayart and I. Z. Ruzsa, Difference sets and frequently hypercyclic weighted shifts, *Ergodic Theory Dynam. Systems* **35**, pp. 691–709 (2015).

[11] H.-P. Beise, Growth of frequently Birkhoff-universal functions of exponential type on rays, *Comput. Methods Funct. Theory* **13**, pp. 21–35 (2013).

[12] H.-P. Beise, On the intersection of the spectrum of frequently hypercyclic operators with the unit circle, *J. Operator Theory* **72**, pp. 329–342 (2014).

[13] H.-P. Beise and J. Müller, Growth of (frequently) hypercyclic functions for differential operators, *Studia Math.* **207**, pp. 97–115 (2011).

[14] L. Bernal-González, On hypercyclic operators on Banach spaces, *Proc. Amer. Math. Soc.* **127**, pp. 1003–1010 (1999).

[15] L. Bernal-González and A. Bonilla, Compositional frequent hypercyclicity on weighted Dirichlet spaces, *Bull. Belg. Math. Soc. Simon Stevin* **17**, pp. 1–11 (2010).

[16] L. Bernal González and A. Montes Rodríguez, Non-finite-dimensional closed vector spaces of universal functions for composition operators, *J. Approx. Theory* **82**, pp. 375–391 (1995).

[17] L. Bernal-González, D. Pellegrino and J. B. Seoane-Sepúlveda, Linear subsets of

nonlinear sets in topological vector spaces, *Bull. Amer. Math. Soc. (N.S.)* **51**, pp. 71–130 (2014).

[18] J. Bès, Dynamics of composition operators with holomorphic symbol, *Rev. R. Acad. Cienc. Exactas Fís. Nat. Ser. A Math. RACSAM* **107**, pp. 437–449 (2013).

[19] J. Bès, Dynamics of weighted composition operators, *Complex Anal. Oper. Theory* **8**, pp. 159–176 (2014).

[20] J. Bès and Q. Menet, Existence of common and upper frequently hypercyclic subspaces, *J. Math. Anal. Appl.* **432**, pp. 10–37 (2015).

[21] J. Bès, Q. Menet, A. Peris and Y. P. de Dios, Recurrence properties of hypercyclic operators, *ArXiv e-prints* (October 2014); *Math. Ann.* To appear.

[22] O. Blasco, A. Bonilla and K.-G. Grosse-Erdmann, Rate of growth of frequently hypercyclic functions, *Proc. Edinb. Math. Soc. (2)* **53**, pp. 39–59 (2010).

[23] J. Bonet and A. Bonilla, Chaos of the differentiation operator on weighted Banach spaces of entire functions, *Complex Anal. Oper. Theory* **7**, pp. 33–42 (2013).

[24] J. Bonet, F. Martínez-Giménez and A. Peris, A Banach space which admits no chaotic operator, *Bull. London Math. Soc.* **33**, pp. 196–198 (2001).

[25] A. Bonilla and K.-G. Grosse-Erdmann, On a theorem of Godefroy and Shapiro, *Integral Equations Operator Theory* **56**, pp. 151–162 (2006).

[26] A. Bonilla and K.-G. Grosse-Erdmann, A problem concerning the permissible rates of growth of frequently hypercyclic entire functions, in *Topics in complex analysis and operator theory*, pp. 155–158 (Univ. Málaga, Málaga, 2007).

[27] A. Bonilla and K.-G. Grosse-Erdmann, Frequently hypercyclic operators and vectors, *Ergodic Theory Dynam. Systems* **27**, pp. 383–404 (2007).

[28] A. Bonilla and K.-G. Grosse-Erdmann, Frequently hypercyclic operators and vectors—Erratum, *Ergodic Theory Dynam. Systems* **29**, pp. 1993–1994 (2009).

[29] A. Bonilla and K.-G. Grosse-Erdmann, Frequently hypercyclic subspaces, *Monatsh. Math.* **168**, pp. 305–320 (2012).

[30] P. S. Bourdon, Invariant manifolds of hypercyclic vectors, *Proc. Amer. Math. Soc.* **118**, pp. 845–847 (1993).

[31] M. De la Rosa, L. Frerick, S. Grivaux and A. Peris, Frequent hypercyclicity, chaos, and unconditional Schauder decompositions, *Israel J. Math.* **190**, pp. 389–399 (2012).

[32] M. de la Rosa and C. Read, A hypercyclic operator whose direct sum $T \oplus T$ is not hypercyclic, *J. Operator Theory* **61**, pp. 369–380 (2009).

[33] D. Drasin and E. Saksman, Optimal growth of entire functions frequently hypercyclic for the differentiation operator, *J. Funct. Anal.* **263**, pp. 3674–3688 (2012).

[34] E. Flytzanis, Unimodular eigenvalues and linear chaos in Hilbert spaces, *Geom. Funct. Anal.* **5**, pp. 1–13 (1995).

[35] M. González, F. León-Saavedra and A. Montes-Rodríguez, Semi-Fredholm theory: hypercyclic and supercyclic subspaces, *Proc. London Math. Soc. (3)* **81**, pp. 169–189 (2000).

[36] W. T. Gowers and B. Maurey, The unconditional basic sequence problem, *J. Amer. Math. Soc.* **6**, pp. 851–874 (1993).

[37] S. Grivaux, A probabilistic version of the frequent hypercyclicity criterion, *Studia Math.* **176**, pp. 279–290 (2006).

[38] S. Grivaux, A new class of frequently hypercyclic operators, *Indiana Univ. Math. J.*

60, pp. 1177–1201 (2011).
[39] S. Grivaux and É. Matheron, Invariant measures for frequently hypercyclic operators, *Adv. Math.* **265**, pp. 371–427 (2014).
[40] K.-G. Grosse-Erdmann, On the universal functions of G. R. MacLane, *Complex Variables Theory Appl.* **15**, pp. 193–196 (1990).
[41] K.-G. Grosse-Erdmann, Hypercyclic and chaotic weighted shifts, *Studia Math.* **139**, pp. 47–68 (2000).
[42] K.-G. Grosse-Erdmann, Dynamics of linear operators, in *Topics in complex analysis and operator theory*, pp. 41–84 (Univ. Málaga, Málaga, 2007).
[43] K.-G. Grosse-Erdmann and A. Peris, Frequently dense orbits, *C. R. Math. Acad. Sci. Paris* **341**, pp. 123–128 (2005).
[44] K.-G. Grosse-Erdmann and A. Peris Manguillot, *Linear chaos*, Universitext, pp. xii+386 (Springer, London, 2011).
[45] D. A. Herrero, Limits of hypercyclic and supercyclic operators, *J. Funct. Anal.* **99**, pp. 179–190 (1991).
[46] V. E. Kim, Complete systems of partial derivatives of entire functions and frequently hypercyclic operators, *J. Math. Anal. Appl.* **420**, pp. 364–372 (2014).
[47] Q. Menet, Existence and non-existence of frequently hypercyclic subspaces for weighted shifts, *Proc. Amer. Math. Soc.* **143**, pp. 2469–2477 (2015).
[48] Q. Menet, Linear chaos and frequent hypercyclicity, *ArXiv e-prints* (October 2014); *Trans. Amer. Math. Soc.* To appear.
[49] A. Montes-Rodríguez, Banach spaces of hypercyclic vectors, *Michigan Math. J.* **43**, pp. 419–436 (1996).
[50] T. K. S. Moothathu, Two remarks on frequent hypercyclicity, *J. Math. Anal. Appl.* **408**, pp. 843–845 (2013).
[51] A. Mouze and V. Munnier, On random frequent universality, *J. Math. Anal. Appl.* **412**, pp. 685–696 (2014).
[52] M. Murillo-Arcila and A. Peris, Strong mixing measures for linear operators and frequent hypercyclicity, *J. Math. Anal. Appl.* **398**, pp. 462–465 (2013).
[53] S. Muro, D. Pinasco and M. Savransky, Strongly mixing convolution operators on Fréchet spaces of holomorphic functions, *Integral Equations Operator Theory* **80**, pp. 453–468 (2014).
[54] M. Nikula, Frequent hypercyclicity of random entire functions for the differentiation operator, *Complex Anal. Oper. Theory* **8**, pp. 1455–1474 (2014).
[55] C. J. Read, The invariant subspace problem for a class of Banach spaces. II. Hypercyclic operators, *Israel J. Math.* **63**, pp. 1–40 (1988).
[56] S. A. Shkarin, On the growth of D-universal functions, *Vestnik Moskov. Univ. Ser. I Mat. Mekh.* **1994**, no. 6, pp. 80–83 (1993); translation in *Moscow Univ. Math. Bull.* **48**, no. 6, pp. 49–51 (1993).
[57] S. Shkarin, On the spectrum of frequently hypercyclic operators, *Proc. Amer. Math. Soc.* **137**, pp. 123–134 (2009).

Classical and new aspects of the domination and factorization of multilinear operators

M. Mastyło
*Faculty of Mathematics and Computer Science,
Adam Mickiewicz University in Poznań,
Umultowska 87,
61-614 Poznań, Poland
E-mail: mastylo@amu.edu.pl*

E. A. Sánchez Pérez
*Instituto Universitario de Matemática Pura y Aplicada,
Universitat Politècnica de València,
Camino de Vera s/n,
46022 Valencia, Spain
E-mail: easancpe@mat.upv.es*

General aspects of the domination and factorization properties of multilinear operators from quasi-Banach lattices to quasi-Banach spaces are presented.

Using a modern approach that allows a full generality, we analyze some classical and new results, showing when certain vector valued norm inequalities provide factorization through some well-known Banach function lattices, as for instance Orlicz spaces. For example, we show that our framework allows to prove a multilinear variant of the Grothendieck factorization theorem for products of $C(K)$-spaces.

Keywords: Multilinear operators, factorization, domination, quasi-Banach function lattice, Banach envelope, Orlicz space.

1. Introduction

Domination and factorization theorems for (linear and continuous) operators between Banach spaces are nowadays standard tools in Functional Analysis.

For example, the rewriting of the classical summability problems in terms of operators allows to understand in the modern framework of the Functional Analysis these problems as domination properties for an operator. The solution of the problems appearing in this context is sometimes due to a factorization theorem that is obtained as a consequence of the domination: The properties of the factorization space; for example, L^p-spaces, is very often the key for solving the original problems.

This is for example what happens in the case of the p-summing operators, whose powerful characterization by means of the factorization through the identification map $i\colon C(K) \to S \subseteq L^p(\mu)$ getting its values in a subspace S of an L^p-space [12, 27], allows to solve a lot of summability problems of the Banach space theory. Also, p-integral operators present other interesting case of a class of operators that can be characterized by a nice factorization property (see [8] or [27]); absolutely continuous operators allow similar characterizations by means of factorizations (see [19, 22]). The abstract version that unifies all of these cases has been also recently obtained, providing a new general point of view for understanding all these problems (see [4]). Other class of examples is given by the so-called Maurey-Rosenthal factorization theorem for Banach function spaces, that shows the requirements on p-convexity of the original space and q-concavity of the operator – a domination property –, in order to factor it through an L^p-space (see [7, 10, 11] and [23, Ch. 6]). Recently, the authors of the present paper have shown that the main arguments that prove the results in these references – that are already a general reformulation of the classical results of Maurey and Rosenthal on factorization of operators through L^p-spaces, can in fact be written for domination and factorizations of operators through Orlicz spaces. Other interesting case of factorization of operators allowing to obtain relevant results on $(q,1)$-summing operators, is the Pisier's Theorem of factorization through Lorentz spaces ([29], see also [15]).

The aim of this work is to show some aspects of the tools for transferring this general theory on domination and factorization of operators to the setting of the multilinear maps, mainly the ones related with the multilinear versions of the Maurey-Rosenthal type theorems of factorization through Orlicz spaces and the theorem and Pisier of factorization through Lorentz spaces.

Recently, some effort has been made in order to find suitable versions of the classical factorization theorems for linear operators to the case of the multilinear map, and also to the case of polynomials ([2, 6, 9, 13, 21, 24–26]). A classical relevant example of factorization of multilinear operator is given by the Grothendieck's theorem, that asserts that every bounded bilinear functional on $C(K_1) \times C(K_2)$ is canonically extendable to a bounded bilinear functional on $L^2(\mu_1) \times L^2(\mu_2)$, where μ_1 and μ_2 are probability Borel measures on K_1 and K_2 (see [28, Theorem 5.5]). A general multilinear extension can be found in [3]. This is not an isolated result on factorization of multilinear maps. Actually, in recent times a big effort has been made in order to find the multilinear versions of a lot of the most important operator ideals. For example, the reader can find a lot of information about the different extension to this setting of the operator ideals of p-summing and (q,p)-summing operators for example in [4, 25]; some extensions of this notion, for example the operator ideal of the p,σ-absolutely continuous multilinear operators, has also been recently studied in [1].

In this work we are interested in a more specific class of multilinear operators: The ones defined in the product of Banach function lattices, having values in a Banach space and factoring through Orlicz spaces or some other class of well-known lattices. The techniques used here are similar to the ones that proved the Pietsch factorization theorem — that is, mainly Hahn-Banach theorem applied to a class of functions defined by the summability inequalities—, but is more related to lattice type geometric inequalities as p-convexity and q-concavity, extending the arguments that prove the Maurey-Rosenthal theorem in the linear case. Also, some results regarding Pisier's theorem will be explained; it is relevant to say that the arguments used for obtaining these results are essentially different to the ones given for the results explained before.

After some introductory material, we will explain *two main methods* that can be considered when we are facing the problem of factorization of multilinear maps.

1) The first proposed factorization method consists on a modification/improvement of the domain of the n-linear map, in a way that it still is n-linear but defined in spaces with better known structure. This provides factorization schemes as

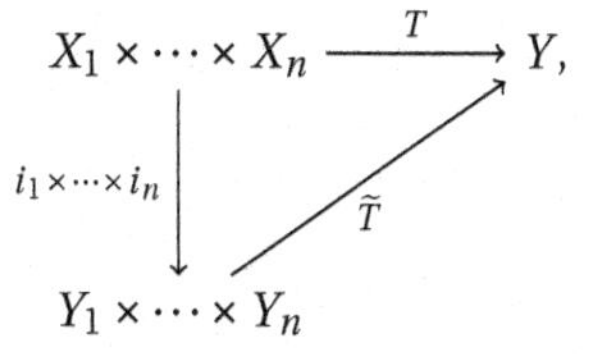

where $i_1, \ldots, i_n$ are some sort of canonical linear operators and the right hand side map closing the diagram is a multilinear operator.

2) The second factorization that we propose is related to the linearization of the original bilinear map, factoring through a Banach space by means of a multilinear map I that must be in a sense canonical, as the a.e. pointwise product in the case of Banach lattices of functions, or the tensor product map. In this case, the right hand side operator $\hat{T}$ closing the diagram,

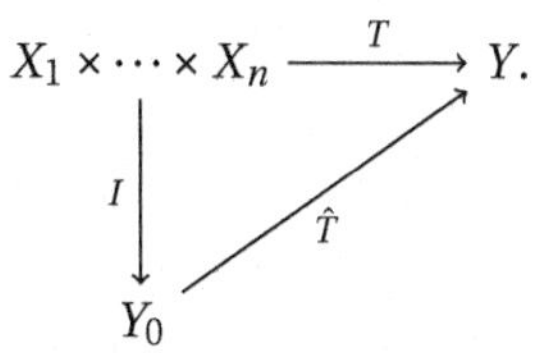

is a linear map.

In this work we will explain the main arguments and results that can be obtained in both directions.

In the first section we will explain the multilinear version of the Maurey-Rosenthal theorem. In the next section, we will give some ideas on the suitable results regarding Pisier's theorem on factorization through Lorentz spaces of $(q,1)$-summing operators. We will explain the case of bilinear operators: However, we must say that the procedure that we use can be extended to the multilinear setting in a more or less obvious way.

2. Preliminaries

We begin by reminding some notations and definitions. A vector space X equipped with a quasi-norm $\|\cdot\|_X$ is called a quasi-Banach space provided that it is complete. The Mackey semi-norm $\|\cdot\|_X^c$ on such a space X is defined as the Minkowski functional of the convex hull of the unit ball $B_X := \{x \in X;\ \|x\|_X \le 1\}$, that is

$$\|x\|_X^c = \inf\{\lambda > 0;\ x \in \lambda\,\mathrm{conv}(B_X)\}, \quad x \in X.$$

Clearly, $\|\cdot\|_X^c$ is a norm on X if and only if the dual X^* separates the points of X. We note that X^* is a Banach space under the norm

$$\|x^*\|_{X^*} = \sup_{x \in B_X} |x^*(x)|,$$

and we have $X^* = (X, \|\cdot\|_X^c)^*$ with equality of norms.

If X is a quasi-Banach space whose dual X^* separates the points of X, then the completion of $(X, \|\cdot\|_X^c)$ is called the *Banach envelope* of X and is denoted by $\widehat{X}$. We observe that for $\kappa\colon X \to X^{**}$ being the canonical embedding defined by $\kappa x(x^*) = x^*(x)$ for all $x \in X$ and $x^* \in X^*$, we have

$$\|\kappa x\|_{X^{**}} = \sup_{\|x^*\|_{X^*} \le 1} |x^*(x)| = \|x\|_X^c.$$

Throughout the paper $(\Omega, \mu) := (\Omega, \Sigma, \mu)$ stands for a complete σ-finite measure space and $L^0(\mu) := L^0(\Omega, \Sigma, \mu)$ denotes the space of (equivalence classes of μ-a.e. equal) complex-valued measurable functions on Ω with the topology of convergence in measure on μ-finite sets. As usual the order $|f| \le |g|$ means that $|f(t)| \le |g(t)|$ for μ-almost all $t \in \Omega$. A subset X of $L^0(\mu)$ is said to be *solid* if conditions $|f| \le |g|$, $f \in L^0(\mu)$, and $g \in X$ imply $f \in X$. If in addition X is a quasi-normed (resp., quasi-Banach) space, then X is said to be a quasi-normed (resp., quasi-Banach) lattice on (Ω, μ).

A quasi-normed function lattice X on (Ω, μ) is said to be order continuous provided $0 \le x_n \downarrow 0$ a.e. implies $\|x_n\|_X \to 0$. The Köthe dual space X' of a normed function lattice X on (Ω,μ) is defined to be the space of all $x \in L^0(\mu)$ such that $\int_\Omega |xy|\,d\mu < \infty$ for every $y \in X$. It is a Banach function lattice on (Ω,μ) when equipped with the norm

$$\|x\|_{X'} = \sup_{\|y\|_X \le 1} \int_\Omega |xy|\,d\mu.$$

It is well-known that a normed function lattice X on (Ω,μ) is order continuous if and only if the map $X' \ni y \mapsto x_y^* \in X^*$ given by the definition of the elements of X' as elements of X^*, i.e.,

$$x_y^*(x) := \int_\Omega xy\,d\mu, \quad x \in X,$$

is an order isometrical isomorphism of X' onto X^* (see, e.g., [17]); it can be proved that if $(X, \|\cdot\|_X)$ is a quasi-normed function lattice on (Ω,μ) such that the topological dual X^* separates the points of X, then $(X, \|\cdot\|_X^c)$ is a normed function lattice on (Ω,μ), which is order continuous provided X is order continuous.

We will use the symbol Φ to denote the set of all increasing and continuous functions $\varphi\colon [0,\infty) \to [0,\infty)$ such that $\varphi(0) = 0$. As usual, a function $\varphi \in \Phi$ is said to satisfy the Δ_2-condition ($\varphi \in (\Delta_2)$ for short) provided there exists $C > 0$ such that $\varphi(2t) \le C\varphi(t)$ for all $t > 0$.

Let X be a quasi-normed lattice on (Ω,μ) and $\varphi \in \Phi$. We define an order ideal in $L^0(\mu)$ as

$$X_\varphi = \{f \in L^0(\mu);\ \exists \lambda > 0, \varphi(\lambda|f|) \in X\},$$

and a functional $\|\cdot\|_{X_\varphi}\colon X_\varphi \to [0,\infty)$ by

$$\|f\|_{X_\varphi} = \inf\{\lambda > 0;\ \|\varphi(|f|/\lambda)\|_X \le 1\}, \quad f \in X_\varphi.$$

It is easy to see that $\|f\|_{X_\varphi} = 0$ if and only if $f = 0$ and $\|\lambda f\|_{X_\varphi} = |\lambda|\,\|f\|_{X_\varphi}$ for all $\lambda \in \mathbb{R}$, $f \in X_\varphi$.

There is a large class of functions $\varphi \in \Phi$ for which $\|\cdot\|_{X_\varphi}$ is a quasi-norm on X_φ. For instance, if φ is a convex function and X is a Banach function lattice, X_φ is a Banach function lattice. In the case when $X = L_1(\mu)$ and $\varphi \in \Phi$, X_φ is the Orlicz space denoted as usual by $L_\varphi(\mu)$. This fact motivates to call X_φ the generalized Orlicz space provided X_φ is a quasi-Banach space.

An important case is given by the function $\varphi(t) = t^p$ for a fixed $0 < p < \infty$, that is defined for all $t \ge 0$. The space X_φ is known as the *p-convexification* X_p of X. Its quasi-norm is given by

$$\|f\|_{X_p} = \||f|^p\|_X^{1/p}, \quad f \in X_p.$$

A relevant fact on these spaces is related to the following classical lattice geometric definition. A quasi-Banach lattice X is p-convex for $0 < p < \infty$ if there is a constant $C > 0$ such that

$$\Big\| \Big(\sum_{k=1}^{n} |x_k|^p \Big)^{1/p} \Big\|_X \le C \Big(\sum_{k=1}^{n} \|x_k\|^p \Big)^{1/p}, \quad x_1, \ldots, x_n \in X.$$

Then, X is p-convex if and only if $X_{1/p}$ is normable.

A Banach space valued operator $T\colon E \to Y$ is said to be *p-concave* for $1 < p < \infty$ if there is a constant $C > 0$ so that

$$\Big(\sum_{k=1}^{n} \|T(x_k)\|_E^p \Big)^{1/p} \le C \Big\| \Big(\sum_{k=1}^{n} |x_k|^p \Big)^{1/p} \Big\|_E, \quad x_1, \ldots, x_n \in E;$$

if T is the identity in E, we say that E is *p-concave.*

We need same technical definition, that will play a crucial role in our construction. For a given function $\varphi \in \Phi$ and a quasi-Banach function lattice X, we will say that X is *φ-admissible* if $\|\cdot\|_{X_\varphi}$ is a quasi-norm on X_φ. If the topological dual $(X_\varphi)^*$ separates the points of X_φ, then we will say that X is *strongly φ-admissible.* For the case $\varphi(t) = t^p$, we simply say that the space X is strongly p-admissible. Order continuity of X_φ is often inherited from X; if an order continuous Banach function lattice X is φ-admissible and φ satisfies the Δ_2-condition, then $\|f_n\|_{X_\varphi} \to 0$ if and only if $\|\varphi(|f_n|)\|_X \to 0$. This implies that in this case X_φ is order continuous if and only if X is order continuous.

3. Factoring through products of Banach function lattices: The first factorization scheme

The main result that provides the tool for this factorization is the following separation theorem. The proof is based on Ky-Fan's Lemma (see, e.g., [12, Lemma 9.10]); a similar proof can be given using the Hahn-Banach theorem; for details we refer to [20].

Theorem 3.1. *Consider scalar functions $\phi, \varphi_1, \varphi_2 \in \Phi$ and let X, Y be quasi-Banach function lattices on (Ω_1, μ_1) and (Ω_2, μ_2) such that X is strongly φ_1^{-1}-admissible and Y is strongly φ_2^{-1}-admissible. Suppose that T is a bilinear operator from $X \times Y$ into a quasi-Banach space E. Assume $0 < C_1, C_2 < \infty$ and that $A \subset X$, $B \subset Y$ are non-empty sets. Consider the following statements:*

(i) *For any finite set of positive scalars $\{\alpha_k\}_{k=1}^n$ with $\sum_{k=1}^n \alpha_k = 1$ and any finite sets $\{f_k\}_{k=1}^n$ in A and $\{g_k\}_{k=1}^n$ in B, $n \in \mathbb{N}$, the following inequality holds:*

$$\sum_{k=1}^{n} \alpha_k \phi(\|T(f_k, g_k)\|_E) \le C_1 \Big\| \sum_{k=1}^{n} \alpha_k \varphi_1(|f_k|) \Big\|_{X_{\varphi_1^{-1}}}^c + C_2 \Big\| \sum_{k=1}^{n} \alpha_k \varphi_2(|g_k|) \Big\|_{Y_{\varphi_2^{-1}}}^c .$$

(ii) *There exist positive functionals* $x^* \in (X_{\varphi_1^{-1}})^*$ *and* $y^* \in (Y_{\varphi_2^{-1}})^*$ *such that*

$$\phi(\|T(f,g)\|_E) \le C_1\, x^*(\varphi_1(|f|)) + C_2\, y^*(\varphi_2(|g|)), \quad (f,g) \in A \times B.$$

(iii) *There exist* $0 \le u \in B_{(X_{\varphi_1^{-1}})'}$ *and* $0 \le v \in B_{(Y_{\varphi_2^{-1}})'}$ *such that*

$$\phi(\|T(f,g)\|_E) \le C_1 \int_{\Omega_1} \varphi_1(|f|)\,u\,d\mu_1 + C_2 \int_{\Omega_2} \varphi_2(|g|)\,v\,d\mu_2,$$

for $(f,g) \in A \times B$.

Then (i) is equivalent to (ii). If $X_{\varphi_1^{-1}}$ *and* $Y_{\varphi_2^{-1}}$ *are order continuous, then all three statements are equivalent.*

The requirements on the functions φ and ϕ appearing in the theorem above can be rewritten in easier terms if they are related to the classical p-convexity/q-concavity inequalities. Using the same arguments, we get the following

Theorem 3.2. *Let* $0 < p, q < \infty$. *Let* X *and* Y *be quasi-Banach lattices such that the corresponding dual spaces* $(X_{1/p})^*$ *and* $(Y_{1/q})^*$ *separate the points of* $X_{1/p}$ *and* $Y_{1/q}$. *Assume* $\phi \in \Phi$, $0 < C_1, C_2 < \infty$ *and that* $A \subset X$, $B \subset Y$ *are non-empty sets. The following are equivalent statements about a bilinear operator* T *from* $X \times Y$ *to a quasi-Banach space* E*:*

(i) *For any set of positive scalars* $\{\alpha_k\}_{k=1}^n$ *with* $\sum_{k=1}^n \alpha_k = 1$ *and any sets* $\{f_k\}_{k=1}^n$ *in* A *and* $\{g_k\}_{k=1}^n$ *in* B, $n \in \mathbb{N}$,

$$\sum_{k=1}^n \alpha_k \phi(\|T(f_k,g_k)\|_E) \le C_1 \Big\| \sum_{k=1}^n \alpha_k |f_k|^p \Big\|_{X_{1/p}}^c + C_2 \Big\| \sum_{k=1}^n \alpha_k |g_k|^q \Big\|_{Y_{1/q}}^c.$$

(ii) *There exist positive functionals* $x^* \in B_{(X_{1/p})^*}$ *and* $y^* \in B_{(Y_{1/q})^*}$ *such that*

$$\phi(\|T(f,g)\|_E) \le C_1\, x^*(|f|^p) + C_2\, y^*(|g|^q), \quad (f,g) \in A \times B.$$

An example of the inequalities that appear in Theorem 3.2 is given by Grothendieck's Theorem: For a pair K_1, K_2 of compact Hausdorff spaces and a bilinear functional T on $C(K_1) \times C(K_2)$ there are probability Borel measures μ_1, μ_2, on K_1, K_2 such that for each pair of functions f_1 and f_2,

$$|T(f_1,f_2)| \le K_G \|T\| \Big(\int_{K_1} |f_1|^2 d\mu_1\Big)^{1/2} \Big(\int_{K_2} |f_2|^2 d\mu_2\Big)^{1/2},$$

where K_G is the Grothendieck's constant. By Young's inequality, this is equivalent to

$$|T(f_1,f_2)| \le \frac{1}{2} K_G \|T\| \Big(\int_{K_1} |f_1|^2 d\mu_1 + \int_{K_2} |f_2|^2 d\mu_2\Big),$$

that gives an example of what we have in Theorem 3.2.

A reformulation of a well-known result regarding bilinear maps on spaces with non-trivial convexity is available by using again this theorem (see [7, Theorem 1]).

Theorem 3.3. *Let $0 < p, q < \infty$ and let $1/r = 1/p + 1/q$. Assume that X and Y are quasi-Banach lattices of functions such that they are strongly $1/p$-admissible and strongly $1/q$-admissible, respectively. Assume $0 < C < \infty$. The following are equivalent statements about a bilinear operator T from $X \times Y$ to a quasi-Banach space E:*

(i) *For each couple of finite sets $\{f_k\}_{k=1}^n$ and $\{g_k\}_{k=1}^n$ of elements of X and Y, respectively,*

$$\Big(\sum_{k=1}^n \|T(x_k, y_k)\|_E^r\Big)^{1/r} \le C\Big(\Big\|\sum_{k=1}^n |x_k|^p\Big\|_{X_{1/p}}^c\Big)^{1/p}\Big(\Big\|\sum_{k=1}^n |y_k|^q\Big\|_{Y_{1/q}}^c\Big)^{1/q}.$$

(ii) *There are positive functionals $x^* \in B_{(X_{1/p})^*}$ and $y^* \in B_{(Y_{1/q})^*}$ such that*

$$\|T(x,y)\|_E \le C\big(x^*(|x|^p)\big)^{1/p}\big(y^*(|y|^q)\big)^{1/q}, \quad (x,y) \in X \times Y.$$

Well-known classical results can be obtained as a consequence of Theorem 3.2. The following one is an extension of Grothendieck's theorem, that is due to Blei (see [3, Theorem 3.2]). The second one is a consequence of the arguments given above for p-convex lattices if no order continuity is assumed.

Corollary 3.1. *Suppose that $K_1, \dots, K_n$ are compact Hausdorff spaces and let $1 \le p_j < \infty$, for each $1 \le j \le n$. Let $0 < C < \infty$ and $\sum_{j=1}^n 1/p_j = 1$. The following are equivalent statements about an n-linear functional U on $C(K_1) \times \cdots \times C(K_n)$:*

(i) *For any set $\{f_j^{(k)}\}_{j=1}^m$ in $C(K_k)$ with $k = 1, \dots, n$,*

$$\sum_{j=1}^m \big|U(f_j^{(1)}, \dots, f_j^{(n)})\big| \le C\Big\|\Big(\sum_{j=1}^m |f_j^{(1)}|^{p_1}\Big)^{1/p_1}\Big\|_{C(K_1)} \cdots \Big\|\Big(\sum_{j=1}^m |f_j^{(n)}|^{p_n}\Big)^{1/p_n}\Big\|_{C(K_n)}.$$

(ii) *There exist probability Borel measures $\mu_1, \dots, \mu_n$ on $K_1, \dots, K_n$, respectively, so that*

$$|U(f_1, \dots, f_n)| \le C\Big(\int_{K_1} |f_1|^{p_1}\, d\mu_1\Big)^{1/p_1} \cdots \Big(\int_{K_n} |f_n|^{p_n}\, d\mu_n\Big)^{1/p_n}$$

for all $f_1 \in C(K_1), \dots, f_n \in C(K_n)$.

Corollary 3.2. *Let $1 < p < \infty$, $1 < p_k < \infty$ for each $1 \le k \le n$ be such that $1/p = 1/p_1 + \cdots + 1/p_n$. Assume that X_k is a p_k-convex Banach lattice for each $1 \le k \le n$, and let E be a p-concave Banach lattice. For any n-linear positive operator*

$T: X_1 \times \cdots \times X_n \to Y$ there exist a constant $C > 0$ and positive functionals $x_k^* \in B_{(X_{1/p_k})^*}$ so that for all $x_1 \in X_1, \dots, x_n \in X_n$,

$$\|T(x_1,\dots,x_n)\|_E \le C\big(x_1^*(|x_1|^{p_1})\big)^{1/p_1} \cdots \big(x_n^*(|x_n|^{p_n})\big)^{1/p_n}.$$

Let us finish this section by centering the attention on bilinear forms. Following a classical way of understanding the linear operators $T: X \to Y$ by means of the bilinear form $(x, y') \mapsto \langle T(x), y'\rangle$ for every $x \in X$ and $y' \in Y^*$.

Theorem 3.4. *Let $\varphi_1, \varphi_2 \in \Phi$ with $\varphi_1, \varphi_2 \in (\Delta_2)$, and let X, Y be order continuous Banach function lattices on (Ω_1, μ_1) and (Ω_2, μ_2), respectively, such that X is strongly φ_1^{-1}-admissible and Y is strongly φ_2^{-1}-admissible. Suppose $0 < C_1, C_2 < \infty$ and consider an operator $S: X \to Y'$, where Y' is the Köthe dual of Y. Consider the following statements about the bilinear form $T: X \times Y \to \mathbb{R}$ defined by $T(f,g) = \int_{\Omega_2} g S(f)\, d\mu_2$ for all $(f,g) \in X \times Y$:*

(i) For every finite sequence of positive scalars $\{\alpha_k\}_{k=1}^n$ with $\sum_{k=1}^n \alpha_k = 1$ and every finite sequences $\{f_k\}_{k=1}^n$ in X and $\{g_k\}_{k=1}^n$ in Y,

$$\sum_{k=1}^n \alpha_k |T(f_k, g_k)| \le C_1 \Big\| \sum_{k=1}^n \alpha_k \varphi_1(|f_k|) \Big\|_{X_{\varphi_1^{-1}}}^c + C_2 \Big\| \sum_{k=1}^n \alpha_k \varphi_2(|g_k|) \Big\|_{Y_{\varphi_2^{-1}}}^c.$$

(ii) There exist functions $0 \le u \in B_{(X_{\varphi_1^{-1}})'}$ and $0 \le v \in B_{(X_{\varphi_2^{-1}})'}$ such that

$$|T(f,g)| \le C_1\Big(\int_{\Omega_1} \varphi_1(|f|)\, u\, d\mu_1\Big) + C_2\Big(\int_{\Omega_2} \varphi_2(|g|)\, v\, d\mu_2\Big),$$

for $(f,g) \in X \times Y$.

(iii) There exist $0 \le u \in (X_{\varphi_1^{-1}})'$ and $0 \le v \in (X_{\varphi_2^{-1}})'$ such that S admits the following factorization:

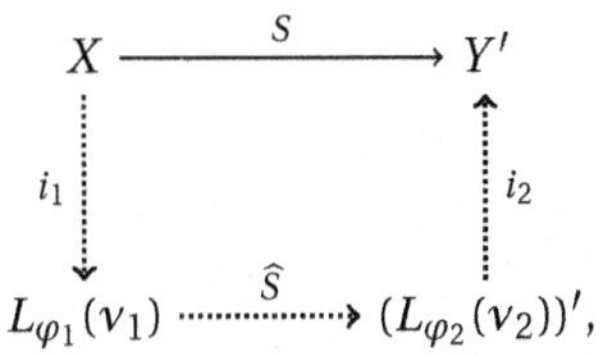

where $L_{\varphi_1}(v_1)$ and $L_{\varphi_1}(v_2)$ are Orlicz spaces on (A_1, Σ_{A_1}, v_1) with $A_1 = \operatorname{supp} u$, $dv_1 = u\, d\mu_1$ and on (A_2, Σ_{A_2}, v_2) with $A_2 = \operatorname{supp} v$, $dv_2 = v\, d\mu_2$, $\widehat{S}$ is a continuous operator, and $i_1: X \to L_{\varphi_1}(v_1)$ and $i_2: (L_{\varphi_2}(v_2))' \to Y'$ are operators given by $i_1(f) = f\chi_{A_1}$ for all $f \in X$ and $i_2(g) = g\chi_{A_2}$ for all $g \in (L_{\varphi_2}(v_2))'$.

Then (i) is equivalent to (ii) and both of them imply (iii).

4. Factorization through tensor products of Banach function lattices

Each multilinear map can be factored through the projective tensor product of the spaces in which the operator acts. This is sometimes called the linearization of the original bilinear map, since the operator that closes the diagram from the tensor product is linear, and the multilinear operator in left hand side of the scheme is the tensor product. That is, we get that T can be factored as

$$\begin{array}{ccc} X_1 \times \cdots \times X_n & \xrightarrow{\quad T \quad} & Y. \\ {\scriptstyle \otimes}\big\downarrow & \nearrow {\scriptstyle \widetilde{T}} & \\ X_1 \hat{\otimes}_\pi \cdots \hat{\otimes}_\pi X_n & & \end{array}$$

However, it is well known, that in general this space does not have a Banach lattice structure, even if the components $X_1, \ldots, X_n$ are Banach lattices. This opens a door for studying some interesting questions related to factorization of multilinear operators. For example, is it possible to factor each multilinear map from a product of Banach lattices through a Banach lattice by putting a standard multilinear map in the left part of the factorization? That is, if $X_1(\mu_1), \ldots, X_n(\mu_n)$ are for instance Banach function spaces, the question is if we can find a factorization through a multiplication map $(f_1, \ldots, f_n) \mapsto f_1 \cdots f_n \in Z(\mu_1 \otimes \cdots \otimes \mu_n)$, for a Banach function space over the product measure Z, as

$$\begin{array}{ccc} X_1(\mu_n) \times \cdots \times X_n(\mu_n) & \xrightarrow{\quad T \quad} & Y. \\ {\scriptstyle \otimes}\big\downarrow & \nearrow {\scriptstyle \widetilde{T}} & \\ Z(\mu_1 \otimes \cdots \otimes \mu_n) & & \end{array}$$

The answer to this general question is no. For example, take the tensor product bilinear operator $\ell^2 \times \ell^2 \to \ell^2 \otimes_\pi \ell^2$; if there is a factorization as the one given above, then we find a Banach lattice through which we can factor the tensor product. By factoring again through this projective tensor product we get the final result that it is isomorphic to a Banach lattice, which is false.

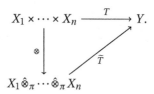

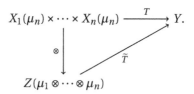

Some research has been made recently on this subject. For example, using vector measures and Banach function spaces techniques, and under the assumption of a factorization as the one above, in [5] it has been studied if it is possible to characterize the maximal Banach function space to which the bilinear map can be extended by continuity. Using integration with respect to bimeasures (see [18]), new attempts to find this space are being developed at the moment.

We will center our attention in a special factorization that has a previous requirement a factorization through the ε tensor product of the bilinear map. This will leave to find a new version of Pisier's factorization theorem on factorization through Lorentz spaces for bilinear maps.

Recall that the factorization theorem due to Pisier [29] establishes that an operator T from a $C(K)$-space to a Banach space Y is (q,p)-summing with $1 \le p < q < \infty$ if and only if there is a probability Borel measure μ on K such that T factors as $T = S \circ j$, where j is the identification map $j\colon C(K) \to L_{q,1}(\mu)$, where the second space is the Lorentz space on the Borel measure space $(K, \mathscr{B}(K), \mu)$ (see, e.g., [12, pp. 326–345]).

As we explained before, two factorization schemes could be available. Let us explain the two cases separately in the following sections.

4.1. *The linearization of the bilinear map through the ε-tensor product*

Let $X_1,\ldots,X_n$, Y be Banach spaces, and let $1 \le q, p < \infty$. We say that an n-linear operator $T\colon X_1 \times \cdots \times X_n \to Y$ is factorable (q,p)-summing if there is a constant $C > 0$ such that for all matrices $\big(x_{jk}^{(1)}\big)_{j=1,k=1}^{M,N},\ldots,\big(x_{jk}^{(n)}\big)_{j=1,k=1}^{M,N}$ in $X_1,\ldots,X_n$, we have

$$\Big(\sum_{j=1}^{M}\Big\|\sum_{k=1}^{N} T\big(x_{jk}^{(1)},\ldots,x_{jk}^{(n)}\big)\Big\|_Y^q\Big)^{1/q} \le C \sup_{x_1^*\in B_{X_1^*},\ldots,x_n^*\in B_{X_n^*}} \Big(\sum_{j=1}^{M}\Big|\sum_{k=1}^{N}\langle x_{jk}^{(1)},x_1^*\rangle\cdots\langle x_{jk}^{(n)},x_n^*\rangle\Big|^p\Big)^{1/p},$$

(see also [30]). Let us show that this definition implies automatically a factorization for the multilinear map. In fact, this definition implies a previous factorization for the multilinear map. If T is a factorable (q,p)-summing n-linear operator $T\colon X_1 \times \cdots \times X_n \to Y$, then T can be factored as

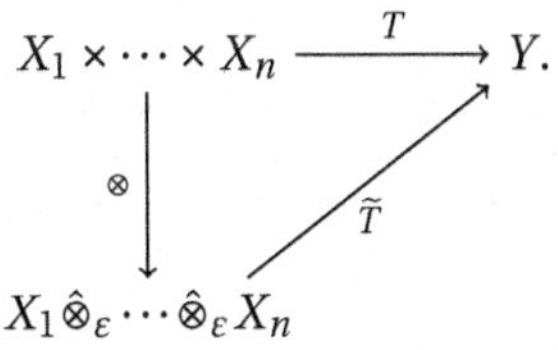

Here $\otimes$ is the tensor product n-linear mapping and $\widetilde{T}$ is a bounded linear operator. Let L be a space of scalar functions defined on the product $K_1 \times \cdots \times K_n$

of compact Hausdorff spaces $K_1,\dots,K_n$ ($n \geq 2$). We write $\odot$ for the map from $C(K_1 \times \cdots \times K_n)$ into L given by

$$\odot(f_1,\dots,f_n)(t_1,\dots,t_n) := f_1(t_1)\cdots f_n(t_n)$$

for all $f_i \in C(K_i)$, $t_i \in K_i$ and each $1 \leq i \leq n$. With these tools, and using the linear Pisier's theorem, we can obtain the following result.

Theorem 4.1. *Let $1 \leq p < q < \infty$. The following statements are equivalent for a bilinear map $T\colon C(K_1) \times C(K_2) \to Y$:*

(i) T is factorable (q,p)-summing.
(ii) There is a constant $C > 0$ such that for all matrices (f_{jk}) and (g_{jk}) in $C(K_1)$ and $C(K_2)$,

$$\Big(\sum_{j=1}^{M}\Big\|\sum_{k=1}^{N} T(f_{jk},g_{jk})\Big\|^q\Big)^{1/q} \leq C\Big\|\Big(\sum_{j=1}^{M}\Big|\sum_{k=1}^{N} \odot(f_{jk},g_{jk})\Big|^p\Big)^{1/p}\Big\|_{C(K_1\times K_2)}.$$

(iii) There is a probability Borel measure μ on $K_1 \times K_2$ such that T allows a factorization as:

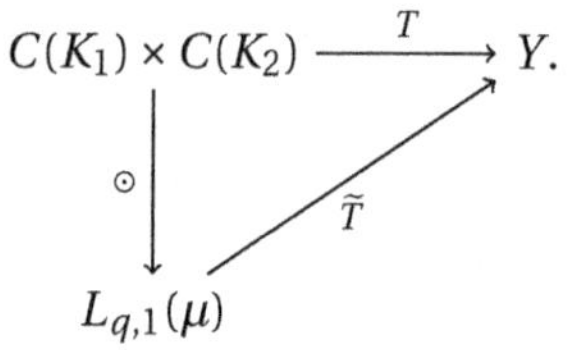

4.2. *The composition factorization for* $(r;p,q)$*-summing operators*

Let $1 \leq r,p,q < \infty$ satisfying that $1/r \leq 1/p + 1/q$. A Banach space valued bilinear operator $T\colon C(K_1) \times C(K_2) \to Y$ is $(r;p,q)$-summing if there is a constant $C > 0$ such that for any choice of finitely many elements $f_1,\dots,f_n$ in $C(K_1)$ and $g_1,\dots,g_n$ in $C(K_2)$,

$$\Big(\sum_{k=1}^{n}\|T(f_k,g_k)\|^r\Big)^{1/r} \leq C\Big\|\Big(\sum_{k=1}^{n}|f_k|^p\Big)^{1/p}\Big\|_{C(K_1)}\Big\|\Big(\sum_{k=1}^{n}|g_k|^q\Big)^{1/q}\Big\|_{C(K_2)}.$$

It is well known that for $1/r = 1/p + 1/q$, $(r;p,q)$-summing bilinear operators satisfy a Pietsch's type factorization theorem. A bilinear operator $T\colon C(K_1) \times C(K_2) \to Y$ is $(r;p,q)$-summing if and only if it factors as:

$$\begin{array}{ccc} C(K_1)\times C(K_2) & \xrightarrow{\quad T\quad} & Y, \\ {\scriptstyle i\times i}\big\downarrow & \nearrow_{S} & \\ L^p(\mu_1)\times L^q(\mu_2) & & \end{array}$$

for probability measures μ_1 and μ_2 on K_1 and K_2 (see [14, 21]). This corresponds to the case that we have called the first factorization scheme in Section 3. In the linear case, Pisier's Theorem states that for $p < q$, a (q,p)-summing operator from $C(K)$ factors through a Lorentz space. So, the question is: For $p/r < 1$ and $q/r < 1$, a bilinear $(r;p,q)$-summing operator T from the product $C(K_1)\times C(K_2)$ to a Banach space Y factors as follows:

$$\begin{array}{ccc} C(K_1)\times C(K_2) & \xrightarrow{\quad T\quad} & Y, \\ {\scriptstyle i\times i}\big\downarrow & \nearrow_{S} & \\ L_{s,1}(\mu_1)\times L_{t,1}(\mu_2) & & \end{array}$$

for two probability Borel measures μ_1 and μ_2 and any $1\le s,t<\infty$? The answer to this question is no in general. It is easy to find counterexamples. In fact, it is possible to show that there are bilinear maps from $C(K_1)\times C(K_2)$ which cannot be factored through any Banach lattice: It is enough to consider a bilinear operator $C(K)\times C(K)\ni(f,g)\longmapsto\langle g,\varphi\rangle P(f)$, where P is any positive non weakly compact operator from $C(K)$ to a Banach lattice Y and φ a non-trivial functional.

However, we can find a substitute of Pisier's Theorem in this setting, that provides a characterization of $(r;p,q)$-summability in our context. Let us define a new space. Let $p,q,r\in[1,\infty)$ be such that $p/r\le 1$ and $q/r\le 1$. Take two compact Hausdorff spaces K_1 and K_2 and two probability Borel measures μ_1 and μ_2 on these spaces. We define a seminorm $\pi_{r;p,q}$ on the tensor product $C(K_1)\otimes C(K_2)$ by the formula:

$$\pi_{r;p,q}(z):=\inf\Big\{\sum_{k=1}^{n}\big(\|f_k\|_{L^p(\mu_1)}^{p/r}\|f_k\|_{C(K_1)}^{1-p/r}\|g_k\|_{C(K_2)}+\|g_k\|_{L^q(\mu_2)}^{q/r}\|g_k\|_{C(K_2)}^{1-q/r}\|f_k\|_{C(K_1)}\big)\Big\},$$

where the infimum is taken over all representations $z=\sum_{k=1}^n f_k g_k$ of $z\in C(K_1)\otimes C(K_2)$. Our space $C(K_1)\hat{\otimes}_{\pi_{r;p,q}}C(K_2)$ is defined as the completion of the quotient space given by the tensor product $C(K_1)\otimes_{\pi_{r;p,q}}C(K_2)$ and the seminorm $\pi_{r;p,q}$.

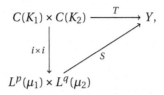

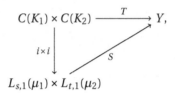

Using these spaces, we obtain the following result, in which the reasonable factorization for $(r; p, q)$-summing bilinear maps is established, although it becomes a characterization only when the bilinear map is positive, that is the image of the products of the positive cones lies in the positive cone of Y.

Theorem 4.2. *Let $1 \leq p, q, r < \infty$ such that $p/r < 1$ and $q/r < 1$. Let $T\colon C(K_1) \times C(K_2) \to Y$ be a bilinear operator. The following statements are equivalent:*

(i) There are probability Borel measures μ_1 and μ_2 on K_1 and K_2 and a constant $C > 0$ such that for every $f \in B_{C(K_1)}$ and $g \in B_{C(K_2)}$,

$$\|T(f,g)\|^r \leq C\Big(\int_{K_1} |f|^p \, d\mu_1 + \int_{K_2} |g|^q \, d\mu_2\Big).$$

(ii) For every finite sequence of positive scalars $\{\alpha_k\}_{k=1}^n$ with $\sum_{k=1}^n \alpha_k = 1$ and every finite sequences $\{f_k\}_{k=1}^n$ in $B_{C(K_1)}$ and $\{g_k\}_{k=1}^n$ in $B_{C(K_2)}$, the following inequality holds

$$\sum_{k=1}^n \alpha_k \|T(f_k, g_k)\|^r \leq C_1 \Big\| \sum_{k=1}^n \alpha_k |f_k|^p \Big\|_{C(K_1)} + C_2 \Big\| \sum_{k=1}^n \alpha_k |g_k|^q \Big\|_{C(K_2)}.$$

(iii) T allows the following factorization:

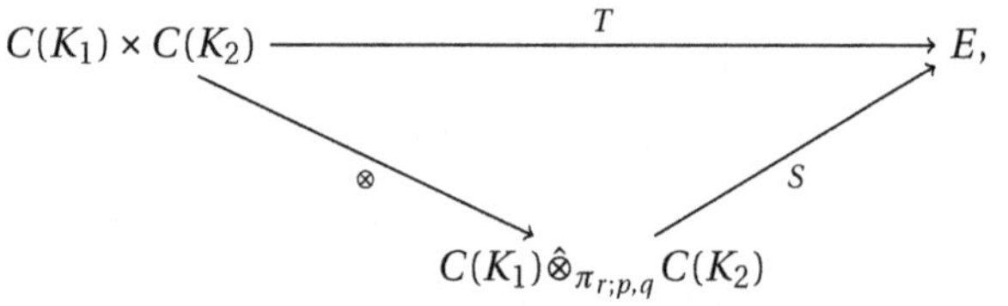

where S is a bounded linear operator.

Moreover, each of the above conditions implies

(iv) T is $(r; p, q)$-summing operator.

All conditions are equivalent whenever T is a positive operator.

References

[1] D. Achour, E. Dahia, P. Rueda and E.A. Sánchez Pérez, Factorization of strongly (p,σ)-continuous multilinear operators, *Linear Multilinear Algebra* **62**, pp. 1649–1670 (2014).

[2] R. Alencar and M. Matos, Some classes of multilinear mappings between banach spaces, in *Publicaciones del Departamento de Análisis Matemático,* (Univ. Complutense de Madrid, Madrid, 1989).

[3] R. C. Blei, Multilinear measure theory and the Grothendieck factorization theorem, *Proc. London Math. Soc. (3)* **56**, pp. 529–546 (1988).

[4] G. Botelho, D. Pellegrino and P. Rueda, A unified Pietsch domination theorem, *J. Math. Anal. Appl.* **365**, pp. 269–276 (2010).

[5] J. M. Calabuig, M. Fernández Unzueta, F. Galaz-Fontes and E. A. Sánchez-Pérez, Extending and factorizing bounded bilinear maps defined on order continuous Banach function spaces, *Rev. R. Acad. Cienc. Exactas Fís. Nat. Ser. A Math. RACSAM* **108**, pp. 353–367 (2014).

[6] E. Dahia, D. Achour and E. A. Sánchez Pérez, Absolutely continuous multilinear operators, *J. Math. Anal. Appl.* **397**, pp. 205–224 (2013).

[7] A. Defant, Variants of the Maurey-Rosenthal theorem for quasi Köthe function spaces, *Positivity* **5**, pp. 153–175 (2001).

[8] A. Defant and K. Floret, *Tensor norms and operator ideals*, North-Holland Mathematics Studies, Vol. 176, pp. xii+566 (North-Holland Publishing Co., Amsterdam, 1993).

[9] A. Defant and M. Mastyło, Interpolation of Fremlin tensor products and Schur factorization of matrices, *J. Funct. Anal.* **262**, pp. 3981–3999 (2012).

[10] A. Defant and E. A. Sánchez Pérez, Maurey-Rosenthal factorization of positive operators and convexity, *J. Math. Anal. Appl.* **297**, pp. 771–790 (2004), Special issue dedicated to John Horváth.

[11] A. Defant and E. A. Sánchez Pérez, Domination of operators on function spaces, *Math. Proc. Cambridge Philos. Soc.* **146**, pp. 57–66 (2009).

[12] J. Diestel, H. Jarchow and A. Tonge, *Absolutely summing operators*, Cambridge Studies in Advanced Mathematics, Vol. 43, pp. xvi+474 (Cambridge University Press, Cambridge, 1995).

[13] V. Dimant, Strongly p-summing multilinear operators, *J. Math. Anal. Appl.* **278**, pp. 182–193 (2003).

[14] S. Geiss, Ideale multilinearer abbildungen, PhD thesis, Jena Universität1984.

[15] N. J. Kalton and S. J. Montgomery-Smith, Set-functions and factorization, *Arch. Math. (Basel)* **61**, pp. 183–200 (1993).

[16] A. Kamińska and M. Mastyło, Abstract duality Sawyer formula and its applications, *Monatsh. Math.* **151**, pp. 223–245 (2007).

[17] L. V. Kantorovich and G. P. Akilov, *Functional analysis*, second edn., pp. xiv+589 (Pergamon Press, Oxford-Elmsford, N. Y., 1982), Translated from the Russian by Howard L. Silcock.

[18] S. Karni and E. Merzbach, On the extension of bimeasures, *J. Analyse Math.* **55**, pp. 1–16 (1990).

[19] J. A. López Molina and E. A. Sánchez Pérez, On operator ideals related to (p,σ)-absolutely continuous operators, *Studia Math.* **138**, pp. 25–40 (2000).

[20] M. Mastyło and E. A. Sánchez Pérez, Domination and factorization of multilinear operators, *J. Convex Anal.* **20**, pp. 999–1012 (2013).

[21] M. C. Matos, On multilinear mappings of nuclear type, *Rev. Mat. Univ. Complut. Madrid* **6**, pp. 61–81 (1993).

[22] U. Matter, Factoring through interpolation spaces and super-reflexive Banach spaces, *Rev. Roumaine Math. Pures Appl.* **34**, pp. 147–156 (1989).

[23] S. Okada, W. J. Ricker and E. A. Sánchez Pérez, *Optimal domain and integral extension of operators*, Operator Theory: Advances and Applications, Vol. 180, pp. xii+400 (Birkhäuser Verlag, Basel, 2008), Acting in function spaces.

[24] D. Pellegrino and J. Santos, Absolutely summing multilinear operators: A panorama, *Quaest. Math.* **34**, pp. 447–478 (2011).
[25] D. Pellegrino and J. Santos, A general Pietsch domination theorem, *J. Math. Anal. Appl.* **375**, pp. 371–374 (2011).
[26] D. Pérez-García and I. Villanueva, Multiple summing operators on $C(K)$ spaces, *Ark. Mat.* **42**, pp. 153–171 (2004).
[27] A. Pietsch, *Operator ideals*, North-Holland Mathematical Library, Vol. 20, pp. 451 (North-Holland Publishing Co., Amsterdam-New York, 1980), Translated from German by the author.
[28] G. Pisier, *Factorization of linear operators and geometry of Banach spaces*, CBMS Regional Conference Series in Mathematics, Vol. 60, pp. x+154 (Published for the Conference Board of the Mathematical Sciences, Washington, DC; by the American Mathematical Society, Providence, RI, 1986).
[29] G. Pisier, Factorization of operators through $L_{p\infty}$ or L_{p1} and noncommutative generalizations, *Math. Ann.* **276**, pp. 105–136 (1986).
[30] P. Rueda and E. A. Sánchez Pérez, Factorization of p-dominated polynomials through L^p-spaces, *Michigan Math. J.* **63**, pp. 345–353 (2014).

Toeplitz products on the Bergman space

María Carmen Reguera*
School of Mathematics, University of Birmingham,
Birmingham, B15 2TT, UK
E-mail: m.reguera@bham.ac.uk
Matematikcentrum, University of Lund,
Lund, 223 62, Sweden
E-mail: mreguera@maths.lth.se

This article reviews the recent developments on the boudedness of products of Toeplitz operators. Special attention is paid to the solution of a conjecture posted by D. Sarason in the 90's on the product of Toeplitz operators in the Bergman space recently achieved by Aleman, Pott and the author in [1].

Keywords: Bergman spaces, Toeplitz products, two-weight inequalities.

1. Introduction

The purpose of this paper is to review some of the recent advances in the study of boundedness of the product of Toeplitz operators in the Bergman space. Much of the effort will be devoted to explain and complement the recent results by Aleman, Pott and the author on the Sarason Conjecture in [1].

Throughout the paper we will denote by $\mathbb{D}$ the unit disc and by $\mathbb{T}$ its boundary. We denote the normalized Lebesgue measure on $\mathbb{D}$ as dA, while σ will denote normalized surface measure on $\mathbb{T}$. We consider the Bergman space, $A^2(\mathbb{D})$, defined as the closed subspace of analytic functions in $L^2(\mathbb{D}, dA)$. We also consider the operator that takes functions in $L^2(\mathbb{D}, dA)$ to return functions in $A^2(\mathbb{D})$, this operator is known as the Bergman projection and we denote it by P_B. The integral form of the Bergman projection is given by:

$$P_B f(z) = \int_{\mathbb{D}} \frac{f(\zeta)}{(1-\bar{\zeta} z)^2} dA(\zeta).$$

This formula and other classical aspects of the Bergman space can be found for instance in [9]. We will also be interested in the Hardy space, $H^2(\mathbb{T})$, the closed subspace of $L^2(\mathbb{T})$ consisting of analytic functions. Associated to the

* Supported partially by the Vinnova VINNMER Marie Curie Incoming project number 2014-01434

Hardy space is the Riesz projection, that we will denote by P_R and that has integral representation:

$$P_R f(z) = \int_{\mathbb{T}} \frac{f(\zeta)}{(1-\overline{\zeta}z)} d\sigma(\zeta).$$

For each function $f \in L^2(\mathbb{D})$ ($f \in L^2(\mathbb{T})$ respectively) we consider the Toeplitz operator in the Bergman space $A^2(\mathbb{D})$ (Hardy space $H^2(\mathbb{T})$ respectively) with symbol f, T_f ($\mathcal{T}_f$ respectively), given by $T_f u = P_B f u$ for all $u \in A^2(\mathbb{D})$ ($\mathcal{T}_f v = P_R f v$ for all $v \in L^2(\mathbb{T})$ respectively).

When the symbol f is analytic, boundedness of the Toeplitz operators T_f in the Bergman and $\mathcal{T}_f$ in the Hardy space is characterized by the boundedeness of the function f in $\mathbb{D}$. In this paper we discuss the boundedness of products of Toeplitz operators in the Bergman space, and more precisely we study a conjecture by D. Sarason [20] that has been recently solved in the negative by Aleman, Pott and the author [1]. The conjecture reads as follows:

Conjecture 1.1 (Sarason Conjecture for the Bergman space). *Let $f, g \in A^2(\mathbb{D})$. Then $T_f T_g^*$ is bounded on $A^2(\mathbb{D})$, if and only if*

$$[|f|^2, |g|^2]_{B_2} := \sup_{z\in\mathbb{D}} B(|f|^2)(z) B(|g|^2)(z) < \infty, \tag{1}$$

where B denotes the Berezin transform,

$$Bf(z) = \int_{\mathbb{D}} \frac{f(\zeta)(1-|z|^2)^2}{|1-\overline{\zeta}z|^4} dA(\zeta).$$

The study of the product of Toeplitz operators originated in the early nineties from Sarason's work on exposed points of H^1, see [23] and [24]. The conjecture was first stated for the Hardy space, with the Berezin condition (1) replaced by an anologous condition involving the Poisson instead of the Berezin transforms. Treil proved the necessity of such condition. In spite of the efforts that provided support for Sarason Conjecture on the Hardy space (see Cruz-Uribe [5], Zheng [31] for instance), Nazarov found a counterexample [17].

The question in the Bergman space however has been open until very recently. Necessity and stronger sufficient conditions than the Berezin condition (1) have been studied by Stroetthof and Zheng [28] and Michalska, Nowak and Sobolewski [16]. The recent work of Aleman, Pott and the author [1] provides a counterexample to the Sarason Conjecture in the Bergman space, as well as a characterization of the boundedness of products of Toeplitz operators in the Bergman space in terms of testing conditions.

Although the results on the Bergman space might seem expected, let us recall that the Bergman and the Hardy spaces present striking differences. Cancellation does not play a big role in the Bergman space, making positive results easier to find, but at the same time, making the search for a counterexample a harder task.

There is an interesting connection between the Sarason Conjecture and the two weight problem for the Riesz and Bergman projections. This connection was first observed by Cruz-Uribe in the case of the Hardy space [5]. By definition of Toeplitz operator one can see $T_f T_g^* = M_f P_B M_{\bar{g}}$, where M_ψ stands for the operator multiplication by ψ. Taking into account that $T_f T_g^* u = T_f T_g^* P_B u$ for any $u \in L^2(\mathbb{D}, dA)$, one can easily see that the boundedness of $T_f T_g^*$ in $A_2(\mathbb{D})$ is equivalent to the two weight problem:

$$P_B : L^2\Big(\frac{1}{|g|^2}, \mathbb{D}\Big) \longrightarrow A^2(|f|^2, \mathbb{D}), \tag{2}$$

or equivalently to

$$P_B(|g|^2 \cdot) : L^2(|g|^2, \mathbb{D}) \longrightarrow A^2(|f|^2, \mathbb{D}). \tag{3}$$

Solving Sarason Conjecture in the Hardy space is equivalent to the two weight problem for the Riesz projection (or the Hilbert transform) due to the general form of the weights when $f, g \in H^2(\mathbb{T})$. This is no longer the case when we consider the Sarason problem in the Bergman space. The weights $|f|^2, |g|^2$ have an extra subharmonic structure in $\mathbb{D}$ that will be essential to the solution of the problem. Characterizing the two weight boundedness for the Bergman projection for general weights is still an open problem. One that promises to be very delicate, as one has to understand the cancellation of the kernel in the Bergman projection in the presence of two very general weights.

In the next section, Section 2, we will study some of the aspects of the counterexample found by Aleman, Pott and Reguera as well as their characterization of the boundeness of Toeplitz products in terms of testing conditions. We sill also inclued a simple proof of the characterization of two weight boundedness for positive dyadic operator by Lacey, Sawyer and Uriarte-Tuero [12]. The last section is dedicated to the discussion of open problems around Sarason Conjecture on the Bergman space.

2. Product of Toeplitz operators on the Bergman space

The Berezin condition (1) is necessary, this was proved by Stroethoff and Zheng in [28]. In fact, (1) is essentially the result of the action of $T_f T_{\bar{g}}$ on the reproducing kernels for the Bergman space. But the Berezin condition (1) can also be seen as a strenghtening of the Békollé-Bonami condition B_2 for the boundedness of the classical Bergman projection on one weighted spaces, see [2, 3]. This justifies our notation in (1).

First sufficient results are due to Stroethoff and Zheng [28], who consider a bumped condition (1), to be precise $[|f|^{2+\epsilon}, |g|^{2+\epsilon}]_{B_2} < \infty$. And a stronger sufficient result using logarithmic bumps is due to Michalska, Nowak and Sobolewski [16], this result also goes beyond the unit disc $\mathbb{D}$ and considers the unit ball of $\mathbb{C}^n$.

2.1. *A counterexample for Sarason Conjecture on the Bergman space*

We discuss the counterexample to the Sarason Conjecture of Aleman, Pott and the author [1]. Their result reads as follows:

Theorem 2.1. *There exist functions $f, g \in A^2(\mathbb{D})$ such that $[|f|^2, |g|^2]_{B_2} < \infty$, but $T_f T_g^*$ is not bounded on $A^2(\mathbb{D})$.*

For the proof of this result we refer the reader to [1]. However we would like to comment on some of the ideas behind it. For that we need to recall the definition of the Dirichlet space D of analytic functions h in $\mathbb{D}$ whose derivative belongs to $A^2(\mathbb{D})$, and the norm is defined by $\|h\|_D^2 = |h(0)|^2 + \|h'\|_2^2$. A positive measure μ on $\mathbb{D}$ is called a Carleson measure for the Dirichlet space D if D is continuously embedded in $L^2(\mu)$, i.e., if

$$[\mu]_{CM} := \sup_{\|h\|_D \le 1} \int_{\mathbb{D}} |h|^2 d\mu < \infty. \tag{4}$$

Given $g \in A^2(\mathbb{D})$ we call $\gamma(g)$ the infimum of constants C such that

$$\int_{Q_I} |g|^2 dA \le C \frac{1}{\log \frac{2\pi}{|I|}} \quad \text{for all arcs } I \subset \mathbb{T}, \tag{5}$$

where Q_I denotes the Carleson box based on the arc $I \subset \mathbb{T}$ and $|I|$ is the normalized length of I. It is well known and easy to prove that $\gamma(g) \lesssim [|g|^2]_{CM}$. But these quantities are not comparable, the counterexample is due to Stegenga [27].

We now go back to the product of Toeplitz operators $T_f T_{\bar{g}}$. Let us fix function $f(z) = 1 - |z|^2$, $z \in \mathbb{D}$, we see that $f \notin A^2(\mathbb{D})$, but it will allow us to illustrate the ideas in [1].

Given $g \in L^2(\mathbb{D})$, and taking into account that $B(|f|^2)$ is subharmonic, one can see that the Berezin condition (1) is equivalent to

$$\sup_{z \in \mathbb{D}} (1 - |z|^2)^2 \log \frac{2}{1 - |z|^2} B(|g|^2)(z) < \infty. \tag{6}$$

Using standard techniques, one can prove the following result:

Proposition 2.1. *Let* $f(z) = 1 - |z|^2$, $z \in \mathbb{D}$, *then*

(1) A function $g \in L^2(\mathbb{D})$ *satisfies* (6) *if and only if*

$$\int_{Q_I} |g|^2 dA \lesssim \frac{1}{\log \frac{2}{|I|}},$$

for all arcs $I \subset \mathbb{T}$.

(2) If $g \in L^2(\mathbb{D})$ *then* $T_f T_g^*$ *is bounded on* $L^2(\mathbb{D})$ *if and only if* $|g|^2 dA$ *is a Carleson measure for the Dirichlet space.*

From the proposition, we conclude that Stegenga's counterexample would give a negative answer to the Sarason Conjecture on the Bergman space if it wasn't for the fact that $f(z) = 1 - |z|^2$, $z \in \mathbb{D}$, is not a function in $A^2(\mathbb{D})$. To fix this inconvenience Aleman et al. [1] find a function $f \in A^2(\mathbb{D})$, that is also Lipschitz and to a certain extent behaves like $1 - |z|^2$. To find such a function the work of Dyn'kin is crucial, [7]. The interested reader is encouraged to read a full account of Aleman et al.'s counterexample to Sarason Conjecture on the Bergman space in [1].

2.2. *A characterization of boundedness of Toeplitz products on the Bergman space*

In this subsection we present the characterization of the boudedness of Toepliz products in the Bergman space in terms of testing conditions proved by Aleman, Pott and the author in [1]. Let us first recall the definition of the maximal Bergman projection P_B^+,

$$P_B^+(f) := \int_{\mathbb{D}} \frac{f(\zeta)}{|1 - \overline{\zeta} z|^2} dA(\zeta). \tag{7}$$

Their result is the following:

Theorem 2.2. *Let* $P_B^+(\cdot)$ *be the maximal Bergman projection on the disc* $\mathbb{D}$, *and let* $f, g \in A^2(\mathbb{D})$. *The following are equivalent*

(1) $T_f T_g^* : A^2(\mathbb{D}) \mapsto A^2(\mathbb{D})$ *is bounded;*
(2) $P_B(|g|^2 \cdot) : L^2(\mathbb{D}, |g|^2) \to L^2(\mathbb{D}, |f|^2)$ *bounded;*
(3) $P_B^+(|g|^2 \cdot) : L^2(\mathbb{D}, |g|^2) \to L^2(\mathbb{D}, |f|^2)$ *bounded;*
(4) (a) $\|P_B^+(|g|^2 1_{Q_I})\|_{L^2(\mathbb{D},|f|^2)} \leq C_0 \|1_{Q_I}\|_{L^2(\mathbb{D},|g|^2)}$,
(b) $\|P_B^+(|f|^2 1_{Q_I})\|_{L^2(\mathbb{D},|g|^2)} \leq C_0 \|1_{Q_I}\|_{L^2(\mathbb{D},|f|^2)}$,
for all intervals $I \in \mathbb{T}$ *and with constant* C_0 *uniform on* I.

Testing conditions were first used by Sawyer in the context of weighted inequalities [25], [26]. Implicitly and independently, they were also present in the celebrated $T1$ theorem of David and Journé [6]. They are called testing conditions because one only needs to test on a special class of functions to deduce boundedness of the operator at hand. We have already shown the connection of the boundedness of Toeplitz products with the two weight problem for the Bergman projection, see (2), (3). The strategy of Aleman et al. [1] is to exploit the fact that f and g are analytic to show that, as in the case of Lebesgue weights or the case of Békollé-Bonami weights, boundedness of the maximal Bergman and the Bergman projections are equivalent. In fact, this is their major contribution, and it reads as follows:

Theorem 2.3. *Let $f, g \in A^2(\mathbb{D})$. Then $T_f T_g^*$ is bounded on $A^2(\mathbb{D})$ if and only if the operator $P^+_{f,g}$, defined by*

$$P^+_{f,g}u(z) = |f(z)| \int_{\mathbb{D}} \frac{|g(\zeta)| u(\zeta)}{|1-\overline{\zeta} z|^2} dA(\zeta),$$

is bounded on $L^2(\mathbb{D})$.

We skip the proof of Theorem 2.3, which can be found in [1].

The characterization of the two weight boundedness for a positive operator in $\mathbb{R}^n$ is well understood, specially if such operator is discrete, see [26], [12]. Aleman et al. proved the maximal Bergman projection P^+_B can be controlled from above and also from below, thanks to its positive nature, by a discrete dyadic operator, see [1], [21]. Therefore one can use the existing two weight dyadic results to establish the result for the maximal Bergman projection. The use of dyadic models to establish results in the continuos setting is not new, see other instances in [4, 8, 15]. We will use the rest of the section to give details on Aleman et al.'s approach.

Definition 2.1. Let $\mathcal{D}$ be a dyadic grid in $\mathbb{T}$. For all $z, \xi \in \mathbb{D}$, we define the positive dyadic operator

$$P^{\mathcal{D}} f(z) := \sum_{I \in \mathcal{D}} \langle f, \frac{1_{Q_I}}{|I|^2} \rangle 1_{Q_I}(z), \tag{8}$$

where Q_I is the Carleson box associated to I, and $|I|$ stands for the normalized length of the interval.

The following proposition compares the maximal Bergman projection with the dyadic operator $P^{\mathcal{D}}$.

Proposition 2.2. *There exist constants C and $\tilde{C}$ and two systems of dyadic grids $\mathscr{D}^j$, $j = 1,2$, such that for every $i \in \{1,2\}$, every $f \in L^1_{loc}$ and $z \in \mathbb{D}$,*

$$\tilde{C}P^{\mathscr{D}^i} f(z) \leq P^+_B f(z) \leq C \sum_{j\in\{1,2\}} P^{\mathscr{D}^j} f(z). \tag{9}$$

We spend the rest of this section recalling a theorem by Sawyer, Lacey and Uriarte-Tuero [12] on the characterization of two weight boundedness for dyadic positive operators in terms of testing conditions. The desired two weight result for the maximal Bergman projection is a consequence of the dyadic result and inequalities (9). The proof of the general two weight result for positive dyadic operators below is essentially the beautiful simplification of Treil [30], with the difference that it avoids appealing to the Carleson embbeding theorem, a practice that has become crucial in the study of two weight problems for cancellative operators. We include it here hoping that the reader will find it interesting and beautiful.

A weight function will be an nonnegative measurable function on $\mathbb{R}^n$, not necessarily locally integrable. Let w, v be weight functions in $\mathbb{R}^n$, let $1 < p < \infty$ and p' its dual exponent. We define $\sigma := v^{1-p'}$, which is usually called the dual weight of v. Let T be an operator. Then the following are equivalent:

$$T : L^p(v) \mapsto L^p(w),$$

$$T(\sigma\cdot) : L^p(\sigma) \mapsto L^p(w), \tag{10}$$

$$w^{1/p} T(\sigma^{1/p'}\cdot) : L^p \mapsto L^p. \tag{11}$$

Below we will use (10) with general weights σ and w. For the Sarason problem, (11) is more natural and will also appear.

Throughout this section, we will denote the expectation of a function f over a cube Q by $\mathbb{E}_Q f$, and the expectation of a function f over a cube Q with respect to a weight σ will be denoted by $\mathbb{E}^\sigma_Q f$.

We consider a dyadic grid in $\mathbb{R}^n$ and denote it by $\mathscr{D}$. The class of operators we are interested in are dyadic positive operators of the form

$$T(f) := \sum_{Q\in\mathscr{D}} \tau_Q(\mathbb{E}_Q f)1_Q, \tag{12}$$

where τ_Q is a sequence of nonnegative scalars and 1_E indicates the characteristic function on the set E.

Given two weights w and σ, we aim to characterize the boundedness of the operator T in the two-weight setting. More precisely, we state the question as follows:

Question 2.1. Characterize the pairs of weights w and σ for which

$$T(w\cdot): L^p(w) \mapsto L^p(\sigma) \text{ is bounded.}$$

The following theorem provides an answer to this question. In this precise form, it is due to Lacey, Sawyer and Uriarte-Tuero [12]. We present a simplified version of their original proof. Our proof can also be adapted to the disc, with the Carleson cubes associated to a dyadic grid in $\mathbb{T}$ as the dyadic family.

Theorem 2.4. *Let w, σ be two weights and let T be a dyadic positive operators as in* (12). *Then*

$$T(w\cdot): L^p(w) \mapsto L^p(\sigma)$$

is bounded, if and only if

$$\|T(w1_Q)\|_{L^p(\sigma)}^p \leq C_0\, w(Q), \tag{13}$$

and

$$\|T^*(\sigma 1_Q)\|_{L^p(w)}^p \leq C_0^*\, \sigma(Q), \tag{14}$$

for all Q dyadic cube in $\mathcal{D}$, and constants C_0 and C_0^* independent of the cubes Q. Moreover, there exists a constant c independent of T and w, σ, such that

$$\|T(w\cdot)\|_{L^p(w)\to L^p(\sigma)} \leq c(C_0 + C_0^*).$$

Remark 2.1. In fact, one needs only weaker testing conditions in order to get boundedness of the operator, namely, (13) and (14) can be replaced by

$$\|T_{in,Q}(w1_Q)\|_{L^p(\sigma)}^p \leq C_0\, w(Q), \tag{15}$$

and

$$\|T_{in,Q}^*(\sigma 1_Q)\|_{L^p(w)}^p \leq C_0^*\, \sigma(Q), \tag{16}$$

respectively, where $T_{in,Q} := \sum_{\substack{P\in\mathcal{D}\\ P\subset Q}} \tau_P(\mathbb{E}_P f)1_P$. The use of these weaker testing conditions (15) and (16) can be traced in the proof of Theorem 2.4 below.

2.2.1. *A Corona decomposition*

For now and throughout this section, we will assume without loss of generality that the function f is positive.

Definition 2.2. Let Q_0 be a cube in $\mathscr{D}$ and let $\mathscr{D}_0$ be a family of cubes contained in Q_0. Let w be a weight in $\mathbb{R}^n$ and let f be a positive locally integrable function. We define

$$\mathscr{L}(Q_0) = \{Q \in \mathscr{D}_0 : Q \text{ is a maximal cube in } \mathscr{D}_0 \text{ such that } \mathbb{E}_Q^w f > 4\mathbb{E}_{Q_0}^w f\}.$$

We define $\mathscr{L}_0 := \{Q_0\}$, and, recursively,

$$\mathscr{L}_i := \bigcup_{L \in \mathscr{L}_{i-1}} \mathscr{L}(L).$$

We will denote the union of all the stopping cubes by $\mathscr{L} := \bigcup_{i \geq 0} \mathscr{L}_i$. We notice that we could also define the starting family $\mathscr{L}_0$ as a union of disjoint maximal cubes and repeat the above construction in each one of the cubes in $\mathscr{L}_0$. Given $Q \in \mathscr{D}_0$, we define $\lambda(Q)$ as the minimal cube $L \in \mathscr{L}$ such that $Q \subset L$ and $\mathscr{D}(L) := \{Q \in \mathscr{D}_0 : \lambda(Q) = L\}$.

We consider now the dyadic Hardy-Littlewood maximal function in its weighted form. For a weight w, we define

$$M_w f(x) = \sup_{Q \in \mathscr{D}} \frac{1_Q}{w(Q)} \int_Q |f| w dm, \tag{17}$$

where dm stands for the Lebesgue measure in $\mathbb{R}^n$. The following result is a well-known classical theorem.

Theorem 2.5.

$$\|M_w f\|_{L^p(wdm)} \leq C \|f\|_{L^p(wdm)}, \tag{18}$$

where the constant C is independent of the weight w.

The stopping cubes in Definition 2.2 provide the right collection of sets to linearise the dyadic Hardy Littlewood maximal function described in (17), i.e., we have the following pointwise estimate:

$$\sum_{\mathscr{L}} (\mathbb{E}_L^w |f|) 1_L(x) \lesssim M_w f(x), \quad \text{for all } x \in \mathbb{R}^n. \tag{19}$$

The proof of (19) is an exercise.

An application of (19) and Theorem 2.5 provide the following useful inequality:

$$\sum_{L \in \mathscr{L}} (\mathbb{E}_L^w |f|)^p w(L) \lesssim \|f\|_{L^p(wdm)}. \tag{20}$$

Proof of Theorem 2.4. We are now ready to prove the main theorem in this section. We will assume there is a finite collection of dyadic cubes $\mathscr{Q}$ in the definition of the operator T, and we will prove the operator norm is independent of the chosen collection. So from now on

$$Tf = \sum_{Q\in\mathscr{Q}} \tau_Q(\mathbb{E}_Q f)1_Q.$$

It is enough to prove boundedness of the bilinear form $\langle T(wf), g\sigma\rangle$, where $0 \le f \in L^p(w)$ and $0 \le g \in L^{p'}(\sigma)$. Following the argument in [30], we seek an estimate of the form

$$\langle T(wf), g\sigma\rangle \le A\|f\|_{L^p(w)}\|g\|_{L^{p'}(\sigma)} + B\|f\|^p_{L^p(w)}. \tag{21}$$

We first divide the cubes in $\mathscr{Q}$ into two collections $\mathscr{Q}_1$ and $\mathscr{Q}_2$ according to the following criterion. A cube Q will belong to $\mathscr{Q}_1$, if

$$\left(\mathbb{E}^w_Q f\right)^p w(Q) \ge \left(\mathbb{E}^\sigma_Q g\right)^{p'} \sigma(Q), \tag{22}$$

and it will belong to $\mathscr{Q}_2$ otherwise. This reorganisation of the cubes allows us to write $T = T_1 + T_2$, where

$$T_i f = \sum_{Q\in\mathscr{Q}_i} \tau_Q(\mathbb{E}_Q f)1_Q,\ i = 1,2.$$

The idea of writing T as the sum of T_1 and T_2 was already present in the work of Treil [30] and previously in the work of Nazarov, Treil and Volberg [18].

We prove the boundedness of T_1 using the testing condition (13). For the boundedness of T_2, we proceed analogously, but using (14) this time.

$$\begin{aligned}\langle T_1(wf), g\sigma\rangle &= \sum_{Q\in\mathscr{Q}_1} \tau_Q \mathbb{E}_Q(fw)\langle g\sigma, 1_Q\rangle \\ &= \sum_{L\in\mathscr{L}}\sum_{Q\in\mathscr{D}(L)} \tau_Q \mathbb{E}_Q(fw)\langle g\sigma, 1_Q\rangle = \sum_{L\in\mathscr{L}} \langle T_L(wf), g\sigma\rangle,\end{aligned}$$

where $\mathscr{L}$ is a collection of stopping cubes in the family $\mathscr{Q}_1$, to be specified below, and $T_L f = \sum_{Q\in\mathscr{D}(L)} \tau_Q \mathbb{E}_Q(f)1_Q$. To find the collection of stopping cubes $\mathscr{L}$, we define $\mathscr{L}_0$ as the collection of maximal cubes in the family $\mathscr{Q}_1$, and follow the Definition 2.2 for given f and w to define $\mathscr{L}$, with $\mathscr{Q}_1$ as our family of dyadic cubes.

We are going to estimate the bilinear form

$$\sum_{L\in\mathscr{L}} \langle T_L(wf), g\sigma\rangle, \tag{23}$$

but before doing this, let us look at the norm of T_L. We claim that

$$\|T_L(wf)\|_{L^p(\sigma)}^p \le C_0 4^p \left(\mathbb{E}_L^w(f)\right)^p w(L). \tag{24}$$

This is easily verified by

$$\begin{aligned}
\|T_L(wf)\|_{L^p(\sigma)}^p &= \Big\| \sum_{Q\in\mathcal{D}(L)} \tau_Q(\mathbb{E}_Q f w) 1_Q \Big\|_{L^p(\sigma)}^p \\
&= \Big\| \sum_{Q\in\mathcal{D}(L)} \frac{w(Q)}{|Q|} \tau_Q(\mathbb{E}_Q^w f) 1_Q \Big\|_{L^p(\sigma)}^p \\
&\le 4^p \left(\mathbb{E}_L^w f\right)^p \Big\| \sum_{Q\in\mathcal{D}(L)} \frac{w(Q)}{|Q|} \tau_Q 1_Q \Big\|_{L^p(\sigma)}^p \\
&\le 4^p \left(\mathbb{E}_L^w f\right)^p \|T(w1_Q)\|_{L^p(\sigma)}^p \le 4^p C_0 \left(\mathbb{E}_L^w f\right)^p w(L),
\end{aligned}$$

where in the first inequality we have used that $Q \in \mathcal{D}(L)$ are not stopping cubes, and in the last inequality, the testing condition (13).

We now estimate (23).

$$\sum_{L\in\mathcal{L}} \langle T_L(wf), g\sigma\rangle = \sum_{L\in\mathcal{L}} \int T_L(wf)(x) g(x)\sigma(x)\,dx = (I) + (II),$$

where

$$(I) = \sum_i \sum_{L\in\mathcal{L}_i} \int_{L \setminus \bigcup_{\substack{L'\in\mathcal{L}_{i+1} \\ L'\subset L}} L'} T_L(wf)(x) g(x)\sigma(x)\,dx,$$

and

$$(II) = \sum_i \sum_{L\in\mathcal{L}_i} \int_{\bigcup_{\substack{L'\in\mathcal{L}_{i+1} \\ L'\subset L}} L'} T_L(wf)(x) g(x)\sigma(x)\,dx.$$

We proceed to estimate (I),

$$\begin{aligned}
(I) &\le \sum_i \sum_{L\in\mathcal{L}_i} \|T_L(fw)\|_{L^p(\sigma)} \|g 1_{L\setminus \bigcup_{\substack{L'\in\mathcal{L}_{i+1} \\ L'\subset L}} L'}\|_{L^{p'}(\sigma)} \\
&\le \Big(\sum_i \sum_{L\in\mathcal{L}_i} \|T_L(fw)\|_{L^p(\sigma)}^p \Big)^{1/p} \Big(\sum_i \sum_{L\in\mathcal{L}_i} \|g 1_{L\setminus \bigcup_{\substack{L'\in\mathcal{L}_{i+1} \\ L'\subset L}} L'}\|_{L^{p'}(\sigma)}^{p'} \Big)^{1/p'} \\
&\le 4 C_0^{1/p} \Big(\sum_{L\in\mathcal{L}} \left(\mathbb{E}_L^w f\right)^p w(L) \Big)^{1/p} \|g\|_{L^{p'}(\sigma)} \lesssim C_0 \|f\|_{L^p(w)} \|g\|_{L^{p'}(\sigma)},
\end{aligned}$$

where in the first two inequalities, we have used Hölder's inequality, and in the third one we have used the testing condition (13) and the fact that $\bigcup_i \bigcup_{L\in\mathcal{L}_i} L\setminus$

$\bigcup_{\substack{L'\in\mathcal{L}_{i+1}\\ L'\subset L}} L'$ forms a partition of the maximal cubes in $\mathcal{L}_0$. For the last inequality, we have used (20).

We now turn to (II). Before we proceed with the estimate, let us note the following remark.

Remark 2.2. Let $L \in \mathcal{L}$ be fixed, then the operator $T_L(fw)$ is constant on L', where $L' \in \mathcal{L}$, $L' \subsetneq L$. We will denote this constant by $T_L(fw)(L')$.

Taking this remark into account, we get the following estimates for fixed $L \in \mathcal{L}_i$:

$$
\begin{aligned}
\int_{\bigcup_{\substack{L'\in\mathcal{L}_{i+1}\\ L'\subset L}} L'} T_L(wf)(x)g(x)\sigma(x)dx &= \sum_{\substack{L'\in\mathcal{L}_{i+1}\\ L'\subset L}} T_L(fw)(L')\int_{L'} g\sigma dx\\
&= \sum_{\substack{L'\in\mathcal{L}_{i+1}\\ L'\subset L}} \int_{L'} T_L(fw)(x)\Big(\mathbb{E}^{\sigma}_{L'}g\Big)\sigma(x)dx\\
&= \int_L T_L(fw)(x)\Big(\sum_{\substack{L'\in\mathcal{L}_{i+1}\\ L'\subset L}} (\mathbb{E}^{\sigma}_{L'}g)1_{L'}(x)\Big)\sigma(x)dx\\
&\leq \Big\|T_L(fw)\Big\|_{L^p(\sigma)}\Big\|\sum_{\substack{L'\in\mathcal{L}_{i+1}\\ L'\subset L}} (\mathbb{E}^{\sigma}_{L'}g)1_{L'}\Big\|_{L^{p'}(\sigma)}\\
&= \|T_L(fw)\|_{L^p(\sigma)}\Big(\sum_{\substack{L'\in\mathcal{L}_{i+1}\\ L'\subset L}} (\mathbb{E}^{\sigma}_{L'}g)^{p'}\sigma(L')\Big)^{1/p'}\\
&\leq 4C_0(\mathbb{E}^{w}_{L}f)w(L)^{1/p}\Big(\sum_{\substack{L'\in\mathcal{L}_{i+1}\\ L'\subset L}} (\mathbb{E}^{w}_{L'}f)^{p}w(L')\Big)^{1/p'},
\end{aligned}
$$

where we have used Remark 2.2, Hölder's inequality, (24) and the hypothesis (22). We now proceed to sum the previous estimates in L to obtain the desired bound for (II).

$$
\begin{aligned}
(II) &\lesssim \sum_i \sum_{L\in\mathcal{L}_i} (\mathbb{E}^{w}_{L}f)w(L)^{1/p}\Big(\sum_{\substack{L'\in\mathcal{L}_{i+1}\\ L'\subset L}} (\mathbb{E}^{w}_{L'}f)^{p}w(L')\Big)^{1/p'}\\
&\lesssim \Big(\sum_{L\in\mathcal{L}} (\mathbb{E}^{w}_{L}f)^{p}w(L)\Big)^{1/p}\Big(\sum_i\sum_{L\in\mathcal{L}_i}\sum_{\substack{L'\in\mathcal{L}_{i+1}\\ L'\subset L}} (\mathbb{E}^{w}_{L'}f)^{p}w(L')\Big)^{1/p'}\\
&\lesssim \|f\|_{L^p(w)}\|f\|^{p/p'}_{L^p(w)} \lesssim \|f\|^{p}_{L^p(w)}.
\end{aligned}
$$

Adding (I) and (II), we get the desired estimate (21). □

We now turn to the two weight characterization for the case of the maximal Bergman projection P_B^+ and its associated dyadic model $P^{\mathscr{D}}$. We start with $P^{\mathscr{D}}$. One can state the following theorem.

Theorem 2.6. *Let $\mathscr{D}$ be a fixed dyadic grid in $\mathbb{T}$ and let $P^{\mathscr{D}}$ as defined in* (8). *Then*

$$P^{\mathscr{D}}(w\cdot): L^p(w) \to L^p(\sigma)$$

is bounded, if and only if

$$\Big\| \sum_{\substack{I\in\mathscr{D}\\ I\subset I_0}} \langle w 1_{Q_{I_0}}, \frac{1_{Q_I}}{|I|^2}\rangle 1_{Q_I} \Big\|_{L^p(\sigma)}^p \le C_0\, w(Q_{I_0}), \tag{25}$$

and

$$\Big\| \sum_{\substack{I\in\mathscr{D}\\ I\subset I_0}} \langle \sigma 1_{Q_{I_0}}, \frac{1_{Q_I}}{|I|^2}\rangle 1_{Q_I} \Big\|_{L^{p'}(w)}^{p'} \le C_0^*\, \sigma(Q_{I_0}), \tag{26}$$

for all I dyadic interval in $\mathscr{D}$, where Q_I represents the Carleson box associated to I and the constants C_0 and C_0^* are independent of the intervals I.
Moreover, there exists a constant $c > 0$ independent of the weights, such that

$$\Big\| P^{\mathscr{D}}(w\cdot) \Big\|_{L^p(w)^p \to L^p(\sigma)} \le c(C_0 + C_0^*).$$

The proof of Theorem 2.6 in the disc $\mathbb{D}$ is identical to the one we describe in Theorem 2.4. In the case of the disc, our dyadic system will be described by the Carleson cubes associated to the intervals in the dyadic grid $\mathscr{D}^\beta$ in $\mathbb{T}$. The boundedness of the weighted Hardy-Littlewood maximal function over dyadic Carleson cubes in the disc will be used instead of Theorem 2.5. We will also consider the testing conditions (15) and (16). The details of the proof are left to the reader.

We obtain the following corollary, which presents a two weight characterization for the maximal Bergman projection.

Corollary 2.1. *Let P_B^+ be the maximal Bergman projection in the disc $\mathbb{D}$, let $1 < p < \infty$ and p' its dual exponent and let w, σ be two weight functions. Then*

$$M_{w^{1/p}} P_B^+ M_{\sigma^{1/p'}} : L^p(\mathbb{D}) \to L^p(\mathbb{D})$$

is bounded, if and only if

$$\|M_{w^{1/p}} P_B^+ M_{\sigma^{1/p'}}(1_{Q_I}\sigma^{1/p})\|_{L^p(\mathbb{D})} \le C_0 \|1_{Q_I}\sigma^{1/p}\|_{L^p(\mathbb{D})}, \tag{27}$$

and

$$\|M_{\sigma^{1/p'}} P_B^+ M_{w^{1/p}}(1_{Q_I} w^{1/p'})\|_{L^{p'}(\mathbb{D})} \le C_0^* \|1_{Q_I} w^{1/p'}\|_{L^{p'}(\mathbb{D})}, \tag{28}$$

for any interval I in $\mathbb{T}$, where the constants C_0 and C_0^* are independent of the choice of interval. Moreover, there exists a constant $c > 0$ independent of the weights, such that

$$\left\| M_{w^{1/p}} P_B^+ M_{\sigma^{1/p'}} \right\|_{L^2 \to L^2} \leq c(C_0 + C_0^*).$$

As in the introduction, the operators M_h stand for the operator of multiplication by the symbol h.

Proof. We only have to prove one direction. By the first inequality in (9), the testing condition (27) and (28) imply the corresponding testing condition for each P^β, and therefore the uniform boundedness of all $P_{\mathcal{D}}$ by Theorem 2.6. The second inequality in (9) now implies the boundedness of $M_{w^{1/p}} P_B^+ M_{\sigma^{1/p'}}$ with the required norm bounds. □

We note that the positivity of P_B^+ and the left hand-side of (9) are crucial here to recover the non-dyadic case from the dyadic one. This advantage is not present in the case of cancellative operators such as the Bergman projection itself.

3. Open problems

In this section we list some of the open problems in the area.

One of the main results in [1], states that the two weight boundedness of the maximal Bergman projection is equivalent to the two weight boundedness of the Bergman projection when the weights are the modulus of analytic functions squared, i.e., for $f, g \in A^2(\mathbb{D})$, $w = |f|^2$ and $\sigma = |g|^2$,

$$P_B(\sigma \cdot) : L^2(\mathbb{D}, \sigma) \to L^2(\mathbb{D}, w) \text{ iff } P_B^+(\sigma \cdot) : L^2(\mathbb{D}, \sigma) \to L^2(\mathbb{D}, w). \tag{29}$$

Question 3.1. Does (29) hold for any pair of weights w and σ?

The general belief is that this should not be the case when the two weights σ, w are allowed to be general weights in $\mathbb{D}$. Candidates of weights for which the maximal Bergman projection is unbounded exist, but proving a two weight estimate for the Bergman projection even when particular weights are given is not an easy task.

The current proof of (29) requires that we consider $w = |f|^2$ and $\sigma = |g|^2$ for some $f, g \in A^2(\mathbb{D})$, this is due to an application of Stoke's theorem.

Question 3.2. Would (29) hold for any pair of subharmonic weights?

The most important question would be a complete characterization of the two weight problem for the Bergman projection in terms of testing conditions. We state it precisely:

Question 3.3. For positive locally integrable weights w and σ in $\mathbb{D}$, could we prove the equivalence of the following conditions?

$$P_B(\sigma\cdot): L^2(\mathbb{D},\sigma) \to L^2(\mathbb{D},w), \tag{30}$$

if and only if

$$\|P_B(\sigma 1_{Q_I})\|_{L^2(\mathbb{D},w)} \le C_0 \|1_{Q_I}\|_{L^2(\mathbb{D},\sigma)}, \tag{31}$$

$$\|P_B^*(w 1_{Q_I})\|_{L^2(\mathbb{D},\sigma)} \le C_0 \|1_{Q_I}\|_{L^2(\mathbb{D},w)}. \tag{32}$$

So far the characterization provided in Theorem 2.2 is only valid for weights $w = |f|^2$ and $\sigma = |g|^2$ for $f, g \in A^2(\mathbb{D})$. This is expected to be a very hard question, in fact, the two weight problem for the Hilbert transform has only been solved recently in a celebrated series of papers by Lacey, Sawyer, Shen and Uriarte-Tuero [11] and Lacey [10]. This is the first instance of a complete solution to a two weight problem for a cancellative operator.

Due to the difficulty of the previous question, there are intermediate questions one could first try to understand.

Question 3.4. Let $Mf(x) := \sup_{\substack{I\in\mathbb{T}\\ x\in Q_I}} \frac{1}{|Q_I|}\int_{Q_I} |f| dA$, under the hypothesis that $M(\sigma\cdot): L^2(\mathbb{D},\sigma) \to L^2(\mathbb{D},w)$ and $M(w\cdot): L^2(\mathbb{D},w) \to L^2(\mathbb{D},\sigma)$, could one prove Question 3.3?

This question is motivated by Nazarov, Treil and Volberg's attempt to prove the two weight problem for the Hilbert transform in [19]. In the case of the Hilbert transform, the two weight boundedness of the maximal function is known to be an unnecessary condition, see [13], [22].

Question 3.5. Let w,σ be a pair of subharmonic weights, could one prove Question 3.3?

The set of questions above are all described in the unit disc $\mathbb{D}$ of $\mathbb{C}$ but the Sarason Conjecture can also be formulated in the unit ball $\mathbb{B}^n$ of $\mathbb{C}^n$. Some partial results can already be found in [29], [14] and [16].

Question 3.6. Extending the results of [1] to the Bergman space defined in the unit ball of $\mathbb{C}^n$.

Acknowledgement

The author is thankful to the organizers of the $6th$ International Course of Mathematical Analysis in Andalucía, held in Antequera, Spain, September 8-12, 2014 for their kind invitation, which resulted in the composition of this paper. The author would also like to thank her collaborators A. Aleman and S. Pott for allowing the use of joint work in this manuscript.

References

[1] A. Aleman, S. Pott and M. C. Reguera, Sarason Conjecture on the Bergman space, *ArXiv e-prints* (April 2013).

[2] D. Bekollé, Inégalité à poids pour le projecteur de Bergman dans la boule unité de $\mathbf{C}^n$, *Studia Math.* **71**, pp. 305–323 (1981/82).

[3] D. Bekollé and A. Bonami, Inégalités à poids pour le noyau de Bergman, *C. R. Acad. Sci. Paris Sér. A-B* **286**, pp. A775–A778 (1978).

[4] M. Christ, Weak type $(1,1)$ bounds for rough operators, *Ann. of Math. (2)* **128**, pp. 19–42 (1988).

[5] D. Cruz-Uribe, The invertibility of the product of unbounded Toeplitz operators, *Integral Equations Operator Theory* **20**, pp. 231–237 (1994).

[6] G. David and J.-L. Journé, A boundedness criterion for generalized Calderón-Zygmund operators, *Ann. of Math. (2)* **120**, pp. 371–397 (1984).

[7] E. M. Dyn'kin, Free interpolation sets for Hölder classes, *Mat. Sb. (N.S.)* **109(151)**, pp. 107–128, 166 (1979).

[8] J. B. Garnett and P. W. Jones, BMO from dyadic BMO, *Pacific J. Math.* **99**, pp. 351–371 (1982).

[9] H. Hedenmalm, B. Korenblum and K. Zhu, *Theory of Bergman spaces*, Graduate Texts in Mathematics, Vol. 199, pp. x+286 (Springer-Verlag, New York, 2000).

[10] M. T. Lacey, Two-weight inequality for the Hilbert transform: a real variable characterization, II, *Duke Math. J.* **163**, pp. 2821–2840 (2014).

[11] M. T. Lacey, E. T. Sawyer, C.-Y. Shen and I. Uriarte-Tuero, Two-weight inequality for the Hilbert transform: a real variable characterization, I, *Duke Math. J.* **163**, pp. 2795–2820 (2014).

[12] M. T. Lacey, E. T. Sawyer and I. Uriarte-Tuero, Two weight inequalities for discrete positive operators, *ArXiv e-prints* (November 2009).

[13] M. T. Lacey, E. T. Sawyer and I. Uriarte-Tuero, A two weight inequality for the Hilbert transform assuming an energy hypothesis, *J. Funct. Anal.* **263**, pp. 305–363 (2012).

[14] Y. Lu and C. Liu, Toeplitz and Hankel products on Bergman spaces of the unit ball, *Chin. Ann. Math. Ser. B* **30**, pp. 293–310 (2009).

[15] T. Mei, BMO is the intersection of two translates of dyadic BMO, *C. R. Math. Acad. Sci. Paris* **336**, pp. 1003–1006 (2003).

[16] M. Michalska, M. Nowak and P. Sobolewski, Bounded Toeplitz and Hankel products on weighted Bergman spaces of the unit ball, *Ann. Polon. Math.* **99**, pp. 45–53 (2010).

[17] F. Nazarov, A counterexample to Sarason's conjecture. http://www.math.msu.edu/ fedja/prepr.html, (1997).

[18] F. Nazarov, S. Treil and A. Volberg, The Bellman functions and two-weight inequalities for Haar multipliers, *J. Amer. Math. Soc.* **12**, pp. 909–928 (1999).

[19] F. Nazarov, S. Treil and A. Volberg, Two weight estimate for the Hilbert transform and corona decomposition for non-doubling measures, *ArXiv e-prints* (March 2010).

[20] J. Peetre, Hankel and Toeplitz operators, in *Linear and Complex Analysis problem book 3*, eds. V. P. Havin and N. K. Nikolski. Lecture Notes in Mathematics, Vol. 1573, pp. 293–358 (Springer Berlin Heidelberg, 1994).

[21] S. Pott and M. C. Reguera, Sharp Békollé estimates for the Bergman projection, *J. Funct. Anal.* **265**, pp. 3233–3244 (2013).

[22] M. C. Reguera and J. Scurry, On joint estimates for maximal functions and singular integrals on weighted spaces, *Proc. Amer. Math. Soc.* **141**, pp. 1705–1717 (2013).

[23] D. Sarason, Exposed points in H^1. I, in *The Gohberg anniversary collection, Vol. II (Calgary, AB, 1988)*, Oper. Theory Adv. Appl., Vol. 41, pp. 485–496 (Birkhäuser, Basel, 1989).

[24] D. Sarason, Exposed points in H^1. II, in *Topics in operator theory: Ernst D. Hellinger memorial volume*, Oper. Theory Adv. Appl., Vol. 48, pp. 333–347 (Birkhäuser, Basel, 1990).

[25] E. T. Sawyer, A characterization of a two-weight norm inequality for maximal operators, *Studia Math.* **75**, pp. 1–11 (1982).

[26] E. T. Sawyer, A characterization of two weight norm inequalities for fractional and Poisson integrals, *Trans. Amer. Math. Soc.* **308**, pp. 533–545 (1988).

[27] D. A. Stegenga, Multipliers of the Dirichlet space, *Illinois J. Math.* **24**, pp. 113–139 (1980).

[28] K. Stroethoff and D. Zheng, Products of Hankel and Toeplitz operators on the Bergman space, *J. Funct. Anal.* **169**, pp. 289–313 (1999).

[29] K. Stroethoff and D. Zheng, Bounded Toeplitz products on Bergman spaces of the unit ball, *J. Math. Anal. Appl.* **325**, pp. 114–129 (2007).

[30] S. Treil, A remark on two weight estimates for positive dyadic operators, *ArXiv e-prints* (January 2012).

[31] D. Zheng, The distribution function inequality and products of Toeplitz operators and Hankel operators, *J. Funct. Anal.* **138**, pp. 477–501 (1996).

Semigroups, a tool to develop Harmonic Analysis for general Laplacians

José L. Torrea*
Departamento de Matemáticas and ICMAT,
Universidad Autónoma de Madrid,
28049 Madrid, Spain
E-mail: joseluis.torrea@uam.es

In this note we introduce the semigroup language as a natural tool to develop a Harmonic Analysis associated to general Laplacians. We make a quick walk through the classical cases of the torus and the line in order to motivate the introduction of semigroups. The case of the discrete Laplacian in the integers is considered in more detail.

Keywords: Discrete Laplacian, heat semigroup, Riesz transforms, conjugate harmonic functions, modified Bessel functions.

1. Introduction

In the last century the development of Harmonic Analysis was impulsed mainly by the understanding of different partial differential equations. Several examples could be presented, but probably the most famous and descriptive examples are the Riesz transforms. They were essential in order to prove "a priori" estimates in L^p for Laplace equation. Another fundamental object was the Littlewood-Paley square functions, that can be used among many other results to prove dimension free estimates for the boundedness of certain operators. Needless to say that the presence of the maximal operators (in many forms) are essential in order to show almost everywhere convergence.

An obvious question is as follows, What happens for other second order differential operators? Can we construct a parallel Harmonic Analysis?

Searching in the literature, a series of papers by B. Muckenhoupt and E. Stein, see Refs [8, 9] and [7], can be considered as the pioneers in this line of thought. We shall describe briefly some of their ideas. Consider the Hermite polynomials H_n, given by Rodrigues' formula, $H_n(x) = (-1)^n e^{x^2} \frac{d^n e^{-x^2}}{dx^n}$, $n = 0, 1, 2\dots$. They are orthogonal with respect to the measure $d\gamma(x) = e^{-x^2} dx, x \in \mathbb{R}$. Also if $L = -\Delta + 2x\nabla = -\partial_x^2 + 2x\partial_x$, then $LH_n = 2nH_n$. Muckenhoupt define

* Research partially supported by grant MTM2011-28149-C02-01 from Spanish Government

the so-called Poisson sum of a function $f(x) = \sum_n a_n H_n(x)$, by

$$g(r,x) = P_r f(x) = \sum_n r^n a_n H_n(x). \tag{1}$$

He observed that P_r has parallel properties to the classical Poisson sums, namely almost everywhere convergence for $L^p(d\gamma), 1 \leq p < \infty$. L^p-convergence $1 < p < \infty$, ... However these Poisson sums do not satisfy neither Cauchy-Riemann equations nor an harmonicity property. Alternatively they considered the sums $f(x,t) = \sum_n e^{-(2n)^{1/2}t} a_n H_n(x)$. These last sums seem satisfy the following "harmonicity" property:

$$(\partial_t^2 - L)f = 0, \quad (\text{observe that } \partial_t^2 e^{-(2n)^{1/2}t} H_n = L e^{-(2n)^{1/2}t} H_n).$$

Moreover they defined the "conjugate" function

$$\tilde{f}(t,x) = \sum_n (2n)^{1/2} e^{-(2n)^{1/2}t} a_n H_{n-1}(x).$$

The following Cauchy-Riemann equations are satisfied

$$\partial_t f(t,x) = (-\partial_x + 2x)\tilde{f}(t,x), \quad \partial_x f(t,x) = \partial_t \tilde{f}(t,x).$$

Also $(\partial_t^2 - L - 2)\tilde{f} = 0$. Observe that $L = (\partial_x)^* \partial_x = (-\partial_x + 2x)\partial_x$ and $\partial_x(\partial_x)^* = L+2$, where $(\partial_x)^*$ is the adjoint of ∂_x with respect to $d\gamma$.

The first purpose of this note is to analyze the (apparently) big differences between the "Poisson" sums $g(r,x)$ and $f(t,x)$ defined above. To understand these differences we compare with the classical cases in the line and in the torus. After that, we present the language of semigroups as a path to built Harmonic Analysis associated to more general Laplacians. Our presentation pretends to give the reader a quick guide to the main ideas, in particular we avoid all the technicalities of the subject (it can be rather involved) and in general we assume that our functions are good enough for the verification of the formulas. As an example we present a Harmonic Analysis associated to the discrete Laplacian on the integers.

2. Walking along the torus and the line

A careful reading of Muckenhoupt and Stein papers could be the following. Given $f \sim \sum_n a_n H_n$, define

$$g(r(t),x) = \sum_n r^n a_n H_n \overset{(r=e^{-2t})}{=} \sum_n e^{-t2n} a_n H_n, \qquad \text{(Poisson)},$$

$$\text{and} \qquad f(t,x) = \sum_n e^{-t(2n)^{1/2}} a_n H_n, \qquad \text{("alternative Poisson")}.$$

Then, at least formally,

$$(\partial_t + L)g(r(t),x) = 0.$$

In other words, for the function $g(r(t),x)$, the substantive "Poisson" is appropriated when speaking about sums but not when speaking about (Laplace) equation. However, for the function f we have

$$(\partial_t^2 - L)f(t,x) = 0, \quad \text{Poisson equation, "harmonic" function.}$$

To show the parallelism with the classical case and to infer ideas for other Laplacians we shall do now a brief review of the situation for the Laplacian in the torus. Given a function $f(\theta) = \sum_k a_k e^{ik\theta}$. We have $-\Delta_\theta f = -\partial_\theta^2 f = \sum_k |k|^2 a_k e^{ik\theta}$. The Poisson sums can be defined (formally) by

$$\begin{aligned} P_r f(\theta) &= \sum_k r^{|k|} a_k e^{ik\theta} = \sum_k r^{\sqrt{|k|^2}} a_k e^{ik\theta} \\ &= 1 + \sum_{k>0} r^k a_k e^{ik\theta} + \sum_{k>0} r^k a_{-k} e^{-ik\theta} \\ {\scriptstyle (z=re^{i\theta})} &= 1 + \sum_{k>0} a_k z^k + \sum_{k>0} a_{-k} \bar{z}^k = U(z). \end{aligned} \tag{2}$$

They are harmonic and satisfy

$$\left(r^2\partial_r^2 + r\partial_r + \Delta_\theta\right) P_r f(\theta) = 0, \qquad \partial_z \partial_{\bar{z}} U = 0.$$

On the other hand, if we define the conjugate of the function f as

$$\begin{aligned} Q_r f(\theta) &= -i \sum_{k\neq 0} \operatorname{sign} k\, r^{|k|} a_k e^{ik\theta} = -i \sum_{k>0} r^k a_k e^{ik\theta} + i \sum_{k>0} r^k a_{-k} e^{-ik\theta} \\ &\qquad {\scriptstyle (z=re^{i\theta})} = -i \sum_{k>0} a_k z^k + i \sum_{k>0} a_{-k} \bar{z}^k = V(z). \end{aligned}$$

We have again the harmonicity

$$\left(r^2\partial_r^2 + r\partial_r + \Delta_\theta\right) Q_r f(\theta) = 0, \qquad \partial_z \partial_{\bar{z}} V = 0.$$

Observe that $r^2\partial_r^2 + r\partial_r + \Delta_\theta = -(r\partial_r)^*(r\partial_r) - (\partial_\theta)^*\partial_\theta$, where $(r\partial_r)^* = -r\partial_r$, being $(r\partial_r)^*$ the adjoint with respect to $d\mu(r) = \frac{dr}{r}$ in $[0,1]$, and $(\partial_\theta)^* = -\partial_\theta$, being $(\partial_\theta)^*$ the adjoint with respect to $d\theta$.

Moreover $F(z) = U(z) + iV(z) = 1 + 2\sum_{>0} a_k z^k$ is holomorphic, $\partial_{\bar{z}} F = 0$. The following Cauchy-Riemann equations are satisfied,

$$\begin{aligned} &\partial_\theta (P_r f)(\theta) = -r\partial_r (Q_r f)(\theta), \\ &r\partial_r (P_r f)(\theta) = \partial_\theta (Q_r f)(\theta), \qquad \left(\text{i.e. } (r\partial_r)^* (P_r f)(\theta) = (\partial_\theta)^* (Q_r f)(\theta)\right). \end{aligned}$$

Before giving the appropriate definition (in our opinion) for a general second order differential operator we shall walk along the corresponding objects for the case of the line $\mathbb{R}$. Given a function $f(x)=\int_{\mathbb{R}}\widehat{f}(\xi)e^{ix\xi}d\xi$. We have (formally) $-\Delta f(x)=-\partial_x^2 f(x)=\int_{\mathbb{R}}\widehat{f}(\xi)|\xi|^2 e^{ix\xi}d\xi$. Then its Poisson integral (again formally) is given by

$$
\begin{aligned}
P_t f(x) &= \int_{\mathbb{R}} e^{-t|\xi|}\widehat{f}(\xi)\, e^{i\xi x}d\xi \\
{\scriptstyle (z=t-ix)} &= \int_0^\infty e^{-z\xi}\widehat{f}(\xi)d\xi + \int_0^\infty e^{-\bar{z}\xi}\widehat{f}(-\xi)d\xi = U(z).
\end{aligned}
$$

As in the case of the torus, this function is harmonic, that is

$$\left(\partial_t^2+\Delta_x\right)P_t f(x)=0, \qquad \partial_z\partial_{\bar{z}}U=0.$$

The conjugate harmonic function is defined by

$$
\begin{aligned}
Q_t f(x) &= -i\int_{\mathbb{R}} \operatorname{sign}\xi\, e^{-t|\xi|}\widehat{f}(\xi)\, e^{i\xi x}d\xi \\
{\scriptstyle (z=t-ix)} &= -i\int_0^\infty e^{-z\xi}\widehat{f}(\xi)d\xi + i\int_0^\infty e^{-\bar{z}\xi}\widehat{f}(-\xi)d\xi = V(z).
\end{aligned}
$$

This function is also harmonic:

$$\left(\partial_t^2+\Delta_x\right)Q_t f(x)=0, \qquad \partial_z\partial_{\bar{z}}V=0.$$

Finally, the function $F(z)=U(z)+iV(z)=2\int_0^\infty e^{-z\xi}\widehat{f}(\xi)d\xi$ is holomorphic, $\partial_{\bar{z}}F=0$, and the following Cauchy-Riemann equations are satisfied,

$$
\begin{aligned}
&\partial_x P_t f(x)=\partial_t Q_t f(x), && \left(\text{i.e. } (\partial_x)^* P_t f(x)=(\partial_t)^* Q_t f(x)\right)\\
&\partial_t P_t f(x)=-\partial_x Q_t f(x).
\end{aligned}
$$

Here $(\partial_x)^*=-\partial_x$ with respect to dx, and $(\partial_t)^*=-\partial_t$ with respect to dt. Observe that

$$\Delta_x=-(\partial_x)^*(\partial_x) \text{ and } \partial_t^2+\Delta_x=-\left[(\partial_t)^*(\partial_t)+(\partial_x)^*(\partial_x)\right].$$

3. Heat semigroup as starting point

In order to include into a general framework the operators considered above, we need to introduce the concept of diffusion semigroups.

Definition 3.1. Given $(\mathcal{M},d\mu)$ measure space. A family $\{T_t\}_{t>0}: L^2\to L^2$ is called a symmetric diffusion semigroup if it satisfies:

- $T_{t_1+t_2}=T_{t_1}T_{t_2}$. $T_0=Id$. $\lim_{t\to 0}T_t f=f$ in L^2.
- $\|T_t f\|_p\le\|f\|_p$, $(1\le p\le\infty)$. Contraction.
- T_t selfadjoint in L^2.
- $T_t f\ge 0$ si $f\ge 0$. Positivity.

- $T_t 1 = 1$. Markov.

A typical example of semigroup is the classical Weierstrass operator:

$$T_t f(x) = \frac{1}{(4\pi t)^{n/2}} \int_{\mathbb{R}} e^{-\frac{|x-y|^2}{4t}} f(y)\, dy.$$

For good functions f we have: $\partial_t (T_t f(x)) = \partial_x^2 (T_t f(x)) = \Delta(T_t f(x))$. Formally, we write $T_t f(x) = e^{t\Delta} f(x)$. A useful remark is the fact that $\widehat{T_t f}(\xi) = e^{-t|\xi|^2} \widehat{f}(\xi)$. We observe also that $\widehat{P_t f}(\xi) = e^{-t\sqrt{|\xi|^2}} \widehat{f}(\xi)$.

A second illustrative example of diffusion semigroup is given by orthogonal systems. Let L be a second order differential operator with eigenfunctions $\{\phi_k\}_k$ and eigenvalues $\{\lambda_k\}_k$. Define

$$e^{-tL}\phi_k(x) = e^{-t\lambda_k}\phi_k(x), \qquad e^{-tL}\Big(\sum_k c_k \phi_k\Big)(x) = \sum_k e^{-t\lambda_k} c_k \phi_k(x).$$

Formally, we have the "heat" equation for L:

$$\partial_t (e^{-tL} f(x)) = -L e^{-tL} f(x).$$

The following two cases are included in this framework:

- Fourier series. $-\Delta e^{ik\theta} = |k|^2 e^{-ik\theta}$.
  $e^{-t(-\Delta)}\big(\sum_k a_k e^{-ik\theta}\big) = \sum_k e^{-t|k|^2} a_k e^{-ik\theta}$.
- Hermite polynomials. $L = -\partial_x^2 + 2x\partial_x$, $LH_n = 2nH_n$. Then
  $e^{-tL}\big(\sum_n a_n H_n\big)(x) = \sum_n e^{-t2n} a_n H_n(x)$.

In other words, the so-called Poisson sums by Muckenhoupt in (1) are nothing but the heat semigroup associated to the operator $L = -\partial_x^2 + 2x\partial_x$. In order to put the operator $f \sim \sum a_n H_n \longrightarrow f(t,x) = \sum_n e^{-t(2n)^{1/2}} a_n H_n$ in our line of thoughts we shall recall the reader the following Gamma function formula,

$$e^{-t\sqrt{\lambda}} = \frac{t}{2\sqrt{\pi}} \int_0^\infty \frac{e^{-\frac{t^2}{4s}}}{s^{3/2}} e^{-s\lambda} ds, \quad \lambda > 0. \tag{3}$$

Observe that, for the case of Fourier series, as $e^{-t(-\Delta)}\big(\sum_k a_k e^{-ik\theta}\big) = \sum_k e^{-t|k|^2} a_k e^{-ik\theta}$, we can define

$$\begin{aligned} e^{-t\sqrt{-\Delta}}\Big(\sum_k a_k e^{-ik\theta}\Big) &= \frac{t}{2\sqrt{\pi}} \int_0^\infty \frac{e^{-\frac{t^2}{4s}}}{s^{3/2}} e^{-s(-\Delta)}\Big(\sum_k a_k e^{-ik\theta}\Big) ds \\ &= \frac{t}{2\sqrt{\pi}} \int_0^\infty \frac{e^{-\frac{t^2}{4s}}}{s^{3/2}} \Big(\sum_k e^{-t|k|^2} a_k e^{-ik\theta}\Big) ds \\ &= \sum_k e^{-t|k|} a_k e^{-ik\theta}. \end{aligned}$$

On the other hand, in the case of Hermite polynomials, if we define the heat semigroup as $e^{-tL}(\sum_n a_n H_n)(x) = \sum_n e^{-t2n} a_n H_n(x)$, then by using again the subordination formula (3) , we have

$$\begin{aligned} e^{-t\sqrt{L}}\Big(\sum_n a_n H_n(x)\Big) &= \frac{t}{2\sqrt{\pi}} \int_0^\infty \frac{e^{-\frac{t^2}{4s}}}{s^{3/2}} e^{-s(-\Delta)}\Big(\sum_n e^{-t2n} a_n H_n(x)\Big) ds \\ &= \frac{t}{2\sqrt{\pi}} \int_0^\infty \frac{e^{-\frac{t^2}{4s}}}{s^{3/2}} \Big(\sum_n e^{-t2n} a_n H_n(x)\Big) ds \\ &= \sum_n e^{-t(2n)^{1/2}} a_n H_n(x). \end{aligned}$$

In general given a positive operator L, the subordinated semigroup is given by (Bochner subordination formula),

$$e^{-t\sqrt{L}} = \frac{t}{2\sqrt{\pi}} \int_0^\infty \frac{e^{-\frac{t^2}{4s}}}{s^{3/2}} e^{-sL} ds, \quad \lambda > 0. \tag{4}$$

Observe that formally $(\partial^2_{tt} - L)e^{-t\sqrt{L}} = 0$. That is, $e^{-t\sqrt{L}}$ satisfies a Laplace equation.

Formula (4) can be used to determine the integral form of certain operators as soon as we know the corresponding integral formula for the heat kernel. An example of this idea is presented in the following lines in which we deduce the classical Poisson kernel on $\mathbb{R}$ by plugging the heat kernel into (4):

$$\begin{aligned} e^{-t\sqrt{-\Delta}}(x,y) &= \frac{t}{2\sqrt{\pi}} \int_0^\infty \frac{e^{-\frac{t^2}{4s}}}{s^{3/2}} e^{-s(-\Delta)}(x,y)\, ds \\ &= \frac{t}{2\sqrt{\pi}} \int_0^\infty \frac{e^{-t^2/4s}}{s^{3/2}} \frac{1}{(4\pi s)^{n/2}} e^{-\frac{|x-y|^2}{4s}} ds \\ \Big(\tfrac{t^2+|x-y|^2}{4s} = u\Big) &= \frac{t}{\pi^{(n+1)/2}} \int_0^\infty \frac{u^{(n+1)/2}}{(t^2+|x-y|^2)^{(n+1)/2}} e^{-u} du \\ &= \frac{\Gamma(\frac{n+1}{2})}{\pi^{(n+1)/2}} \frac{t}{(t^2+|x-y|^2)^{(n+1)/2}}. \end{aligned}$$

The use of formulas related with the Gamma function has been of help in the last century to build a kind of functional calculus for positive operators. E. Stein, in [10], proposed the use of that kind of formulas in order to define different operators. Namely "Fractional integrals" associated to a Laplacian can be given via the numerical formula $\lambda^{-\alpha} = \frac{1}{\Gamma(\alpha)} \int_0^\infty t^\alpha e^{-\lambda t} \frac{dt}{t}$, $\lambda, \alpha > 0$. In fact, given ϕ_k, eigenfunction of L with eigenvalue λ_k, we have

$$L^{-\alpha}\phi_k = \frac{1}{\Gamma(\alpha)} \int_0^\infty t^\alpha e^{-tL} \phi_k \frac{dt}{t} = \frac{1}{\Gamma(\alpha)} \int_0^\infty t^\alpha e^{-t\lambda_k} \phi_k \frac{dt}{t} = \frac{1}{\lambda_k^\alpha} \phi_k.$$

In general, for positive L,

$$L^{-\alpha} f = \frac{1}{\Gamma(\alpha)} \int_0^\infty t^\alpha e^{-tL} f \, \frac{dt}{t}.$$

As an illustration of the use of these formulas we compute the kernel of the fractional integral $(-\Delta_x)^{-\alpha/2}$ in $\mathbb{R}^n$.

$$\begin{aligned}
(-\Delta_x)^{-\alpha/2} f(x) &= \frac{1}{\Gamma(\alpha/2)} \int_0^\infty t^{\alpha/2} e^{-t(-\Delta_x)} f(x) \, \frac{dt}{t} \\
&= \frac{1}{\Gamma(\alpha/2)} \int_0^\infty \int_{\mathbb{R}^n} \frac{1}{(4\pi t)^{n/2}} e^{-\frac{|x-y|^2}{4t}} f(y) dy t^{\alpha/2} \frac{dt}{t} \\
&= \frac{1}{\Gamma(\alpha/2)} \int_{\mathbb{R}^n} \int_0^\infty \frac{1}{(4\pi t)^{n/2}} e^{-\frac{|x-y|^2}{4t}} t^{\alpha/2} \frac{dt}{t} f(y) dy \\
&= \frac{1}{(4\pi)^{n/2}} \frac{1}{\Gamma(\alpha/2)} \int_{\mathbb{R}^n} \int_0^\infty \frac{4^{(n-\alpha)/2}}{|x-y|^{n-\alpha}} u^{(n-\alpha)/2} \frac{du}{u} f(y) dy \\
&= \frac{1}{\pi^{n/2} 4^{\alpha/2} \Gamma(\alpha/2)} \Gamma\Big(\frac{n-\alpha}{2}\Big) \int_{\mathbb{R}^n} \frac{1}{|x-y|^{n-\alpha}} f(y) dy \\
&= c_{n,\alpha} \int_{\mathbb{R}^n} \frac{1}{|x-y|^{n-\alpha}} f(y) dy.
\end{aligned}$$

Observe that, in the above calculation, we do not use Fourier transforms.

Assume that we have a Laplacian with a factorization of the type $L = \partial^* \partial$. Then a natural Riesz transform associated to L should be $\partial(L)^{-1/2}$. Hence if $\{\phi_k\}_k$ is the family of orthogonal eigenfunctions we have

$$\begin{aligned}
\int \Big(\partial(L)^{-1/2}\phi_k\Big)\Big(\partial(L)^{-1/2}\phi_\ell\Big) \, d\mu &= \int \Big(\partial^* \partial(L)^{-1/2}\phi_k\Big)\Big((L)^{-1/2}\phi_\ell\Big) d\mu \\
&= \int \Big(L(L)^{-1/2}\phi_k\Big)\Big((L)^{-1/2}\phi_\ell\Big) \, d\mu \\
&= \int \Big((L)^{1/2}\phi_k\Big)\Big((L)^{-1/2}\phi_\ell\Big) \, d\mu \\
&= \lambda_k^{1/2} \lambda_\ell^{-1/2} \int \phi_k \phi_\ell \, d\mu.
\end{aligned}$$

This gives boundedness in $L^2(d\mu)$ of $Rf = \partial L^{-1/2}$.

What about the boundedness in L^p of Riesz transforms? The general procedure in Harmonic Analysis is that once L^2-boundedness is known, the kernel is used to get L^p boundedness. In other words, we shall make the following computation

$$\begin{aligned}\partial_x L^{-1/2} f(x) &= \partial_x \int_0^\infty e^{-tL} f(x)\, t^{1/2} \frac{dt}{t} = \int_0^\infty \partial_x e^{-tL} f(x)\, t^{1/2} \frac{dt}{t} \\ &= \int_0^\infty \partial_x \int_{\mathbb{R}^n} e^{-tL}(x,y) f(y)\, dy\, t^{1/2} \frac{dt}{t} \\ &= \int_{\mathbb{R}^n} \left(\int_0^\infty \partial_x e^{-tL}(x,y)\, t^{1/2} \frac{dt}{t} \right) f(y)\, dy \\ &= \int_{\mathbb{R}^n} K(x,y) f(y)\, dy.\end{aligned}$$

It can be seen that, for good enough functions,

$$\partial_{x_i} (-\Delta)^{-1/2} f(x) = \frac{\Gamma\left(\frac{n+1}{2}\right)}{\pi^{(n+1)/2}} P.V. \int_{\mathbb{R}^n} \frac{x_i - y_i}{|x-y|^{n+1}} f(y) dy.$$

Again we have not used Fourier transforms. Now we arrive to the question of how to introduce the "conjugate operator". For that we mean a harmonic function satisfying some kind of Cauchy-Riemann equations together with the Poisson semigroup, $u(x,t) = e^{-t\sqrt{L}} f(x)$. After some seconds of reflexion it seems natural to consider $v(x,t) = \partial_x (L)^{-1/2} e^{-t\sqrt{L}} f(x)$. Observe that

$$\begin{aligned}\partial_x^* v(x,t) &= \partial_x \partial_x (L)^{-1/2} e^{-t\sqrt{L}} f \\ &= (L)^{1/2} e^{-t\sqrt{L}} f = -\partial_t e^{-t\sqrt{L}} f = -\partial_t u(x,t),\end{aligned}$$

and

$$\partial_t v(x,t) = -\partial_x (L)^{-1/2} \sqrt{L} e^{-t\sqrt{L}} f = -\partial_x u(x,t).$$

Then

$$\partial_{tt}^2 v(x,t) = -\partial_x \partial_t u = \partial_x \partial_x^* v(x,t) = \tilde{L} v(x,t).$$

4. Discrete Laplacian

Consider the discrete Laplacian given by

$$\Delta_d f(n) = f(n+1) - 2f(n) + f(n-1), \qquad n \in \mathbb{Z}.$$

The solution of

$$\begin{cases} \frac{\partial}{\partial t} u(n,t) = u(n+1,t) - 2u(n,t) + u(n-1,t), \\ u(n,0) = \delta_{nm}, \text{ for every fixed } m \in \mathbb{Z}, \end{cases}$$

is given by $u(n,t) = e^{-2t}I_{n-m}(2t)$, where $I_k(t)$ is the Bessel function of imaginary argument, see [4] and [6]. We consider the formal series

$$e^{-t(-\Delta_d)}f(n) = W_t f(n) = \sum_{m\in\mathbb{Z}} e^{-2t}I_{n-m}(2t)f(m). \tag{5}$$

Formally, $u(n,t) = W_t f(n)$ is the solution to the heat equation

$$\begin{cases} \frac{\partial}{\partial t}u(n,t) = u(n+1,t) - 2u(n,t) + u(n-1,t), \\ u(n,0) = f(n), \end{cases}$$

where u is the unknown function and the sequence $f = \{f(n)\}_{n\in\mathbb{Z}}$ is the initial datum at time $t = 0$. By using some formulas related with Bessel functions, we have the following results, see [3].

Proposition 4.1. *Let $f \in \ell^\infty$. The family $\{W_t\}_{t\ge 0}$ satisfies*

(i) $W_0 f = f$.
(ii) $W_{t_1}W_{t_2}f = W_{t_1+t_2}f$.
(iii) If $f \in \ell^2$ then $W_t f \in \ell^2$ and $\lim_{t\to 0} W_t f = f$ in ℓ^2.
(iv) (Contraction property) $\|W_t f\|_{\ell^p} \le \|f\|_{\ell^p}$ for $1 \le p \le +\infty$.
(v) (Positivity preserving) $W_t f \ge 0$ if $f \ge 0$, $f \in \ell^2$.
(vi) (Markovian property) $W_t 1 = 1$.

Proposition 4.2. *Let $f \in \ell^\infty$. Then,*

$$u(n,t) = \sum_{m\in\mathbb{Z}} e^{-2t}I_{n-m}(2t)f(m), \quad t>0, \quad n\in\mathbb{Z},$$

is a solution of the heat equation.

$$\partial_t u(n,t) - \Delta_d u(n,t) = 0.$$

The proof of the last two propositions involved several formulas related with Bessel functions. We list here the more important for our needs.

$$I_k(t) = i^{-k}J_k(it) = \sum_{m=0}^{\infty} \frac{1}{m!\,\Gamma(m+k+1)}\left(\frac{t}{2}\right)^{2m+k}, \quad k\in\mathbb{Z}.$$

The function I_k is defined in the whole real line, even in the whole complex plane, where I_k is an entire function. Moreover $I_{-k}(t) = I_k(t)$ and

$$I_r(t_1+t_2) = \sum_{k\in\mathbb{Z}} I_k(t_1)I_{r-k}(t_2), \quad \text{for} \quad r\in\mathbb{Z}. \quad \text{Neumann's identity.}$$

On the other hand $I_k(t) \ge 0$ for every $k\in\mathbb{Z}$ and $t \ge 0$, and

$$\sum_{k\in\mathbb{Z}} e^{-2t}I_k(2t) = 1, \qquad I_k(t) = Ce^t t^{-1/2} + R_k(t),$$

with $|R_k(t)| \le C_k e^t t^{-3/2}$, for $t \to \infty$.

Also $\frac{\partial}{\partial t} I_k(t) = \frac{1}{2}(I_{k+1}(t) + I_{k-1}(t))$, and from this it follows immediately

$$\frac{\partial}{\partial t}(e^{-2t} I_k(2t)) = e^{-2t}(I_{k+1}(2t) - 2I_k(2t) + I_{k-1}(2t)).$$

The following (Schläfli) integral representation is essential in our reasonings

$$I_\nu(z) = \frac{z^\nu}{\sqrt{\pi}\, 2^\nu \Gamma(\nu + 1/2)} \int_{-1}^{1} e^{-zs}(1-s^2)^{\nu-1/2}\, ds, \quad |\arg z| < \pi, \quad \nu > -\frac{1}{2}.$$

Also the following consequences

$$I_{\nu+1}(z) - I_\nu(z) = -\frac{z^\nu}{\sqrt{\pi}\, 2^\nu \Gamma(\nu + 1/2)} \int_{-1}^{1} e^{-zs}(1+s)(1-s^2)^{\nu-1/2}\, ds,$$

$$I_{\nu+2}(z) - 2I_{\nu+1}(z) + I_\nu(z) = \frac{z^\nu}{\sqrt{\pi}\, 2^\nu \Gamma(\nu + 1/2)}$$
$$\times \left(\frac{2}{z} \int_{-1}^{1} e^{-zs} s(1-s^2)^{\nu-1/2}\, ds + \int_{-1}^{1} e^{-zs}(1+s)^2(1-s^2)^{\nu-1/2}\, ds \right).$$

Remark 4.1. As a consequence of the general picture described above the operator

$$P_t f(n) = \frac{1}{\sqrt{\pi}} \int_0^\infty \frac{e^{-u}}{\sqrt{u}} W_{t^2/(4u)} f(n)\, du = \frac{t}{2\sqrt{\pi}} \int_0^\infty \frac{e^{-t^2/(4v)}}{\sqrt{v}} W_v f(n) \frac{dv}{v}$$

satisfies formally the "Laplace" equation

$$\partial_{tt}^2 P_t f(n) + \Delta_{\mathrm{d}} P_t f(n) = 0.$$

But we prove that it satisfies the equation for $f \in \ell^\infty$.

To define the "Discrete Riesz transforms", we consider the "first" order difference operators

$$Df(n) = f(n+1) - f(n) \quad \text{and} \quad \tilde{D}f(n) = f(n) - f(n-1),$$

that allow factorization of the discrete Laplacian as $\Delta_{\mathrm{d}} = \tilde{D}D$. Then the Discrete Riesz transforms are defined as

$$\mathcal{R} = D(-\Delta_{\mathrm{d}})^{-1/2} \quad \text{and} \quad \tilde{\mathcal{R}} = \tilde{D}(-\Delta_{\mathrm{d}})^{-1/2}.$$

The operator $D(-\Delta_{\mathrm{d}})^{-1/2}$ is given by the multiplier $-ie^{-i\theta/2}$ and also described as the convolution with the kernel $\frac{1}{\pi(n+\frac{1}{2})}$. While $\tilde{D}(-\Delta_{\mathrm{d}})^{-1/2}$ is given by the multiplier $-ie^{i\theta/2}$ and described as the convolution with the kernel $\frac{1}{\pi(n-\frac{1}{2})}$. This gives boundedness in ℓ^2 (also ℓ^p).

We can go further and consider the Discrete Conjugate functions

$$Q_t f = \mathscr{R} P_t f \quad \text{and} \quad \widetilde{Q}_t f = \widetilde{\mathscr{R}} P_t f.$$

We prove that they satisfy Cauchy-Riemann equations and also that the Riesz transforms are the limit when $t \to 0$ of the corresponding Conjugate function.

Proposition 4.3. *Let Q_t and $\widetilde{Q}_t$ be as above and let f be a compactly supported function.*

(i) The operators Q_t, $\widetilde{Q}_t$ and P_t satisfy the Cauchy–Riemann type equations

$$\begin{cases} \partial_t(Q_t f) = -D(P_t f), \\ \widetilde{D}(Q_t f) = \partial_t(P_t f); \end{cases} \qquad \begin{cases} \partial_t(\widetilde{Q}_t f) = -\widetilde{D}(P_t f), \\ D(\widetilde{Q}_t f) = \partial_t(P_t f). \end{cases}$$

Moreover, $\partial^2_{tt} Q_t f(n) + \Delta_d Q_t f(n) = 0$, and $\partial^2_{tt} \widetilde{Q}_t f(n) + \Delta_d \widetilde{Q}_t f(n) = 0$.

(ii) We have, for $n \in \mathbb{Z}$,

$$\lim_{t\to 0} Q_t f(n) = \mathscr{R} f(n), \quad \text{and} \quad \lim_{t\to 0} \widetilde{Q}_t f(n) = \widetilde{\mathscr{R}} f(n).$$

5. Further applications

The numerical formula for the Gamma function,

$$\lambda^\alpha = \frac{1}{\Gamma(-\sigma)} \int_0^\infty (e^{-t\lambda} - 1) \frac{dt}{t^{1+\alpha}}, \quad 0 < \alpha < 1,$$

can be used in order to define the positive powers of an operator L as

$$L^\alpha f(x) = \frac{1}{\Gamma(-\sigma)} \int_0^\infty (e^{-tL} f(x) - f(x)) \frac{dt}{t^{1+\alpha}}, \quad 0 < \alpha < 1.$$

This formula gives a precise definition, with exact constants, for the "Fractional Laplacian". In the last 10 years this topic has been a kind of central object in PDE's. See [2], [11].

Recently an application to the study of the wave equation has been carried out in [5]. The main difficulty in this case is the presence of oscillatory integrals. To give an idea to the reader we can say that the starting (non trivial) point is to show that the oscillatory integral has sense

$$u(\cdot, t) = \frac{i^\sigma t^{2\sigma}}{4^\sigma \Gamma(\sigma)} \int_0^\infty e^{-i\frac{t^2}{4s}} e^{is\Delta} f \frac{ds}{s^{1+\sigma}},$$

and satisfies the problem

$$\begin{cases} \partial_t^2 u + \frac{1-2\sigma}{t} \partial_t u = \Delta u, \\ u(\cdot, 0) = f. \end{cases}$$

The ideas developed above can also be applied to the study of one-sided operators, see [1].

Acknowledgments

I thank the organizers of the VI International Course of Mathematical Analysis in Andalucía for the invitation to present these ideas in such a high mathematical and delightful atmosphere.

References

[1] A. Bernardis, F. J. Martín-Reyes, P. R. Stinga and J. L. Torrea, Maximum principles, extension problem and inversion for nonlocal one-sided equations, *J. Differential Equations* To appear.

[2] L. Caffarelli and L. Silvestre, An extension problem related to the fractional laplacian, *Comm. Partial Differential Equations* **32**, pp. 1245–1260 (2007).

[3] Ó. Ciaurri, T. A. Gillespie, L. Roncal, J. L. Torrea and J. L. Varona, Harmonic analysis associated with a discrete laplacian, *J. Anal. Math.* To appear.

[4] F. A. Grünbaum and P. Iliev, Heat kernel expansions on the integers, *Math. Phys. Anal. Geom.* **5**, pp. 183–200 (2002).

[5] M. Kemppainen, P. Sjögren and J. L. Torrea, Wave extension problem for the fractional laplacian, *Discrete Contin. Dyn. Syst.* **35**, pp. 4905–4929 (2015).

[6] N. N. Lebedev, *Special functions and their applications*, Revised English edition. Translated and edited by Richard A. Silverman, pp. xii+308 (Prentice-Hall, Inc., Englewood Cliffs, N.J., 1965).

[7] B. Muckenhoupt and E. M. Stein, Classical expansions and their relation to conjugate harmonic functions, *Trans. Amer. Math. Soc.* **118**, pp. 17–92 (1965).

[8] B. Muckenhoupt, Hermite conjugate expansions, *Trans. Amer. Math. Soc.* **139**, pp. 243–260 (1969).

[9] B. Muckenhoupt, Poisson integrals for Hermite and Laguerre expansions, *Trans. Amer. Math. Soc.* **139**, pp. 231–242 (1969).

[10] E. M. Stein, *Topics in harmonic analysis related to the Littlewood-Paley theory.*, Annals of Mathematics Studies, No. 63, pp. viii+146 (Princeton University Press, Princeton, N.J.; University of Tokyo Press, Tokyo, 1970).

[11] P. R. Stinga and J. L. Torrea, Extension problem and Harnack's inequality for some fractional operators, *Comm. Partial Differential Equations* **35**, pp. 2092–2122 (2010).

Author index